American Institute of Chemical Engineers

20th International Conference of the Center for Chemical Process Safety

AA003219

CCPS 2005

Held at the 2005 AIChE Spring National Meeting

April 10-14, 2005
Atlanta, Georgia, USA

Printed from e-media with permission by:

Curran Associates, Inc.
57 Morehouse Lane
Red Hook, NY 12571
www.proceedings.com

ISBN: 978-1-60423-464-0

Some format issues inherent in the e-media version may also appear in this print version.

American Institute of Chemical Engineers

20th International Conference of the Center for Chemical Process Safety
2005

TABLE OF CONTENTS

Preventing Incidents at Newly Acquired Facilities – Implementation of Lessons Learned

Brian R. Dunbobbin, C. Lorn Paxton, Glen A. Peters
Air Products & Chemicals Inc.
Allentown, PA 18051 USA
dunbobbr@apci.com

Martin A. Dennehy
Air Products plc.
Walton on Thames,
England KT12 4RZ
dennehma@apci.com

The large number of acquisitions taking place in the chemical industry has increased the need to efficiently assess the process safety risks at such facilities so that they meet the requirements of the acquiring company. This can be difficult if the technology issues are not fully disclosed during the "Due Diligence" phase of an acquisition.

This paper describes a tiered approach to ensuring that the acquiring company identifies as quickly as possible key process safety risks so that employee, community and company interests are adequately protected. In particular, three aspects of the work process for identifying process safety risks at an acquired facility are shown to be important. The first is the method used to identify process safety risks during the "Due Diligence" phase. This involves the selection of appropriate team members, and development of comprehensive checklists that the acquisition team can use to identify potential risks at an early stage. The second is the ability of the acquiring company to focus its own technical and process safety experience on the technologies being acquired; this experience is often in the form of standards that are used in the design and operation of similar facilities. Whether or not to retrofit these standards becomes an important issue when the newly acquired facility is operated by the new company. Finally, a Process Safety Assessment work process, with defined protocols, has been developed so that senior Process Safety, Operations and Engineering personnel can assess the potential process safety risks at a facility quickly and consistently in the first few months of operation following an acquisition.

1.0 Introduction

Following acquisition of a chemical facility it is important that the acquiring company quickly understand the potential hazards and risks that the facility may present. This ensures that employee, community and company interests are adequately protected. In particular there are three issues that should be understood by the acquiring company:

1) Are there major hazards present in the facility that could cause a serious incident?
2) Are the management systems at the facility adequate to ensure safe operation? Are these management systems compatible with existing company systems?
3) Do the standards and practices utilized at the newly acquired facility meet those of the acquiring company?

Unfortunately getting answers to these questions can be a difficult task, especially if the technology issues are not fully disclosed during the pre-purchase, or initial due diligence, phase of an acquisition.

Environmental issues are often easily identified during the pre-purchase phase of an acquisition since regulations normally require disclosure and public reporting. Process safety issues on the other hand are often tied closely with the technology of the processes being acquired and the selling company has little incentive to give full disclosure of technology "trade secrets" prior to finalization of the sale. The process safety hazards and risks may not be fully identified until the new owner has taken over the facility and reviewed the details of the technology and the hazards that may be present. It can then be an expensive task to bring the acquired facility into compliance with the process safety standards of the new owner.

Air Products utilizes a tiered approach to ensure that potential key process safety risks are identified as quickly as possible. In particular, three aspects of the work process for identifying process safety risks at an acquired facility are shown to be important:

1. **Pre-Purchase Assessment (sometimes referred to as the due diligence phase).** A consistent work process and methodology must be used to identify process safety risks during the pre-purchase phase. This involves the selection of appropriate pre-purchase assessment team members, including experienced process safety practitioners, and development of comprehensive checklists that the acquisition team can use to identify potential risks at an early stage. This checklist must be efficiently prioritized so that it can be used quickly in the normally brief pre-purchase period. At this stage of the acquisition a transition plan is also put in place. This transition plan identifies key elements that must be in place on the first day of ownership.
2. **Technology Issues.** The ability to focus technical and process safety experience on the technologies being acquired; this experience is often in the form of standards that are used in the design and operation of similar facilities. Whether or not to retrofit these standards becomes an important issue which can often be complicated by the fact that the newly acquired facility safety record may be excellent, but operates to different standards and work processes.
3. **Post Purchase Assessment.** A process safety assessment work process, with defined protocols that build upon the checklists used in pre-purchase phase, has

been developed so experienced process safety, operations and engineering personnel can assess the potential process safety risks at a facility quickly and consistently in the first few months of operation following an acquisition.

In the following sections of this paper we will describe in detail the major elements of each of these three aspects of the work process.

2.0 Pre-Purchase Assessment Phase of an Acquisition

After the business areas have determined that a potential acquisition is feasible, a small team is usually formed to review documentation and visit the potential acquisition candidate for a pre-purchase assessment to identify potential environmental, health and safety impacts. This is often done in secret by a small, two or three person team, so employees and stakeholders are not alarmed at the prospect of a potential sale of their facility. During the usually brief site visits, sometimes lasting for only several hours, the team will usually walk around the facility and get an overview of the technology, equipment and operating capabilities of the facility. It is normally not meant to give a detailed assessment of the technology since this often remains confidential until later in the acquisition process. Obtaining as much documentation as possible prior to the site visit can be extremely helpful. Documentation normally needed at this early stage includes plot plans, permits, lists of chemicals on-site, chemical inventories, etc.

It is at this stage of a potential acquisition that the acquiring company would like to identify key issues that could impact the economic viability of the acquisition. Within the process safety area these would include major hazards that could cause serious incidents and the possible costs necessary to provide an acceptable level of risk. The pre-purchase assessment team also develops a transition plan that summarizes actions for management and identifies items that must be in place on day one of the transfer of ownership.

It is extremely important that the team contains representatives from the correct functional areas. We will not get into the full team composition here, but rather we will concentrate on the need for adequate experienced process safety representation. This process safety representative must have a level of experience and prestige so representations on the safety of the facility are heard by senior management.

To assist the process safety member of the pre-purchase assessment team we have developed a checklist that can be followed to ensure that the "right" questions are asked both prior to and during the site visit.

The major sections of the checklist include elements of any good process safety management system and are given below:

1. Determination of Regulatory Applicability
2. Process Safety Information
3. Process Hazard Reviews
4. Operating Procedures
5. Training*
6. Pre-startup Safety Reviews (PSSR)*
7. Hot Work*

8. Management of Change (MOC)*
9. Mechanical Integrity
10. Incident Investigation
11. Emergency Planning and Response Procedures
12. Compliance Audits (Internal & External)
13. Trade Secrets*
14. Facility Public Risks
15. Transportation Risk
16. Site Security
17. Safety Systems and Standards*
18. Technology Specific Items*

Items in the checklist marked by an asterisk are often not addressed fully, if at all, during this pre-purchase phase due to time constraints and a lack of information, therefore these are usually treated as low priority items and are only addressed if time allows. In particular, the last two items, Safety Systems and Standards, and Technology Specific Items, may be extremely difficult to assess at this stage if the technology is new to the acquiring company. Also, the site visit assessment team is **often limited to two or three team members** and having all the necessary technology skills on the team can be difficult. All of the items not completely covered in this pre-purchase assessment will be reviewed in detail as part of the transition plan and during the post purchase assessment.

Each section of the checklist contains detailed questions that can be answered by the process safety pre-purchase assessment team member after discussing the questions with facility management.

In the Mechanical Integrity section of the checklist for instance the following would be addressed:

a. *Determine whether there is a mechanical integrity system/procedures for the following equipment:*
   - *Relief devices*
   - *Major process equipment like reactors and distillation columns*
   - *Storage tanks*
   - *Piping*
   - *Pumps and compressors*
   - *Safety instrumented systems (critical interlocks)*

b. *Determine if initial thickness baselines for tanks have been completed.*

c. *Determine the current plant test or inspection frequencies for this equipment.*

d. *Review previous and current year maintenance budgets to identify any significant planned expenditures for equipment replacement.*

e. *Review MI test records for large equipment (reactors, distillation columns, large rotating machinery, storage tanks, etc.) to determine if significant useful life is still available and whether the need for replacement is imminent.*

f. Identify what organization performs the MI inspections (contractor, corporate group, etc.)

g. Determine expenditures for MI inspections for the previous year and budget allocation for current year.

Answers to these questions give good insight into how well maintained the facility is, how close to its expected life the equipment is, and whether or not relief devices have been identified and serviced. If a facility cannot produce a list of relief devices, it is extremely unlikely that relief device testing and inspection is being performed on a routine basis.

After completing as much of the checklist as possible, the pre-purchase assessment team will be in a better position to assess what is known and what is not known about a facility in respect to its hazards and risks. Unfortunately not all of the information required will be freely given to the team during these initial phases of the acquisition. However, one has to determine if there is enough information available to assess the potential costs to perform any necessary upgrades to bring a facility into compliance with the safety requirements of the acquiring corporation as a whole.

The types of problems that can show up during pre-purchase assessment include:

1) Facility siting issues caused by local population centers being too close to the storage of hazardous materials.
2) Chemical reactivity problems with materials being used at the facility e.g. pump seal fluid compatibility with main process fluids.
3) Inadequate documentation on the maintenance and design of relief devices.
4) Lack of process safety information and corresponding inadequate hazard reviews.

Some of these issues are readily handled by putting a contingency on the value of the facility. It is the responsibility of the pre-purchase assessment team to recommend to management which issues require significant contingencies.

2.1 Transition Plan

The major element of the transition plan is the identification of items that must be in place on day one of the transfer of ownership. These may include:

1) Emergency response issues such as emergency contact numbers on documentation. Sometimes these will involve contracting with the selling company to continue providing these services until the acquiring company personnel are adequately trained.

2) Regulatory notifications such as US EPA RMP or European Seveso ownership and registration.

3.0 Technology & Standards Evaluation

During and after the pre-purchase phase of an acquisition it is important to focus technical and process safety experience on the technologies being acquired. This experience is often in the form of standards that are used in the design and operation of similar facilities. For example Air Products has many standards covering the safe operation of air separation plants. When we acquire such a technology, we can quickly assess whether or not the facility in question conforms to our standards.

Whether or not to retrofit these standards onto a newly acquired facility becomes an important issue, but is not always straight forward. Just because a facility operates to different standards does not mean that it is unsafe.

To assess whether or not to retrofit standards onto facilities we utilize a work process that ensures that senior process safety, technology and operational input is given. A list of standards that have required retrofitting in the past has been developed and this is given to the process safety assessment teams when looking at new acquisitions. When questions arise related to the need to retrofit standards, senior management is consulted to ensure that adequate technical resources are dedicated to answering the question.

4.0 Process Safety Assessment

Shortly after the transfer of the assets, a process safety assessment is performed by experienced process safety, operations and engineering personnel to assess the potential process safety risks at a facility in detail. The assessment will generate an integration plan that outlines how the facility will meet the acquiring company safety and technology standards and how the facility plans to resolve any potential safety risks. This plan should be agreed to by all parties. This assessment typically takes about 2-3 days for a large facility and utilizes an assessment protocol that ensures consistency. This assessment protocol is similar to the pre-purchase checklist referred to earlier. It contains the same elements as the pre-purchase checklist, but since more time is available to conduct this assessment, it has more in-depth questions and more guidance for the assessors.

Figure 1 is an extract of questions from the hazard review element of the protocol. These protocol questions range from the basic, "Has a hazard review been conducted for the processes at the site?" to "Are sufficient funds budgeted to correct the deficiencies and complete the action items?" and "Has a system been established to promptly address the team's findings and recommendations?" Note that if documentation cannot be found to support each question then it is assumed that the work hasn't been done. Guidance notes are also included to assist the process safety assessor in the task. These are important since they help to ensure consistency between assessors.

The work process for the process safety assessments consists of three distinct phases:

1) Assessment Pre-Planning
2) On-Site Activities
3) Post Assessment Follow-Up

A brief look at each of these areas gives the major considerations for each phase:

4.1 Assessment Pre-Planning

This phase is probably the most significant since it includes communication with the site, team selection and definition of the protocols to be used. It sets the ground rules for the assessment and ensures that all required documentation is available for review by the assessment team. Several weeks prior to the site visit the team leader will identify the appropriate personnel to conduct the assessment. Key employees will include experienced process safety specialists, operations and engineering personnel. Ideally the people selected should have experience and knowledge of the processes being reviewed at the acquired facility. For example, we utilize persons knowledgeable in air separation technology (operations and engineering) when conducting an assessment of an air separation facility. These knowledgeable persons can quickly ascertain potential design and operations hazards of the acquired facility based on a review of company engineering standards.

Communication with the site will allow the site to understand exactly what will be reviewed, when it will occur, and who will be participating. As part of pre-planning, the team leader will prepare a draft daily schedule showing specific areas the team will review and the morning or afternoon the specific area will be reviewed. This schedule is sent to the site with a request that names of appropriate site people are assigned to review a particular area with the assessor on a particular day. This allows the site manager to ensure sufficient resources are available during the assessment for an adequate review, and gives the assessor a point of contact once on-site. The site is also sent a copy of the Process Safety Assessment Protocol questions so they can begin gathering the documentation required to answer the protocol questions.

This pre-planning phase has been shown to be extremely important since it ensures an efficient on-site assessment process, facility buy-in with the assessment process and leads to true team building once on-site. This team building is important since the assessment is not intended as a simple audit of the site. It is intended as a mechanism to identify potential process safety issues and to develop an integration plan that the newly acquired facility can practically achieve using resources from the site and the acquiring company.

4.2 On-Site Activities

Having communicated the intent of the assessment during the pre-planning phase, the first thing that the assessment team does after arriving onsite is to conduct an opening meeting with the site team. This opening meeting allows the team leader to introduce the assessment team, communicate the activities to be completed on-site, as well as the post assessment follow-up activities. The team leader also reviews the daily schedule agreed upon prior to coming onsite to ensure the appropriate people are still available and no emergencies, either plant or people wise, will cause interference with the assessment.

At the completion of the opening meeting, the assessment team takes a site tour to familiarize the team with site layout and conditions. Once the tour is completed each individual from the assessment team meets with the assigned site contact to begin the assessment. At the end of each day the assessment team meets with the site team to communicate the areas reviewed, and discuss findings and items that may need more detailed follow-up. This daily debriefing is very important. It informs the site of the actions identified during the assessment, and it allows for no surprises at the closing meeting.

At the completion of the on-site assessment, the assessment team prepares written actions and communicates these actions to the site team and facility management at a closing conference. The site should have previously been informed of all actions at the daily site debriefings, and only agreement on the wording of each item is usually necessary.

While performing this assessment one quickly determines the adequacy of the process safety management systems in place at the facility and a determination can then be made as to areas that require attention. In some cases this can mean implementing, for example, a whole new mechanical integrity program, while in other cases only small adjustments may be necessary to bring the facility into compliance with corporate requirements. It is important that the assessment team contain experienced process safety and technology members since they have to be able to assess the hazards and recommend to management whether or not a facility is safe to operate.

4.3 Post-Assessment Follow-up and Integration Plan

Within 30 days of the on-site assessment, facility management, in conjunction with the assessment team, prepares an integration plan that addresses assessment actions and puts forth an action plan for resolving them. The assessment team reviews the plan and recommends modifications as necessary to ensure that it is adequate to closeout all of the assessment findings. Once finalized, the integration plan is entered into a database for follow-up until all items are completed.

An extract from an integration plan is shown in Figure 2. Each item has a corresponding action plan, with the responsible person identified. The plan should also contain an expected and an actual completion date.

Where necessary, a plan for gathering necessary process safety information and for performing hazard reviews should be part of the integration plan. These hazard reviews are performed in addition to the process safety assessments since a hazard review is much more detailed and can take a significant amount of time to prepare and execute, especially if a facility has little in the way of process safety information documented.

5.0 Conclusion

A multi-tier approach to assessing the hazards and risks associated with newly acquired facilities has been used successfully to ensure that employee, community and company interests are quickly addressed and adequately protected.

Adequate process safety representation on the teams working on the early stages of new acquisitions is essential and a checklist outlining key issues to be addressed simplifies the process and ensures consistency. This checklist has to be concise since the pre-purchase phase of an acquisition has to be completed in a very short time period by a small multi-functional team. Identification of all potential hazards and risks is not always possible in these early stages due to confidentiality concerns, but early identification of potential issues can be very beneficial and ensure a smooth transition of ownership. A transition plan, outlining items that need to be in place on day one of transfer of ownership, is prepared by the pre-purchase team if the acquisition is finalized.

Where the technology is known, and existing standards exist within the acquiring company, a determination of whether or not to retrofit these standards onto the newly acquired facility has to be made. This can be a difficult decision since the existing operating procedures and design may be perfectly safe. A list of standards that should be considered for retrofit should be developed for such technologies so that process safety and technology management can determine the retrofit requirements. In general, process safety standards that are not technology specific are usually retrofitted.

After taking over operation of the facility, a more detailed process safety assessment should be performed to obtain a thorough evaluation of the hazards and risks associated with a facility. Since more time is available to conduct this post purchase assessment than is available for the pre-purchase assessment, an assessment protocol containing more guidance and detail is used as a tool to ensure thoroughness. An integration plan is developed showing how and when the newly acquired facility will comply with corporate requirements. The team performing this assessment should be experienced and senior enough to ensure that management gets a good picture of the true risks to the corporation.

Figure 1 - Excerpt from the Process Safety Assessment Protocol

	PROCESS SAFETY ASSESSMENT for ACQUISITIONS/JOINT VENTURES **Hazard Reviews**	
PSM-3	Has a hazard review been conducted for the processes at the site? Assessor Guidance: • Assessor should verify a list of the various processes exists for the site and confirm that a hazard review has been conducted for the specific process.	Y ***N*** NA ***DK***
PSM-4	Determine which of the methodologies listed below were used for the hazard reviews; • What-if/Checklist; • Hazard and Operability Study (HAZOP); • Failure Mode and Effects Analysis (FMEA); • Fault Tree Analysis; or • An appropriate, equivalent methodology.	Y ***N*** NA ***DK***
PSM-5	Does the hazard review address the areas listed below (Assessor should assess the quality of the hazard review)? a) The hazards of the process a) The identification of any previous incident which had a likely potential for catastrophic consequences in the workplace b) Engineering and administrative controls applicable to the hazards and their interrelationships, such as appropriate application of detection methodologies to provide early warning of release c) Consequences of failure of engineering and administrative controls d) Facility siting e) Human factors Assessor Guidance: • Confirm that the hazard review documentation includes a discussion of facility siting, human factors, and previous incidents. • Assessor should review the hazard review to assess the quality of the hazard review. • Facility siting means considering the location of covered processes within the plant property. Facility siting issues can include location of vessels containing HHCs and their proximity to other equipment, control rooms, maintenance shops, and administration buildings • Human factors involve issues and techniques with the functional relationship between human operators and engineered systems. Human factor considerations include potential human error causes of accidents; examining the location of and access to critical safety instruments, alarms, and equipment; or reviewing critical procedures used by operators and maintenance personnel	Y ***N*** NA ***DK***

PSM-6	Was the hazard review performed by a team with expertise in engineering and process operations, including at least one employee with experience and knowledge specific to the process and one member knowledgeable in the specific hazard review methodology used? <u>Assessor Guidance:</u> • Hazard review must identify through documentation who these individuals are. Also, determine what organization(s) performed the hazard reviews (plant only, support from corporate, contractor, etc.)	Y ***N*** NA ***DK***
PSM-7	Has a system been established to promptly address the team's findings and recommendations? Does <u>each</u> resolved hazard review recommendation have documentation to support resolution of the items? <u>Assessor Guidance:</u> • Assessor should determine if any high-risk action items are not complete. How many action items are beyond their due date?	Y ***N*** NA ***DK***
PSM-8	Has a written schedule been established of when the findings and recommendations are to be resolved? <u>Assessor Guidance:</u> • Each item requiring resolution must have a planned completion date and individual responsible for implementing the corrective action. A single completion date for the entire hazard review does not qualify as a "system" to address the recommendations as required above.	Y ***N*** NA ***DK***
PSM-9	Are sufficient funds budgeted to correct the deficiencies and complete the action items? <u>Assessor Guidance:</u> • The assessor should review hazard review recommendations and identify any items that will require significant capital to implement	Y ***N*** NA ***DK***
PSM-10	Are the hazard reviews updated and revalidated periodically by a qualified team? (Assessor should record the date of the last hazard review). <u>Assessor Guidance</u>: • US PSM regulations require a 5-year revalidation frequency for covered sites. • Air Products requires all sites receive a minimum 10-year revalidation frequency when not governed by regulation.	Y ***N*** NA ***DK***
PSM-11	Has the site identified fatal accident rate (FAR) events and estimated capital required to upgrade systems to meet this target, as appropriate?	Y ***N*** NA ***DK***

Figure 2 – Extract from an Integration Plan

Action	ACTION PLAN With Responsible Person	Estimated Comp. Date	Actual Comp. Date	STATUS
A complete P&ID was not available nor a hazard review for Process A. • ***Develop P&ID's for Process A.*** • ***Conduct hazard review for Process A.***	• Complete P&ID's update for Process A (Facility Manager) • Ensure that adequate process safety information is available for the hazard review (Facility Manager) • Complete the hazard review for Process A. (Facility Manager)	Mar. 31 2004 June 30 2004 Sept 31 2004		

Paper 141b

Organizational Factors That Influence Safety

Behavioral Science Technology, Inc.
417 Bryant Circle
Ojai, CA 93023
bstojai@bstsolutions.com

Prepared for Presentation at
American Institute of Chemical Engineers
2005 Spring National Meeting
Process Plant Safety Symposium
Atlanta, Georgia
April 11-13, 2005

UNPUBLISHED

Introduction

For the past 20 years the author and his associates have worked with over 1,600 organizations to improve safety performance. In the course of that work, we have discovered that organizations optimize their efforts when they provide and support strong roles at each of three core roles: the front-line employee, supervisory and senior manager levels. These three roles, when appropriately engaged, make up a cord of three strands that is not easily broken.

Of these three areas, the leadership role has proved to be both the most impactful for safety improvement, as well as the most complex to understand and influence. One striking example of this is seen in the study of sites using the same improvement methodology. Even when compared by industry, site size, and systems, some organizations achieve steady improvement while others consistently struggle. More than any other factor, the quality of the organization's leadership, in particular its influence on the organizational culture, determined the level of success achieved.

While it is generally recognized that leadership is important to performance, the "how" of that leadership is often debated. What makes a leader effective at influencing and improving safety? What are the qualities of such a leader? What are the key tasks that a leader must perform in order to generate desired results?

While this paper will focus on these questions as it pertains to senior leaders, the principles discussed are applicable to any level in the organization.

How Leadership Influences Safety

How is it that safety leadership assures performance improvement?

The primary activity of safety initiatives, whether at the site or corporate level, is to reduce the amount of *exposure* that occurs in the workplace. While not all exposure is equal in terms of severity potential, all incidents result from exposure to hazards and reducing exposure is the primary mechanism of safety improvement. Exposure arises from management and work system problems and the pressures of the organizational climate. These lead to hazards in the workplace and worker exposure. Leaders directly and indirectly impact systems and climate, and thereby hazards and exposure.

In order to understand these connections, the following sections present findings from ongoing research that answer some of the critical questions surrounding safety leadership.

The Four Elements of Successful Safety Leadership

At its most basic, leadership is about establishing a direction and helping people to move in that direction. Practically speaking, leadership comes down to two tasks:

- Getting subordinates to do the right work the right way, and
- Maintaining a successful relationship with the people doing that work.

In many organizations, particularly in the area of safety, the two tasks appear to be at odds with each other. Oftentimes leaders believe that doing one well means sacrificing the other. While there is undeniably a tension that must be balanced between these two tasks, we have found that the leaders in organizations who perform both of these tasks well are able to maintain a healthy and appropriate balance. And these leaders are more effective at fostering high levels of safety performance.

The successful realization of this balance, both on the interpersonal level and on a level that impacts the whole organization, starts with who the leader is and what is important to him or her. It radiates out from there in how the leader goes about influencing others concerning what's important to the organization. Then it expresses itself in what things the leader does day-to-day and especially in how he or she does them. And it ends by shaping nine elements of the organization's culture and safety climate, elements that play out directly in safety performance.

Personality & Values

The leader's personality and values are at the core of who the leader is, and consequently at the core of his or her effectiveness as a safety leader. These two elements strongly influence what the leader will tend to focus on and the likelihood that the leader will favor or use the most effective influencing style for safety and the best safety leadership practices.

How the Leader Influences

Leaders differ in their style of influencing others. The influence style describes how the leader interacts with subordinates to enlist their energy in an enterprise. Research shows that a transformational leadership style is characteristic of the most effective safety leaders.

What Leaders Do

The leader enacts his or her personality, values and leadership style in daily practices. These are the day-to-day behaviors through which the leader effectively (or not) influences the vigor of the safety climate and builds strong safety systems.

Organizational Culture

The safety climate is the readily measurable, more accessible and more rapidly changed aspects of the organizational safety culture. It is the place where the leader has leverage to impact the culture, or "how we do things around here." Research indicates that there are specific aspects of the organization's culture and safety climate that are predictive of high safety performance.

Personality & Values

At the very core of who a leader is, and consequently how he or she acts and responds as a leader, are personality and values. Studies have identified a strong link between a leader's personality and many dimensions of employee performance. Personality describes the stable attributes of the person on five dimensions, known in the research literature as the *Big Five.* The Big Five dimensions are:

- **Extroversion** - involves warmth towards others, outgoingness, assertiveness, level of activity, seeking excitement, optimism and positive emotions.
- **Agreeableness -** involves trusting others, being straightforward, consideration for others, compliance with standards, modesty, and sympathy.
- **Conscientiousness** - involves a sense of competence, orderliness, sense of responsibility, need to achieve, self-discipline, and deliberateness.
- **Emotional Stability** - involves confidence in self and others, an upbeat approach to challenges, and the ability to handle stress well.
- **Openness to Experience** - involves a sense of curiosity, an exploratory approach to challenges and an imaginative mindset.

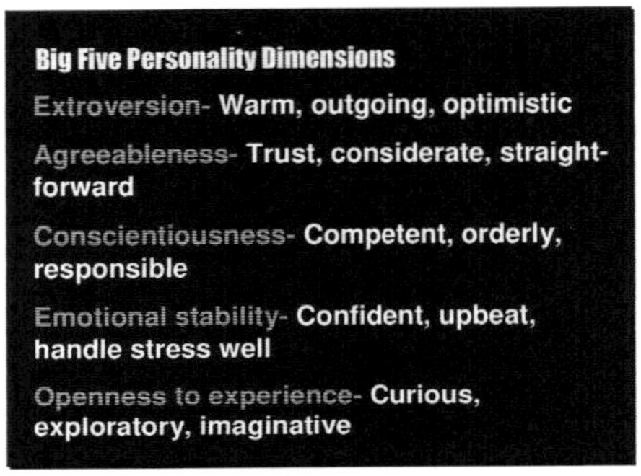

Exhibit 1. Big Five Personality Dimensions.

In addition, the leader's values need to be compatible with those required for the job. In our experience, the most successful safety leaders have a high value for service to others, problem solving and quality as well as a low tolerance for exposing others to risk. The makeup of a leader's personalities and values dispose them to their influence style, and the likelihood that they will engage in the practices identified as optimal for producing a high-performing organizational safety culture.

While a personality is well-formed by adulthood and its Big Five dimensions are stable over time, the important point is that leaders gain flexibility and breadth of options through insight into their own values and personality structure. Effective leaders understand their values and personalities and are able to develop highly effective ways for bringing their assets to the forefront and offsetting their liabilities.

How the Leader Influences

There are a number of influence styles that leaders use. The research shows that leaders who have a strong transformational leadership style typically have groups that perform better in various ways, including safety outcomes.

Transformational leadership has four dimensions. The first is **charisma**. Does the leader provide vision and a sense of mission, instill pride, gain respect and trust and increase optimism? The second is **inspiration**, which is sometimes grouped with charisma. Essentially, it defines whether the leader acts as a model, communicates a vision, sets high standards, and uses symbols to focus efforts. The third is **individual attention**. Does the leader coach, mentor, provide feedback, link the individuals' needs to the organization's mission, and provide personal attention? The fourth dimension is **intellectual challenge**. Does the leader provide subordinates with a flow of challenging new ideas aimed at rethinking old ways of doing things, challenge dysfunctional paradigms, promote rationality and careful problem solving?

Interestingly, leaders who have high levels of transformational leadership are not dependent on their bosses to place a high priority on safety. Their safety best practices are strong whether or not there is an external emphasis on safety. It is likely that this is related to who the leader is: a transformational leader is more likely to demonstrate a value for the well being of his or her subordinates, and this motivation to protect employees comes from a different—and more interior—place than organizational authority. Another finding related to transformational leadership is that its relationship to safety outcomes is entirely mediated by preventative action. In other words, it's not just the leader's influence style that matters, but also what the leader does: supervisors with strong relationships with workers (transformational style) talk and listen to them and take action about safety (preventative action) and that leads to lower injury rates.

What a Leader Does: Best Practices

What a leader does and how he or she does it is a manifestation of the leader's personality, values influence style and level of insight or understanding of his or her own personality. The leader's typical behaviors or practices in turn strongly influence the organizational safety climate and ultimately the culture.

The author and his colleagues have identified several best safety leadership practices. In our work we noted that certain management and leadership practices recur across those organizations with high-performing cultures. Comparing that experience with existing literature on leadership influences on safety and organizational culture the author and his colleagues came to the conclusion that there are at least eight definable leadership practices connected to the development and support of a high-performing safety culture. These are Vision, Credibility, Collaboration, Feedback and Recognition, Accountability, Communication, Values Safety, and Action Oriented.

Vision – The effective leader has identified the strategic role of safety performance to his or her organization, is able to "see" what such performance excellence would look like, and can convey that vision in a compelling way to the organization.

Credibility – The effective leader is credible to other people in the organization, is willing to admit his or her mistakes to others, "goes to bat" for direct reports and the interests of the group, and gives honest information about safety even it if is not well received.

Collaboration – The effective leader works well with other people, promotes cooperation and collaboration in safety, actively seeks input from people on issues that affect them, and encourages others to implement their decisions and solutions for improving safety.

Feedback & Recognition – The effective leader is good at providing feedback and recognizing people for their accomplishments. This person publicly recognizes the contributions of others, uses praise more often than criticism, gives positive feedback and recognition for good performance, and finds ways to celebrate accomplishments in safety.

Accountability – The effective leader gives people a fair appraisal of the efforts and results in safety, clearly communicates people's roles in the safety effort, and fosters the sense that every person is responsible for the level of safety in their organizational unit.

Communication – The effective leader is a great communicator. He or she encourages people to give honest and complete information about safety even if the information is unfavorable. This leader keeps people informed about the big picture in safety, and communicates frequently and effectively up, down, and across the organization

Action-Orientation – The effective leader is proactive rather than reactive in addressing safety issues. This leader gives timely, considered responses for safety concerns, demonstrates a sense of personal urgency and energy to achieve safety results, and demonstrates a performance-driven focus by delivering results with speed and excellence.

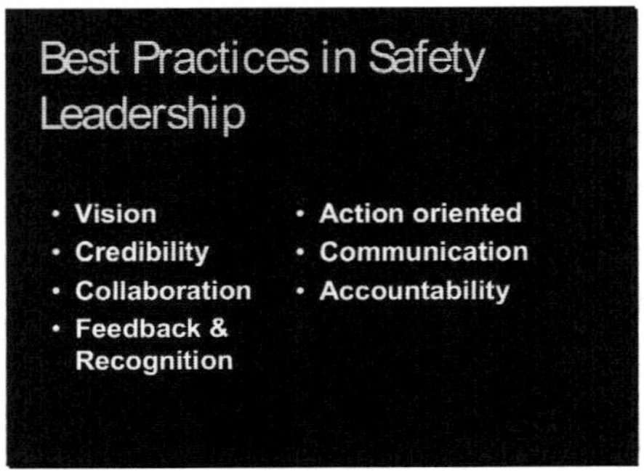

Exhibit 2. Best practices in Safety Leadership.

The Organizational Culture and Safety Climate the Leader Creates

Our experience working with organizations interested in performance improvement, and subsequent ongoing research, have highlighted the fact that far from being an ambiguous facet of organizational life, organizational culture and safety climate are definable and measurable in very practical terms. High-performing organizations consistently show high trust, good communication, management credibility, and organizational value for safety. Low-performing organizations tend to show the opposite. We have identified nine characteristics of organizational culture that are predictive of successful safety outcomes. These are:

1. Teamwork–the effectiveness of workgroups in meeting targets and deadlines. It is clear how the leadership best practice of collaboration and feedback can build teamwork. Transformational leadership can build teamwork when the transformational leader directs attention to the team and uses it to develop and implement safety solutions.

2. Workgroup relations–the degree to which coworkers respect each other. Transformational leadership instills pride, gains respect and trust, increases optimism among members of a workgroup and collaborates practices increasing the likelihood that team members will reciprocate these things with each other.

3. Procedural justice–the level that workers rate the fairness of first-level supervisors. When transformational leadership is strong in a leader, the individual attention is shared across subordinates, not focused on particular favorites, which supports this procedural justice. But procedural justice also requires action-oriented practices: systems and procedures need to be aligned to visibly support fairness.

4. Perceived organizational support–the level to which employees feel the organization is concerned for their overall well being. This factor in particular requires the transformational style and follow-up action oriented practices. Part of the definition of transformational leadership is individual attention, which influences Perceived Organizational Support only when the leader actually does things to demonstrate the concern for employees well being and interests.

5. Leader-member exchange–the strength of the two-way relationship that workers feel they have with their supervisors. Strong perceptions of leader-member exchange require that the leader follow-up his or her individual attention, intellectual challenge, and inspiring words with behaviors and practices, for example actually going to bat for employees when needed. This requires not only action oriented practices, but those needed for good communications and credibility.

6. Management credibility–the perception of consistency and fairness of management in dealing with workers. Part of the definition of transformational leadership is that the leader gains respect and trust of employees, and acts as a model. This dovetails with leadership best practices needed for credibility. When the majority of leaders are successful in this, overall management credibility will be improved.

7. Organizational value for safety–this is often called 'safety climate', the perceived level of the organization's commitment to safety. Interestingly, this factor matters less if the leader has the right values. This means that even when the organization itself is not supporting safety at a high level, the individual transformational leader can effectively support it. Conversely, if the leader's style does not include strong transformational elements, the leader is more dependent on the organization's support for safety performance.

8. Upward communication–the adequacy of upward messages about safety. Listening to subordinates is an action that needs to accompany transformational leadership for it to be effective. But when both are in place, upward communication is positively impacted.

9. Approaching others–the probability that workers will speak to each other about safety issues. In a feedback-rich environment, modeled by the leader, people are more likely to approach each other.

Companies with cultures that are high in these nine characteristics tend to be more successful in initiatives they undertake, generate change more rapidly, and achieve higher performance overall in critical business functions than companies that score low on these factors.

Conclusion

Organizational culture and safety climate are strongly influenced by the quality and type of leadership enacted throughout the organization and this in turn impacts safety performance. Leaders can gain insight into their own values and personality structures and how these play out in their influence style and practices. Understanding how these factors interact enables leaders to use their own behavior as a central leverage point for safety. It shows them what to do to build strong safety performance.

Process Safety Indicators

Paper by:
Ian Travers
Viki Beckett
Health and Safety Executive
Hazardous Installations Directorate
Daniel House, Trinity Road, Bootle, Merseyside, UK. L20 3TW
Ian.travers@hse.gsi.gov.uk Viki.Beckett@hse.gsi.gov.uk

Abstract

HSE has been working closely with chemical and major hazard industries in the UK to develop and implement performance measurement systems for process safety risks. The aim is to embed process safety indicators within safety management systems. Currently there is over reliance on loss time incidents as a means of highlighting a failing system. Personal accidents are rarely caused by the same factors as loss of containment incidents or major accidents. Drawn from best practice within industry, this paper outlines a six-stage process which can be used by companies to develop and tailor leading and lagging indicators covering process risks. To supplement this work HSE and industry have been collaborating to collect and analyse causation data and lessons learnt from loss of containment incidents.

From this work we are hoping that eventually we can identify a small set of process safety indicators which should feature in all chemical processing and major hazard industries.

Background

Process safety indicators can provide ongoing assurance that risks are being adequately controlled. Senior Executives need to monitor the effectiveness of internal controls against business risks. For major hazard installations and chemical manufactures, process safety risks will be a significant aspect of business risk, asset integrity and reputation. Good corporate governance requires systems and arrangements to provide a high level of assurance. However, many organisations do not have good information to show how well they are managing major hazard risks. Discovering weaknesses in control systems by having a major incident is too late and too costly. Knowing that process risks are effectively controlled has a clear link with business efficiency as several indicators can be used to show plant availability and optimise operating conditions.

Measuring and monitoring is an essential part of any safety management system. Employers and operators of major hazard establishments are legally required to have such systems in place. The aim of this initiative is for a greater focus on checking the effectiveness of control of process safety risks. Process safety audits provide a periodic check on the completeness and operation of systems but are often too infrequent to detect deterioration in critical control systems. Carefully selected and targeted performance measurement using a small range of process safety indicators will provide ongoing information on the current status of critical controls. Suitable leading indicators will highlight system deterioration in time to take corrective action whereas lagging indicators show whether the desired outcome is being achieved.

To meet societal concerns about the industry, chemical and major hazard companies collectively need to demonstrate risks are being adequately controlled. The industry is often judged by the worst performer or against the last major incident to gain public attention. Developing process safety indicators will build and maintain trust with stakeholders and contribute towards Sustainable Development Programmes.

The incidence of major incidents and in particular, loss of containment dangerous occurrences[1], in the UK has not improved in recent years. There are approximately 250 process-related serious incidents each year in the chemicals sector. Despite the introduction of the Control of Major Accident Hazards Regulations, 1999, COMAH[2], there are approximately 170 such incidents at COMAH establishments each year.

The report on the BP Grangemouth[3] incidents in 2000 highlighted weaknesses in safety management systems, in that incorrect assumptions about major hazard performance were drawn from measuring lost time accident frequencies. The conclusions from other inquiries into major accidents in the chemical sector have also shown that operators frequently measure the wrong issues and fail to focus on indicators of major hazard performance[4/5].

Pilot Study

To promote the concept of process safety measurement and explore the current knowledge in the UK, HSE and a small group of chemical companies based in Scotland undertook a pilot study. The study concluded that it was too early to formulate a core set of indicators for the whole sector and that given the current knowledge and experience in the UK most benefit was to be gained by focusing on indicators relating to specific process risks on site or within an individual organisation. The study found that there was no accepted system or method companies could follow to develop indicators. The involvement of the site management team in selecting indicators was seen as essential as it gave an opportunity to consider the desired outcomes of the process safety management system. The main conclusion was that we should seek to establish a best practice in the development and implementation of indicators and to share examples and ideas throughout the sector.

The study also piloted the collection of information on underlying causes of loss of containment incidents and sharing the lessons learnt. This information is provided to HSE on a voluntary basis and the anonymised results are published on a restricted access website[6].

Following on from the pilot HSE and industry have been working together researching into best practice followed by companies who have successfully implemented a process safety measurement system. The Chemical Industries Association, CIA and UK Petroleum Industry Association, UKPIA, have adopted process safety performance measurement as key initiatives within Responsible Care and Sustainable Development Programmes.

Best Practice Guidance

This close working between HSE and industry has culminated in a draft 'Step by Step Guide to Developing Process Safety Performance Indicators'. This has just been published for consultation[7].

The 6-stage process is outlined in Diagram 1, **Annex 1**. It is presumed that companies using the Guide already have a process safety management system in place; the emphasis is to provide the means to check whether the risk controls are effective. For organisations that already have process safety indicators it is not suggested that these should be abandoned and the process started over from new. Such enterprises can use the Guide to review their system to see if any important aspects have been overlooked.

Step 1: Establish the Organisation

1.1 Appoint a Steward or Champion - Developing and implementing process safety indicators is frequently a new area of work within an organisation and the need for such a system will not be immediately obvious to many people. Someone within the organisation will need to promote the idea and drive forward the initiative. This may include making a business case for the adoption of performance indicators and securing suitable resources.

1.2 Consider deploying an Implementation Team - In large organisations there will be too much for one person to deal with and it is often more appropriate to form a team to manage the introduction of process safety indicators. This has the benefit of drawing in people from a range of business operations providing the opportunity for pooling of ideas. A steering committee may also be helpful to oversee the implementation programme and to check the indicators match current business priorities.

1.3 Secure Senior Management Involvement - It is vital that corporate presidents, vice presidents and executive managers are committed to adopting meaningful indicators. They have ultimate responsibility for the control of risk and are therefore the main customer of the enhanced information on the control of risk. It is important that executive managers agree that the indicators chosen provide them with the right scope and level of information they require to be satisfied that process safety risks are under control.

Step 2: Decide on the Scope

Setting the scope is about selecting the right indicators to provide just enough information about the adequacy of process safety controls. Busy management teams will quickly loose interest in an extensive raft of indicators and so it is essential to avoid KPI overload. Collection and analysis of data is resource intensive and so the arrangements for monitoring performance have to be cost effective. Even for the largest organisations in the order of 15 to 20 process safety indicators will be sufficient to provide a high degree of assurance across the whole business.

An important consideration is to determine the level at which indicators are to be set, e.g. to measure the performance of:

- a plant or installation
- the whole site,
- the organisation, or
- a combination of these.

To enable a comprehensive picture of performance across an entire enterprise indicators set at differing levels should be compatible with each other. For large or complex sites such as refineries, it is common practice to have a hierarchy of indicators within an organisation where a small selection of key indicators areg collated and reported up to the next management level. This assists benchmarking across an organisation whilst keeping local management information within reasonable limits.

Selecting the Right Indicators – The aim is to select the right indicators to show that the key risk controls are working as designed. The first consideration should therefore be to identify safety critical activities. Focus on the absolutely essential aspects of process safety. As measurement should aim to highlight system deterioration, ensure that ongoing activities and systems are covered first. Think of these as systems, procedures or actions that have to operate correctly and as designed on each and every occasion in order to prevent an incident.

Step 3: Identify the Outcomes and Systems to Deliver Them

After deciding on the scope of the indicators, the main process safety risks and areas where greater assurance is needed, decide what success looks like in these areas. For example, if staff competence is a key area decide what 'competence' means. It may be that staff have all attended induction training, have read and understood any relevant procedures, are aware of the main health and safety risks attached to the process they are carrying out. If corrosion of pipe work is a main process safety risk, then success could be 'no loss of containment incidents due to pipe work corrosion'.

It is essential to identify the outcome and the aspect of the process safety management system aimed at delivering that outcome. Describe the control systems in terms of 'success' otherwise it will be impossible to identify the right indicators to show the desired outcome is being achieved. The outcome should link to the lagging indicators set in Step 4. Organisations will already have key process control systems in place. Setting indicators provides an opportunity to review the suitability and priority given to each aspect of control.

Step 4: Determine the Indicators

Once the preparation has been completed and the areas for measurement identified, the indicators can be determined. For each aspect of the process safety management system to be measured, at least one lagging (outcome) and one leading (activity) indicator should be developed. Leading indicators highlight the effectiveness of the risk control systems in place to deliver the outcome whereas; lagging indicators show whether the outcome has actually been achieved.

Set the Tolerance/Target - A tolerance should be set for each leading indicator. This represents the point at which deviation from the outcome should be flagged up to the management team for attention. For example, for the indicator: "Percentage of overdue safety critical maintenance actions beyond the planned maintenance interval " a company may decide that 100% of actions must be completed to schedule and therefore set a zero tolerance. Another may tolerate slippage of up to 10% beyond the scheduled interval but that beyond this should be brought to the attention of the management team. Lagging indicators don't normally have a set tolerance but are treated as a target and trended against previous figures.

The management team should set the tolerance, not the person responsible for the activity. This enables management to decide whether to intervene when performance has deviated beyond an acceptable level.

Step 5: Establish the Data Collection and Reporting Systems

Once the indicator has been selected and the tolerance set, it is important to ensure that the relevant information is readily available within the organisation. Experience has shown that information and data required to establish process safety indicators are usually already available and collected for other purposes e.g. for quality control or business efficiency. However, it may be necessary to collate, analyse and organise the data differently to ensure it supports the suite of indicators chosen.

Co-ordinate the performance data through one person who will be responsible for collecting all the information, compiling reports for the management team, and raising the alarm if there are any areas of concern requiring immediate action.

Presentation - It is best to keep the presentation of performance data as simple as possible – summarised in a single sheet. It is important to clearly show any deviations from set tolerances or targets and important trends. It is essential the senior management team routinely see key performance information as they are the main customer for this information and will need to make decisions on corrective action.

Deviations from Tolerances - Deviations from tolerances **must** be followed up otherwise there is little point in collecting the information in the first place. The main aim of a performance information system is to indicate where process control management systems have deteriorated or are not delivering the intended outcome.

Step 6: Review

Periodic Review - Indicators should not be developed and then forgotten. Periodically, the scope of the full set of indicators needs to be reviewed to ensure indicators still reflect the main process risks. Indicators may need to be changed because of:

- Introduction of new, high-risk processes,
- Improvement programmes,
- Alteration in plant design,
- Reduction of staff/loss of competence in certain areas.

If reviews are not carried out, the process of using KPIs may become meaningless and the information collected may not give the necessary assurance to senior managers that the major hazard risks are under control.

Review of Leading Measures Against Lagging Measures - There should be a strong link between the leading and lagging measures and the process control in place to prevent an incident.

- *You're measuring in the wrong place!*
  If performance is poor against a group of leading indicators but the associated lagging indicator is satisfactory, it is likely that the leading indicator selected is too far removed from the critical control measure that delivers or maintains the desired outcome. For instance, competence of operators may be measured whereas, more importantly the completion of a particular operation in a set way may be more critical to ensuring the safety of a specific activity.

- *You are doing what you intended but it's the wrong thing!*
  If a group of leading indicators are on target and closely linked to the process control system but the associated lagging indicator shows poor performance, it is likely that the process control system is ineffective in delivering the desired outcome.

Examples of Indicators

Annex 2 gives some examples of process safety indicators derived using the Step by Step Guide. The aim is to have leading and lagging indicators covering the full scope of a process safety management system.

Lagging Indicators

The lagging indicator should directly link to the outcome as this provides the means by which achievement of the outcome is monitored as shown in Diagram 2, Annex 2. Lagging indicators are usually expressed in absolute terms i.e. number of events, as even where tolerances or targets have been set against the lagging indicator each event signals a failure to meet the desired outcome. There should be internal reporting and investigation of each such event to discover the underlying cause and to apply learning across the whole enterprise. The definition of an 'incident' will require careful consideration in order to capture all significant events, especially where a loss of containment incident definition is based upon the quantity of material lost. Better definitions are consequence-based, either for the actual or potential consequence. However, as the aim of such measurement systems is to identify weakness in process safety management, a risk-based approach is to base the definition of an incident on the significance of the management system failure that gave rise to the incident. This will help identify weakness in common management systems where there same controls are applied elsewhere to prevent a major incident, even when the circumstances of the actual incident had no potential or real impact on process safety.

Leading Indicator

Leading indicators should be set as close as possible to the critical controls that deliver the outcome. Where several systems are as seen as critical, then a separate indicator should be set for each. Leading indicators should always be framed in terms of degree of compliance with the tolerance set by the management team. This will simplify reporting and presentation of the information to the management team, for instance by using traffic lights and help focus attention to areas where remedial action is warranted.

Safety Critical?

Frequently indicators are set for a whole range of similar activities or tasks undertaken on site e.g. 'percentage of overdue maintenance actions'. Such broadly based indicators are likely to mask the significance of the overdue action and hinder management decisions about where to intervene to take corrective action. A risk-based approach is to focus on safety critical issues. This will require careful consideration of the items of plant, equipment as well as activities and tasks that are critical to delivering the outcome of the process management system.

Sector Indicators

One of the original aims of our pilot study was to explore the options for a national set of process safety indicators for the UK chemical and major hazard industries. At the time this proved too ambitious. It is clear from our subsequent work that the indicators that suit an individual organisation do not automatically translate into a range of common indicators for the whole sector. Organisational indicators must serve the direct business priorities for that enterprise and these can (and should) change over time whereas sector measures need to be fixed for a longer period to allow for benchmarking and analysis of trends. A further difference is the frequency at which the measurement should take place. For organisational indicators measurement should occur frequently throughout the year – preferably on a monthly basis. Sector indicators require updating annual or at the most bi-annually.

As an interim approach to developing a range of sector indicators we have focused on a very small set of sector leading indicators that show the extent to which operators:

- have adopted process safety performance indicators, and
- whether these cover all key process safety risks, and
- have been adopted according to a systematic approach.

We can combine these with the main lagging indicator:

- the number of major accidents and process safety related dangerous occurrences in the UK.

Core Indicators

Our work has shown that there is a consensus of common or core leading indicators that feature within most company process safety measurement systems. The details of each naturally varies from organisation to organisation to reflect business priorities but the scope encompass the following elements of a process safety management system:

- Hazard analysis and risk assessment / Process reviews

- Staff training / competence / resourcing,
- Safety communication / awareness
- Plant design / plant change
- Process controls / operating procedures
- Planned inspection / maintenance
- Emergency arrangements
- Incident reporting and investigation
- Monitoring / auditing / review, and
- Regulatory compliance.

Therefore, our expectation is that, as performance measurement becomes part of the fabric of process safety management and companies move to a more systematic approach based on outcomes, then a common format of leading and lagging indicators will develop. Moving to an agreed set of sector indicators would then follow as the next logical step.

Causation of Loss of Containment Incidents

To supplement the work on performance indicators we have established a scheme of voluntary reporting of information on causation and lessons learnt from loss of containment incidents. This information is sent to HSE for analysis and is additional to the information provided within formal notifications required by statute. To prevent the adverse use of this data by regulatory inspectors and the risk of disclosure of the company identity we have established a series of safeguards to protect the identity of the organisation submitting the voluntary information.

We use a set framework of causation rather than a free text description of the incident. Over time this will help identify trends and common failings in management systems. The real advantage of this system is that the analysis is done in real time rather than as a retrospective review and that the data and lessons learnt are published quickly[6].

The scheme has been running for almost one year and is already revealing interesting findings i.e. most incidents occur during normal operations and the main failing seem to have a strong element of human factors as an underlying cause. Overtime this analysis should help pinpoint common weakness in process controls and lead to better-targeted programmes to deal with these issues.

Sharing Information and Best Practice

One of the main features of this programme has been the sharing of information and experience amongst operators. To support this we have provided a restricted access web site for safety professionals within the major hazard and chemical industries. To participate in the programme contact: chemicalindustries.pspm@webcommunities.hse.gov.uk The information covered in this paper is routinely posted on the web site. For more information see:
http://webcommunities.hse.gov.uk/ui/inovem.ti/group/chemicalindustries.pspm/grouphome

Conclusion

By close collaboration with industry we have been able to establish a systematic approach to the development of process safety performance indicators. This provides the UK chemical and major hazard industries with the means to gain a high degree of assurance on the management of business risks. These systems contribute directly to good corporate governance and therefore underpin sustainable development.

Sharing information within a secure environment has been a fundamental aspect of this programme. The collective benefits of this approach have overcome initial concerns and sensitivities about more openness with potential competitors and the regulator.

Annex 1:
Diagram 1: Step-by-Step Guide to Setting Performance Indicators

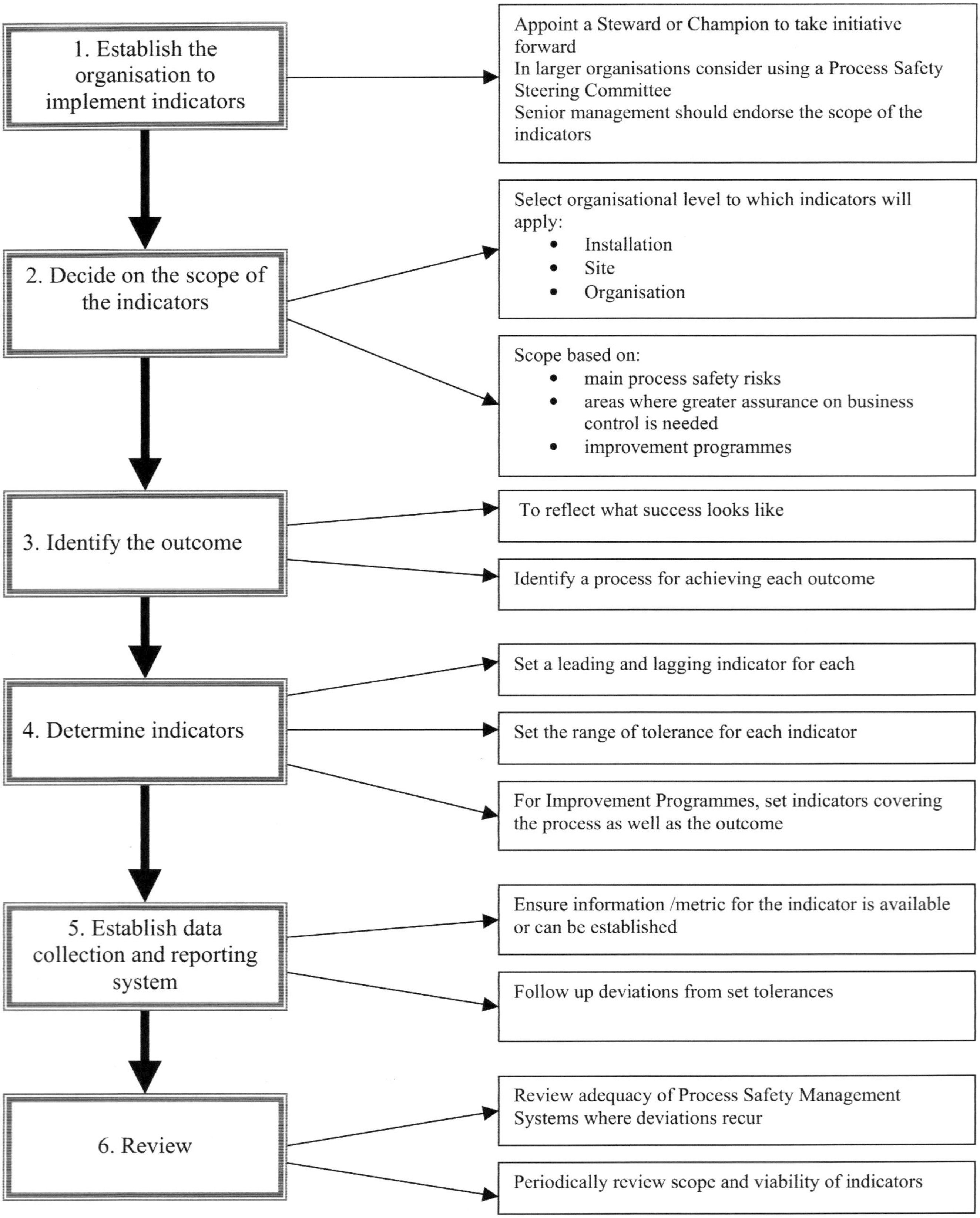

Annex 2: Examples of Performance Indicators

The following examples show how 'Step 3: Identify the Outcome', and 'Step 4: Determine the Indicator', have been applied to some elements of Plant Integrity. For any organisation the aim is to have indicators covering the full scope of plant integrity utilising, wherever possible, a combination of both leading and lagging indicators.

The key stages, as illustrated in Diagram 2, are:

1. Decide on the Outcome – in terms of success for the system,
2. Determine a Leading Indicator – to clearly show when the outcome is not being achieved,
3. Identify the Process Controls that directly deliver the outcome,
4. For each key control system set a leading indicator.

Diagram 2: Linking Indicators to the Outcome and Process Controls

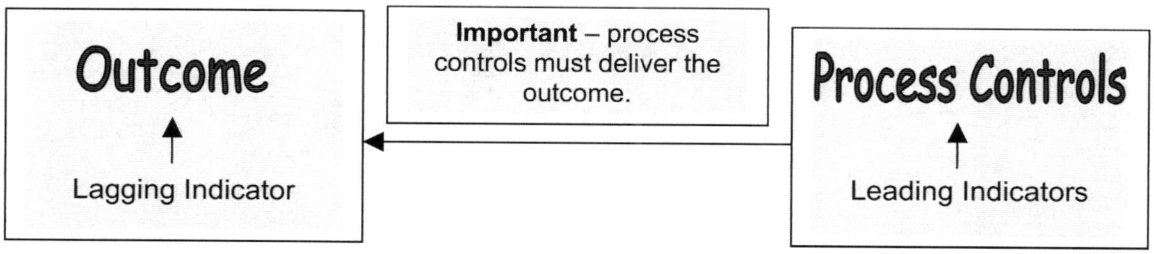

Plant Design

Outcome: Plant operation optimised with equipment running efficiently and reliably with no unexpected breakdown due to deficiencies in the plant design and specification.

Lagging Indicator: Numbers of incidents involving loss of containment of hazardous material or failure of safety critical plant/equipment where deficiency in plant design was found to be a contributory factor.

Process Control: Plant Design – including systems to ensure original design and specification as well as a periodic assessment of plant and equipment against current standards, codes and good practice.

Leading Indicators: **On commissioning** - percentage of safety critical items of plant or equipment that comply with specified design standards.

On a periodic basis - percentage of safety critical items of plant or equipment that comply with current design standards or codes.

Plant Change/Modifications (temporary/permanent)

Outcome: Following plant change/modification, the plant continues to operate in an optimised state, with equipment running efficiently and reliably and without unexpected breakdown.

Lagging Indicator: Numbers of incidents involving loss of containment of hazardous material or failure of safety critical plant/equipment where deficiencies in plant change was found to be a contributory factor.

Process Control: Plant change – including scope, hazard assessment, approval, implementation, post commissioning checks, updating operation procedures and records.

Leading Indicators:
Note: There is no need to constantly measure the scope, this should be a one off initial measure.

Percentage of HAZOP actions completed to schedule.

Percentage of post commissioning checks carried out within specified timescale.

Percentage of plant change authorisations completed retrospectively.

Planned Inspection/Maintenance

Outcome: No unexpected failure of safety critical plant/equipment due to corrosion, wear and damage.

Lagging Indicator: Numbers of incidents involving loss of containment of hazardous material or failure of safety critical plant/equipment where corrosion, wear and damage was found to be a contributory factor.

Process Control: Planned Inspection system
Planned Maintenance system

Leading Indicators: Percentage of at risk (from corrosion, wear or damage) safety critical plant/equipment inspected within the planned timescale.

Note: One aspect, e.g. corrosion, may be the key risk on the plant, so it may be better to separate out the leading and lagging measures to give a greater focus on that specific issue.

Instrumentation/Alarms

Outcome: All safety critical instrumentation and alarms are in good working order to correctly indicate when process conditions exceed safe operating limits.

Lagging Indicator: Percentage of safety critical instrumentation/alarms that fail to operate at the designed level/circumstances - whist in use or during testing.

Process Control: Planned plant inspection/testing.
Planned maintenance.

Leading Indicators: Percentage of safety critical instrumentation/alarm checks completed to schedule.

Percentage of maintenance actions to deal with defective instrumentation/ alarms completed to schedule.

References

1. Reporting of Injuries, Diseases and Dangerous Occurrences Regulations 1995 (RIDDOR), SI 1995 / 3163.

2. The Control of Major Hazard Regulations 1999 (COMAH), SI 1999 / 3246. Implement the Seveso II Directive (96/82/EC) in Great Britain.

3. Major Incident Report BP Grangemouth Scotland: 29th May to 10th June 2000, A Public Report Prepared by the HSE on behalf of the Competent Authority, http://www.hse.gov.uk/comah/bpgrange/index.htm

4. The Lessons from Longford: the Esso gas plant explosion. Andrew Hopkins. CHH Australia, 2000, Sydney, NSW. ISBN 1 86468 422 4.

5. Large Property Damage Losses in the Hydrocarbon-Chemical Industries – A Thirty-Year Review, MARSH Risk Consulting, 2001. www.marsh.com.

6. http://webcommunities.hse.gov.uk/ui/inovem.ti/group/chemicalindustries.pspm/grouphome

7. Step by Step Guide to Developing Process Safety Indicators: Draft for Consultation, HSE, 2004. http://webcommunities.hse.gov.uk/ui/inovem.ti/group/chemicalindustries.pspm/grouphome

Continuously Improving PSM Effectiveness - A Practical Roadmap

Steve Arendt
Vice President
ABS Consulting
10301 Technology Drive
Knoxville, TN 37932
(865) 671-5812
sarendt@absconsulting.com

PSM practices and formal management systems have been in place in many companies for more than 20 years. PSM is widely credited for perceived reductions in major accident risk and in improved chemical industry performance. Nevertheless, many companies continue to be challenged by resource pressures, inadequate management system audit results, and stagnant process safety incident performance.

While most companies claim to advocate Continuous Improvement, few have actually put substance into these words and succeeded in practice. This is partly because there is no clear vision of what PSM performance and continuous improvement are. This paper defines the following function:

PSM Effectiveness = f [PSM performance + PSM efficiency]

Continuous improvement can address performance issues or efficiency issues or both. Companies are seeking new ways to improve process safety management activities based on the following strategies:

- Doing less PSM work, based on risk judgments
- Doing the same PSM activities, but more efficiently
- Using the same resources, use better practices to generate better results
- Getting better PSM results, but with fewer resources
- Extending existing PSM practices into new areas
- Extending existing PSM practices throughout the life-cycle
- Adding new PSM activities to existing PSM elements
- Creating new PSM elements
- Restructuring the PSM system

This paper defines these terms and gives case study examples and lessons of ways that companies are trying to improve performance, efficiency, and effectiveness using the above strategies.

Managing Catastrophic Risk:
Quantitative Methods that Provide Insight and Decision Support

Scott Schiller
ConocoPhillips Company

Jatin Shah
Baker Engineering and Risk Consultants, Inc.

Abstract

Risk exposures often involve potential economic issues that can result in "business interruption" from damage to critical assets or public pressure that can alter or shut down an operation. Strong emotions from stakeholders further complicate risk exposures that involve safety impacts affecting a company's employees and contractors, as well as the general public. Furthermore, risk exposures that have high-impact potential (catastrophic) create another layer of complexity as some stakeholders' lose sight of the frequency side of the risk equation. Management teams are forced to toe a fine line in order to balance the needs of the various stakeholders and economically operate their facilities.

Fortunately, quantitative methods are available to help the risk professional evaluate the complex interactions that make many risk exposures difficult to evaluate using just qualitative methods. The more complicated risk exposures, which are often comprised of the high or catastrophic impacts coupled with low to very low frequency, often require the use of quantitative analytical methodologies like Layer of Protection Analysis (LOPA) and/or Quantitative Risk Assessment (QRA) methodology. In addition to providing a better understanding of the risk exposure, QRA provides an objective means to review and compare the relative benefits of potential risk reduction alternatives and forms the basis for economic comparison analysis between such risk reduction alternatives. Incremental risk analysis is used when one needs to clearly show the point of diminishing return when multiple options are available to address the risk exposure. The objective basis of quantitative techniques and the use of relative comparisons provide key insights needed to make decisions around the management of catastrophic risk exposures.

Case studies are provided to illustrate how quantitative analysis techniques are utilized to evaluate risk exposures from "as-is" operations and how these existing risk exposures are modified by potential risk mitigation alternatives. A case study also illustrates how augmenting the QRA study by utilizing incremental risk analysis provided additional insight for the facility decision-makers by making available to them comparison graphics and the support data which depict the point of diminishing returns relative to the viable risk reduction alternatives under consideration.

Introduction

Making decisions consciously or unconsciously, with good or bad consequences, represents the tool all of us use in facing the opportunities, the challenges, and the uncertainties of life. Some decisions are fairly obvious – "no-brainers", but no-brainers are the exceptions. Most important decisions are tough and complex with no easy or obvious solutions. According to Hammond, Kenny, and Raiffa, "the ability to make smart choices is a fundamental life skill."[1]

Difficult decisions involve high stakes and serious consequences often involving numerous and complex considerations leading to a great deal of discomfort. There is a concern when people react by making the decisions too quickly, too slowly, or too arbitrarily — often resulting in a mediocre choice dependent upon luck for success. Many organizations help their managers make smarter decisions by utilizing an effective decision making process that:

- Focuses on what's important
- Is logical and consistent
- Acknowledges both subjective and objective factors
- Blends analytical and intuitive thinking
- Requires only as much information and analysis needed to resolve a dilemma
- Is straightforward, reliable, easy to use, and flexible

Most of the decisions faced by owners and operators of facilities in the chemical, oil and gas industries are difficult and complex, potentially burdening managers with difficult decision-making dilemmas. Some of the most challenging involve EHS risk issues. The problem does not go away if put in a pile for further consideration. The solutions do not get less expensive if put into a pending file. The issue does not become less of a liability if allowed to age. Some examples of these types of dilemmas are:

- *Dilemma #1* Given the upcoming maintenance shutdown of one of the facility's most hazardous units, a manager wants to take the opportunity to further reduce the overall risk on the unit. A list of potential risk reduction options were developed by a team based upon various suggestions and industry developments.
  What should be done?
  How much will it cost?
  What will be the residual risk?

- *Dilemma #2* A business management team is exploring an opportunity to expand into a major market for their existing product.
  What are the risk exposures?
  Can we manage the additional risk exposure?
  Does the business opportunity offset the risk exposure?

- *Dilemma #3* Another manager is trying to manage an outstanding Process Hazards Analysis (PHA) action item, which has identified the need for an additional shutdown to be added to an existing process. The additional shutdown will require a unit shutdown and modification to a thick-walled pressure vessel.

What is a satisfactory level of risk?
Will the action item reduce risk to that level?
What is the value of the risk reduction proposal?

Critical issues that hold up the decision-making process are the lack of understanding of the problem, lack of sufficient data, or data that contains information contradictory to the decision-making process. Many risk issues have multiple drivers often not realized by stakeholders, and the proposed solutions have inherent trade-offs requiring more objective data that managers can use as justification for the decisions that they make.

Risk management professionals at many organizations have a basket of tools ranging from qualitative to quantitative techniques that they can utilize to ensure a logical and consistent decision-making process. While qualitative tools help them identify and focus on what's important and often resolve the simpler decisions, semi-quantitative and even quantitative tools are brought to bear for more difficult and complex decisions. Sometimes qualitative techniques create more questions than they shed light on, but in most cases these techniques do help separate the wheat from the chaff and allow one to focus on the more important issues at hand. However, to obtain the funding that one may need, more quantitative techniques are often needed to provide the justification required by the organizational approval processes.

While the more quantitative tools provide additional information and insight to help support an informed decision, they do require additional resources (financial and time). Sufficient details are needed to meet the information requirements in the management systems (process) you operate under. The key to a good decision-making process is to follow the axiom of utilizing only as much information and analysis needed to make the decision.

This paper provides an overview of the risk analysis methodologies and how they can be utilized in an effective decision-making process.

Risk Analysis Techniques

Qualitative Analysis

Hazard Identification

Hazard identification involves any number of prescribed "qualitative" techniques or methodologies that most commonly develop into hazard situations (scenarios). A thorough understanding of work activities, processes, and products is required to properly utilize hazard identification techniques. Opportunities for hazard identification include incident and near-miss investigations, process hazard analyses, job planning, job safety analyses, audits, or any other activity where the identification of hazards is desirable.

The hazard identification process varies based on the size and complexity of the activity or process and the magnitude of potential hazards. It is recognized that the hazard associated with more simple activities or processes can be identified using industry popular techniques such as What If Brainstorming, Job Safety Analysis (JSA),

Checklists, and What If Checklists, while other more complex activities or processes may need to use more intricate techniques such as (but not limited to) HAZOP or FEMA [2].

Risk Analysis

Hazard scenarios can be displayed as the two risk components of likelihood and consequence. From these risk components, the qualitative value or measure of risk is determined. The Risk Matrix can be used to rank the risks posed by each of the identified hazard scenarios. This is accomplished by:

- **Determining the Consequence** of the hazard scenario: Using the Consequence table, identify the associated consequence severity for each scenario. This is done by applying a qualitative tabulation or "scale" of consequence severities from which a choice is made. The consequence value should be assigned without any existing safeguards (except those existing "passive" safeguards).
- **Determining the Likelihood** of the event: Using the descriptions in the left column, determine which best fits the hazard scenario being evaluated. Take into account whatever known existing mitigations and systems would tend to decrease the chance of occurrence.
- **Determining the Risk**: On the Risk Matrix, plot the Consequence number and the Likelihood number to obtain the risk ranking. The risk Equation is Risk = Severity x Likelihood. The Risk Equation can be used to obtain the numeric risk value for a particular scenario (see Figure 1 below).

Risk Categories

20 to 25	**High**
10 to 16	**Significant**
5 to 9	**Medium**
1 to 4	**Low**

Risk Matrix

Likelihood

Severity / Frequency		Negligible	Minor	Significant	Major	Catastrophic
Frequent	5	*5*	*10*	*15*	*20*	*25*
Probable	4	*4*	*8*	*12*	*16*	*20*
Rare	3	*3*	*6*	*9*	*12*	*15*
Remote	2	*2*	*4*	*6*	*8*	*10*
Improbable	1	*1*	*2*	*3*	*4*	*5*
		1	**2**	**3**	**4**	**5**

Consequence

Figure 1. Qualitative Risk Matrix Model Example

Typically, such analyses are done in a "team" environment and there may be some variation in ranking between one evaluation team and another. This is primarily due to each team's basic understanding of the risk matrix methodology and its application, particularly with respect to the assigning of likelihood. However, this approach does allow for the generation of a "risk inventory" for each activity or process within an organization. The relative importance of each postulated scenario can then be determined by comparing the risks with other identified scenarios, which will identify the most important risks to be addressed and promote the most efficient use of loss prevention resources.

Describing the scenarios of interest from the above hazard identification techniques in terms of risk allows the risk professional to better categorize and initially start to manage the hazardous scenarios that are generated. Even with little scenario detail, this qualitative assessment process can provide the following:

- Ranking the degree of risk to individual events
- Identifying areas or activities of potential concern such as personnel safety, environmental impacts, and business & asset loss issues
- Prioritizing actions required based on the level of risk presented.

It is important to remember that the intent is to generate an inventory of risk scenarios with sufficient prioritization to identify the issues that need further attention.

Qualitative hazard identification and risk assessment involving the use of a risk matrix are all covered in various hazard risk industry reference literature (books, periodicals, and the Internet). An example reference would be The American Institute of Chemical Engineers (AIChE) Center for Chemical Process Safety (CCPS) publication "Guidelines for Hazard Evaluation Procedures, 2nd Edition with Worked Examples" [2].

When stakeholders are confronted with the management of a number of risk issues, the initial intent would typically be to identify the risks that will cause the most impact to the business and attempt to prioritize the risk data. Here, a qualitative risk assessment is the right initial choice. Qualitative approaches afford the most efficient means of assimilating all of the risks so that the issues of interest can be screened out and identified.

The success of the risk identification process dictates the effectiveness of a Risk Management Program. What one doesn't know, one cannot manage.

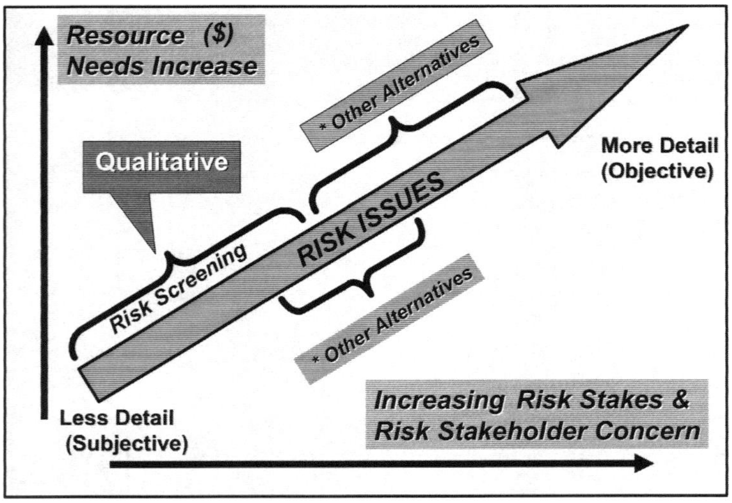

Figure 2. Risk Issue Ramp (1 of 3)

Figure 2, Risk Issue Ramp (1 of 3), shows where the initial risk scenario inventory work fits on the overall risk management picture of a unit process or a facility. Note as risk stakes increase and higher levels of management get involved with the decision process, the subjective nature of qualitative risk assessment may need to be supplemented with other more quantitative methodologies to provide additional clarity and insight to the stakeholders that are involved.

Semi-Quantitative Analysis

As the stakes increase and the decisions involve more complicated interactions, the need for more clarity and objectivity increases. Semi-quantitative risk assessments begin to "sharpen the pencil" and start moving from a subjective perspective to a more objective perspective. Semi-Quantitative Risk Assessment is a viable alternative for looking at existing qualitative risk scenarios and more closely evaluating the effectiveness of the recognized safeguards (or levels of protection) that may exist.

Layer of Protection Analysis (LOPA)

Often the additional clarity required involves the likelihood that the proposed scenario will occur and how to account for the various systems (administrative, engineered) that can affect this likelihood. LOPA, a semi-quantitative methodology, allows one to take into account the independent layers of protection against an undesirable event that are present. LOPA also is used to determine if more independent layers of protection are required, i.e., can the risk be tolerated given the mitigation measures present? [3] This is basically done by first assigning numerical frequency numbers to:

- An initiating cause,
- The possible failure of each protection layer, and
- A universal or company-recognized consequence impact of interest.

Then a comparison is made between these frequency values to determine if the existing protection layers are adequate. These numbers oftentimes are chosen from empirical data consistent with industry. Also, they can be generated and supported from an individual facility's historical data.

Many types of protection layers are recognized, as illustrated in Figure 3, the LOPA Onion. A scenario may have one or many layers depending on the process complexity and potential consequence severity. However, only one layer is required to perform its function successfully and avert the catastrophic event.

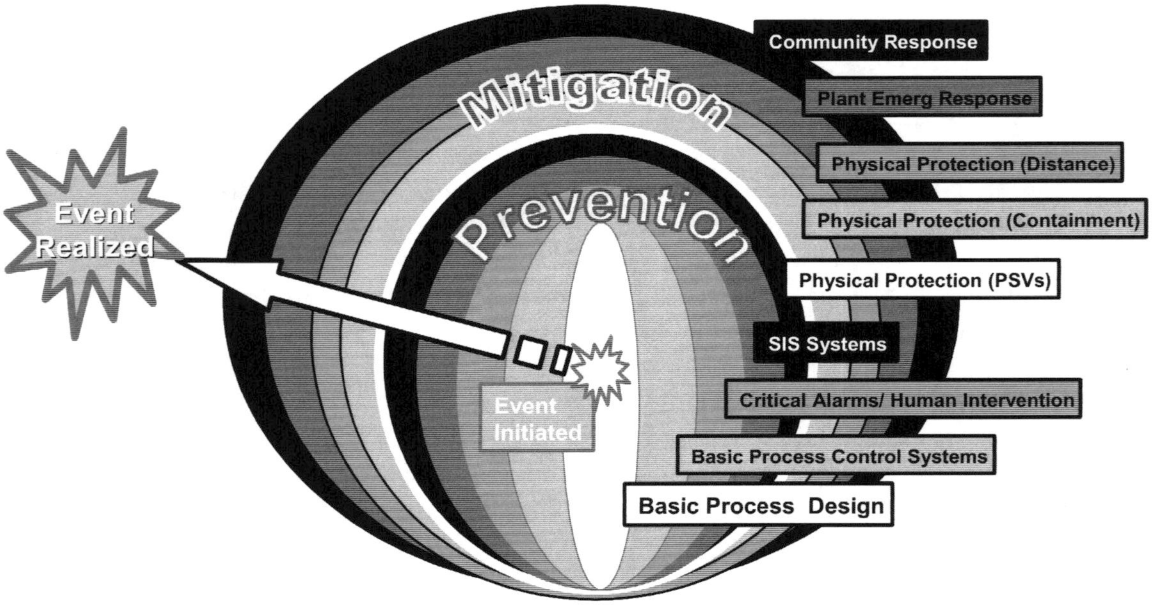

Figure 3. Process Hazards Model, The LOPA Onion

If the estimated risk of a scenario is not acceptable, additional Independent Protection Layers (IPLs) may be added in order to increase the probability that at least one layer will function successfully, since no layer is 100% reliable. LOPA does not suggest which IPLs to add or which design to choose, but it does assist in evaluating the alternative mitigation measures.

A scenario is typically identified during a qualitative hazard evaluation, such as a PHA, management of change evaluation, or design review. LOPA is then applied after an unacceptable consequence (with a credible cause for it) is selected. It then provides an order-of-magnitude approximation of the risk of the scenario given the layers of protection either in place or proposed. The results can then be extended to make risk judgments and to help the risk professional decide how much additional risk reduction may be required to reach a tolerable risk level.

LOPA provides more detail to the risk picture, particularly from the standpoint of the necessary IPLs that may be required for a given scenario. LOPAs, depending on the development of the initiating causal data and the subsequent protection layer data, can fit a rather wide range of detail from lesser detail (subjective in nature) to more detail (objective in nature). Some LOPA work used to reconcile the safeguards listed with a PHA generated scenario may only take a few minutes to develop and complete, while other studies where the description of the existing scenario requires well developed initiating event frequency analysis and protection layer failure frequency verification may take several days to complete.

Risk Comparison

Sometimes there are questions regarding both the likelihood of the scenario as well as the hazard associated with the scenario. Another semi-quantitative risk assessment technique called a risk comparison can be effectively used to address this situation. Here, scenarios of interest probably generated from qualitative hazard identification work are re-worked such that their risk pairs reflect more detail. This is done so the eventual comparison effort becomes more definitive, bringing more insight to the risk issue. In such assessments either one or the other risk pairs (consequence or likelihood) may be upgraded and given more detail. It is also common for both the likelihood and consequence components to be upgraded with more detail. In any of these cases the resulting data can then be compared, using any number of graphical choices for comparison.

In case study 1, contained in this paper, a risk comparison analysis is presented. It involves the quantification of both risk pairs from an initial inventory of existing scenarios, which were initially generated using a qualitative process. The comparison format is a risk matrix, which is upgraded to show more detailed information on both the X and Y coordinate axis. The various scenarios of interest are then listed in their prospective matrix cell or box corresponding to their risk pair values and can be evaluated against each other relative to their location on the matrix. As in the qualitative risk matrix, each of the matrix cells or boxes reflects differing levels of risk.

Figure 4, Risk Issue Ramp (2 of 3) shows where semi-quantitative methodologies, like LOPA or Risk Comparison, would be utilized when risk issues are beginning to generate interest and take up manager's time. Note a noticeable overlap exists as subjective information transitions to objective information.

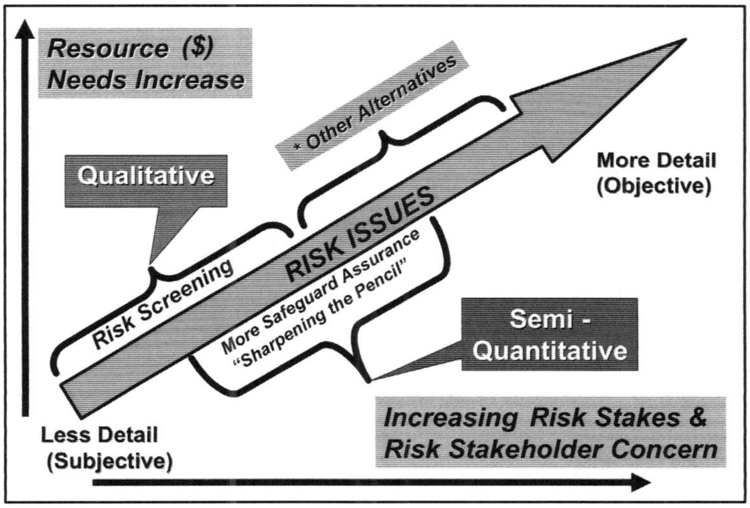

Figure 4. Risk Issue Ramp (2 of 3)

Quantitative Analysis

The American Institute of Chemical Engineers (AIChE) Center for Chemical Process Safety (CCPS) publication "Guidelines for Chemical Process Quantitative Risk Analysis (CPQRA)" [4] provides a comprehensive discussion of how to conduct QRA studies. The underlying basis of a QRA study is simple in concept, and it provides an objective means to answer questions such as:

- What can go wrong?
- How likely is it?
- What are the impacts?
- So what?

QRA is typically the method that is used only after other more subjective alternatives/methods have been rigorously evaluated (or possibly completed), as it is viewed as a resource-intensive exercise. The methodology of the QRA technique is quite flexible and can be used to evaluate many types of risk exposures [5], including:

- Safety (Employee and Public)
- Environment (Chronic and Acute)
- Property Damage
- Business Interruption
- Operational Business Risk Exposures
- Non-Operational Business Risk Exposures
- Enterprise-wide Risk

To illustrate the usefulness of this technique, the focus of this paper will be restricted to a QRA study of an operating facility. The identification of ***"what can go wrong?"*** is typically conducted by a combination of reviewing process drawings and previous Process Hazard Analysis (PHA) studies. The identified events range from minor incidents attributable to small leaks to major accidents involving catastrophic failure scenarios.

The question ***"how likely is it?"*** is typically based on statistical analysis of historical accident data within the given industry and in some cases through the use of Fault Tree Analysis.

To determine ***"what are the impacts?"*** an analysis of the consequences of a release are evaluated through the use of consequence analysis software and superimposed onto the receptors to estimate the level of impacts that each event can generate. The event likelihood together with the impacts is combined to present the results in a format that is easier to understand and digest.

To answer the final question ***"so what?"*** the results are typically compared to risk threshold guidelines (internal or governmental). The risk profiles for each proposed mitigation option are then evaluated to see if the risks are tolerable. In the absence of guidelines, one can use comparative analysis or an incremental risk analysis to add further insight.

Typically there are two types of risk results generated by a QRA study: Geographic Risk and Societal Risk, discussed below.

Geographic risk is expressed in terms of impacts per year, and is represented visually in the form of an iso-risk contour as shown in Figure 5. Geographic risk is independent of the actual population surrounding a given hazard, thus capturing only the potential for impacts. Geographic risk takes into account the equipment failure rate data, meteorological probabilities, impact zones, and the vulnerability probabilities. It is in essence the inherent risk of an impact at a given location from the process.

The result indicates the risk of impact at a

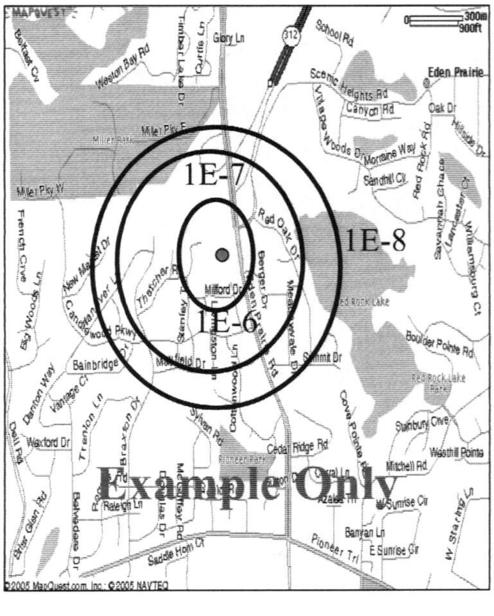

Figure 5. Example of Geographic Risk Contours

given location if that location is continuously occupied by an individual 24 hours a day, 365 days a year. For example, if a person was continually present at a particular location, the geographic risk could be the same as the individual risk.

The geographic risk is calculated by summing the probability of all outcomes that have the potential to cause an impact at that given location. The impact probability is the product of the individual event probability, the atmospheric stability, wind speed and wind direction probability, and the outdoor vulnerability fraction. An x-y coordinate system is used to generate a table of impact probabilities for each coordinate pair. By connecting those points with a common value of impact/year (i.e., 1.0E-05, 1.0E-06) the iso-risk contours are created. These contours are then superimposed on a map of the area or facility plot plan to provide aerial insight as to which areas surrounding the facility are subjected to various levels of risk. Geographic risk provides information that is vital for land use planning and many governments also have established geographic risk criteria that facilities must meet.

Whereas geographic risk measures the risk potential surrounding the facility, societal risk evaluates the predicted number of impacts based upon the location and presence probability of the population surrounding the unit. ***This measure of risk is site-specific*** and is highly dependent upon the physical location of people, the fraction of the time they are present, and whether they are located indoors or outdoors. Societal risk estimates the potential impact and likelihood of each event outcome. The event outcome frequency-impact data (also referred to as F-N pairs) is used to generate two types of societal risk formats. The first is the societal risk index (SRI), also known as the probable loss of life (PLL), and the second is the cumulative frequency versus the number of impact curve (F-N Curve).

The SRI or PLL data is generated by using the F-N (Frequency-Number) pairs as described above and multiplying the number of impacts by its specific event frequency to generate the expected fatality-per-year value for that event outcome. This calculation is performed for each discrete event outcome to generate a table of expected fatality rates, as illustrated in the example below.

Scenario ID	Frequency	Wind Direction	Wind Combination	Impacts	SRI (impacts per year)
Node01_s_day	3.13E-02	N	D5	1	3.13E-02
Node01_s_day	2.35E-02	S	D5	0	0.00E+00
Node01_s_day	1.57E-02	E	D5	2	3.13E-02
Node01_s_day	7.83E-03	W	D5	3	2.35E-02
Node01_m_day	2.41E-03	N	D5	7	1.69E-02
Node01_m_day	7.24E-04	S	D5	4	2.89E-03
Node01_m_day	1.45E-04	E	D5	10	1.45E-03
Node01_m_day	1.45E-05	W	D5	8	1.16E-04
Node01_l_day	1.08E-05	N	D5	40	4.32E-04
Node01_l_day	3.24E-06	S	D5	25	8.10E-05
Node01_l_day	6.48E-07	E	D5	35	2.27E-05
Node01_l_day	6.48E-08	W	D5	50	3.24E-06
				TOTAL	**1.08E-01**

The above example is for a single "node" or grouping of equipment items with similar expected consequence zones, over a range of representative hole sizes ("s"=small, "m"=medium and "l"=large), with releases occurring during the daytime period. Then for each scenario, the SRI is calculated by multiplying the event frequency by the number of impacts. Additionally, these expected rates of impact could be summed for all events to generate the SRI value for the entire process, facility, or node. Specifically for this example, the total SRI for Node01 equals 1.08E-01 impacts per year.

The advantage of this calculation is the ability to develop risk priority tables by relative risk contribution. If the risk assessment was conducted for an entire business unit, the SRI can be used to prioritize entire facilities and separate processes; even down to the level of individual pieces of equipment. One benefit of the SRI metric is that relative contributions can be used to target risk mitigation efforts. The targeting efforts can illustrate the average risk reduction afforded by the various risk mitigation technologies as compared to the Base Case risk exposure.

Another benefit of using SRI is the ability to compare unit specific risks with empirical average individual risk data that reflect more recognizable activities and their associated risk values. The SRI is divided by the exposed population to arrive at the average individual risk value that can be compared to the values in the table below listing activities/situations common to everyday life and their associated average individual risks.

Table 1. Common Life Activities and Associated Individual Risks

Activities/Situations	Avg. Individual Risks
Disease (age 60-65)	1.67E-2
Motorcycling	1.00E-3
Sky Diving	6.67E-4
General Aviation	6.67E-4
Disease (age 20-25)	5.00E-4
Auto Driving	3.33E-4
Home Accidents	2.22E-4
Proposed Building Siting Risk Criteria	1.00E-4
Swimming	3.00E-5
Scheduled Airlines	2.50E-5
Acts of God	3.00E-6

As powerful a tool as SRI is, it has a major limitation in that it equates the risk of small impact/high frequency events with large impact/low frequency events as shown in the example below:

Number of Impacts	Frequency of Impacts	SRI
100	1.00E-05	1.00E-03
1	1.00E-03	1.00E-03
10	1.00E-04	1.00E-03

The trouble with blindly using SRI is that the result removes the risk aversion of large impact/low frequency events. For this reason the F-N curve is also used in conjunction with SRI data to help more fully understand the full risk potential and the magnitude of the largest possible events.

The F-N curve represents the cumulative frequency of all events exceeding a given outcome. To generate the F-N curve, the event probability and impact values (the F-N pairs) are sorted from smallest to largest number of impacts. For the largest impact, the event frequency is the same as the cumulative frequency. We then take the event with next largest impact and add its event frequency to the event frequency of the largest event and obtain the cumulative frequency. This process is repeated for all events to generate a table of cumulative frequency at various levels of impact.

Table 2. Cumulative Frequency Table

Scenario ID	Frequency	Wind Direction	Wind Combination	Impacts	Cumulative Frequency
Node01_s_day	2.35E-02	S	D5	0	8.16E-02
Node01_s_day	3.13E-02	N	D5	1	5.81E-02
Node01_s_day	1.57E-02	E	D5	2	2.68E-02
Node01_s_day	7.83E-03	W	D5	3	1.11E-02
Node01_m_day	7.24E-04	S	D5	4	3.31E-03
Node01_m_day	2.41E-03	N	D5	7	2.58E-03
Node01_m_day	1.45E-05	W	D5	8	1.74E-04
Node01_m_day	1.45E-04	E	D5	10	1.60E-04
Node01_l_day	3.24E-06	S	D5	25	1.48E-05
Node01_l_day	6.48E-07	E	D5	35	1.15E-05
Node01_l_day	1.08E-05	N	D5	40	1.09E-05
Node01_l_day	6.48E-08	W	D5	50	6.48E-08

It is the objective way in which ***"so what?"*** is addressed by the QRA technique that is its strength.

The most powerful argument for using a QRA technique is that it provides a consistent and objective basis to evaluate the risk exposures. As long as the studies are conducted utilizing a consistent approach (data usage, assumptions, and level of detail), management can be assured in the objectivity of the analysis and have the confidence to use the information to support their decisions by utilizing:

- Comparison with guidelines or established risk thresholds
- Prioritization amongst operations
- Identification of the diminishing point of economic return (Cost Comparison Analysis, Incremental Risk analysis)

As already stated, many risk issues have multiple drivers with inherent trade-offs. Often times, these drivers take on an economic or financial perspective, where an economical justification becomes critical to the final decision or business plan. In this way QRA can begin to bring economic clarity to the risk management process.

Figure 6, Risk Issue Ramp (3 of 3) illustrates how QRA provides yet more information and detail to the risk issues faced by stakeholders.

Case Study 2 in this paper illustrates how incremental risk analysis can be utilized to provide additional decision insight to management. The incremental risk analysis approach is useful in helping managers choose between several effective risk reduction options to:

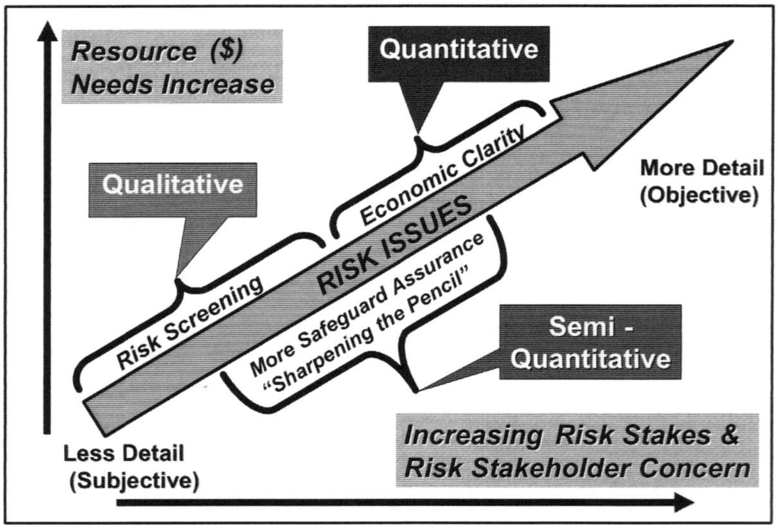

Figure 6. Risk Issue Ramp (3 of 3)

1.) Identify the point of diminishing return.
2.) Prioritize the order of implementation of the risk reduction options.
3.) Identify complementary instead of duplicative options.
4.) Derive the most risk reduction benefit for a given investment.

The analysis uses the QRA data to identify the risk reduction option that provides the greatest risk reduction benefit for a given investment. Then it is assumed that this risk mitigation option is implemented and becomes the new base case against which all remaining risk reduction options are evaluated. The process is repeated until all mitigation options are addressed.

Case Studies

Case Study 1 Risk Comparison

An opportunity to expand market share from the existing base of customers within the U.S. to new customers in Europe was being evaluated by the business leaders of a large chemical manufacturer. The materials being evaluated included both toxic and reactive materials. While still in the proposal stage, the business leaders wanted to evaluate the potential risk exposure associated with the transportation requirements to deliver materials manufactured in the U.S. to European locations. Since many of the details of the proposed requirements for European transport were not known, a quantitative risk assessment could not be used to determine risk significance utilizing internal risk guidelines. However, to avoid investing a lot of resources only to learn at a later stage that the risk exposure would not justify the business opportunity, a high-level risk assessment study was conducted. A qualitative What-if study identified the following issues:

- The transportation risks associated with the European transport operation had the potential for higher risk exposure due to the additional operations required.
- European transport operations required the use of inter-modal containers.
- Inter-modal containers have to be lifted from trucks at U.S. ports onto ships and vice-versa at European ports.
- Some of the materials were water reactive and were expected to have a greater consequence exposure for European routes.
- Truck mileage for U.S. transport is considerably greater than for European transport.

Due to some of the trade-offs between the transportation systems, the project team decided to obtain more objective quantitative information for each chemical transport operation. Since the specific information to calculate the traditional results of QRA study were not available, it was decided to utilize a semi-quantitative approach.

For each product considered for European transport it was decided to model a couple of release scenarios at two weather conditions. The likelihood and consequence analyses were conducted for the existing U.S. operations as well as proposed European options. Although route-specific information was not available, the team was able to utilize generalized averages to calculate the likelihood of the two spill sizes. Consequence analysis utilized chemical and container specific information to estimate the hazard zones for a given end-point.

Results for a water reactive chemical "A" is presented in Figure 7 while the results for a compressed gas "B" are presented in Figure 8.

"A" Risk Comparison
Europe vs US Transport Risk

Likelihood Frequency Per Ton Per Year					
	More than Once in hundred thousand per year (>1E-5)				
	Once in a hundred thousand to once in a million per year (1E-5-1E-6)	US 1D EU-L 1D/RupD		US RupD	
	Once in a million to Once in ten million per year (1E-6-1E-7)	EU-W 1D EU-L 1F/RupF	EU-W RupD	US 1F US RupF	
	Less than once in ten million per year (<1E-7)			EU-W 1F EU-W Rup F	
		Less than 0.5	0.5-1	1-3	Greater than 3

Hazard Area (sqkm)

Consequence Severity

SCENARIO KEY	
Scenario ID Components:	
EU-W	European Transport Water Spills Option
EU-L	European Transport Land Spills Option
US	US Transport Option
1	Small-hole size
Rup	Large Hole size
F	F stability 1.5 m/s wind speed
D	D stability 5 m/s wind speed

Figure 7. Europe vs U.S. Transport Risk Comparison for Chemical "A"

"B" Risk Comparison
Europe vs US Transport Risk

Likelihood Frequency Per Ton Per Year					
	More than Once in hundred thousand per year (>1E-5)		EU-Opt2/3-RupD	EU-All options-1D US-1D	
	Once in a hundred thousand to once in a million per year (1E-5-1E-6)			US-Rupture D EU-Opt1-RupD EU-Opt2/3 RupF	US1F EU-All Options-1F EU-Opt1-RupF US Rupture F
	Once in a million to Once in ten million per year (1E-6-1E-7)				
	Less than once in ten million per year (<1E-7)				
		Less than 0.5	0.5-1	1-3	Greater than 3

Hazard Area (sqkm)

Consequence Severity

SCENARIO KEY	
Scenario ID Components:	
EU	European Transport Option
US	US Transport Option
1	Small-hole size
Rup	Large Hole size
F	F stability 1.5 m/s wind speed
D	D stability 5 m/s wind speed
Container Design Options for BF3:	
Option 1	40' ISO trailer containing 6 tubes
Opiton 2	20' ISO trailer containing 12 tubes
Option 3	20' ISO trailer containing 8 tubes

Figure 8. Europe vs U.S. Transport Risk Comparison for Chemical "B"

Each chemical transport comparison shows that the proposed transport of these chemicals to Europe does not pose a risk exposure different from what is currently tolerated by the business for the existing us operations. A closer look shows that European transport of chemical A actually poses lower risk exposure than the risks within the U.S. Another comparison of the two chemicals to each other is shown in Figure 9.

"A/B" Risk Comparison
Europe vs US Transport Risk

Likelihood Frequency Per Ton Per Year						
	More than Once in hundred thousand per year (>1E-5)		B-EU-Opt2/3-RupD	B-EU-All options-1D B-US-1D		
	Once in a hundred thousand to once in a million per year (1E-5-1E-6)	A-US 1D A-EU-L 1D/RupD		A-US RupD B-US-Rupture D B-EU-Opt1-RupD B-EU-Opt2/3 RupF	B-US1F B-EU-All Options-1F B-EU-Opt1-RupF B-US Rupture F	
	Once in a million to Once in ten million per year (1E-6-1E-7)	A-EU-W 1D A-EU-L 1F/RupF	A-EU-W RupD	A-US 1F A-US RupF		
	Less than once in ten million per year (<1E-7)			A-EU-W 1F A-EU-W Rup F		
		Less than 0.5	0.5-1	1-3	Greater than 3	Hazard Area (sqkm)

Figure 9. Chemical A and B Transport Comparison

This comparison clearly shows that Chemical A has a lower risk exposure than Chemical B based upon the limited data that was used in the analysis.

When making risk comparisons, as was done in this case study, it is important to clearly understand what data was utilized in this study so that the comparisons can be as close to an "apples-to-apples" comparison as possible. Using this logic, management can have more faith in the first two comparisons where the comparison is for the same chemical. The last comparison between the two chemicals has additional potential for one to draw an incorrect conclusion.

The comparison approach utilized in this case study:

1. Works well early in the project phase when key information is not determined or unavailable
2. Helps one gain a better understanding of conflicting qualitative data
3. Utilizes sufficient detail to provide more objective information
4. Provides results in a format that is easy to understand.
5. Requires more resources than a qualitative review but considerably less than a QRA study

When utilizing this approach, the experience of the risk analyst is crucial to ensure proper use of this technique. The following should be clearly understood:

1. The data that is used
2. The data that is not available

3. The Assumptions that were required
4. What comparisons and inferences should not be made

Case Study 2 - Incremental Risk Analysis

An operator of an HF Alkylation unit was preparing for an upcoming turnaround and wanted to evaluate several mitigation options that had been identified from a variety of qualitative studies. Many of the simpler non-intrusive risk reduction options were already implemented, however, management wanted a more objective review of the potential risk reduction benefits associated with several more complicated options as well as when they are combined. It was decided that an update of a previously conducted QRA study was needed.

The traditional results of a QRA study including geographic risk contours, FN curves, and Societal Risk Index (SRI) results were calculated for the As-Is unit as well as each of the proposed mitigation options. Figure 10 provides a comparison of how the geographic risk is affected by the different mitigation options.

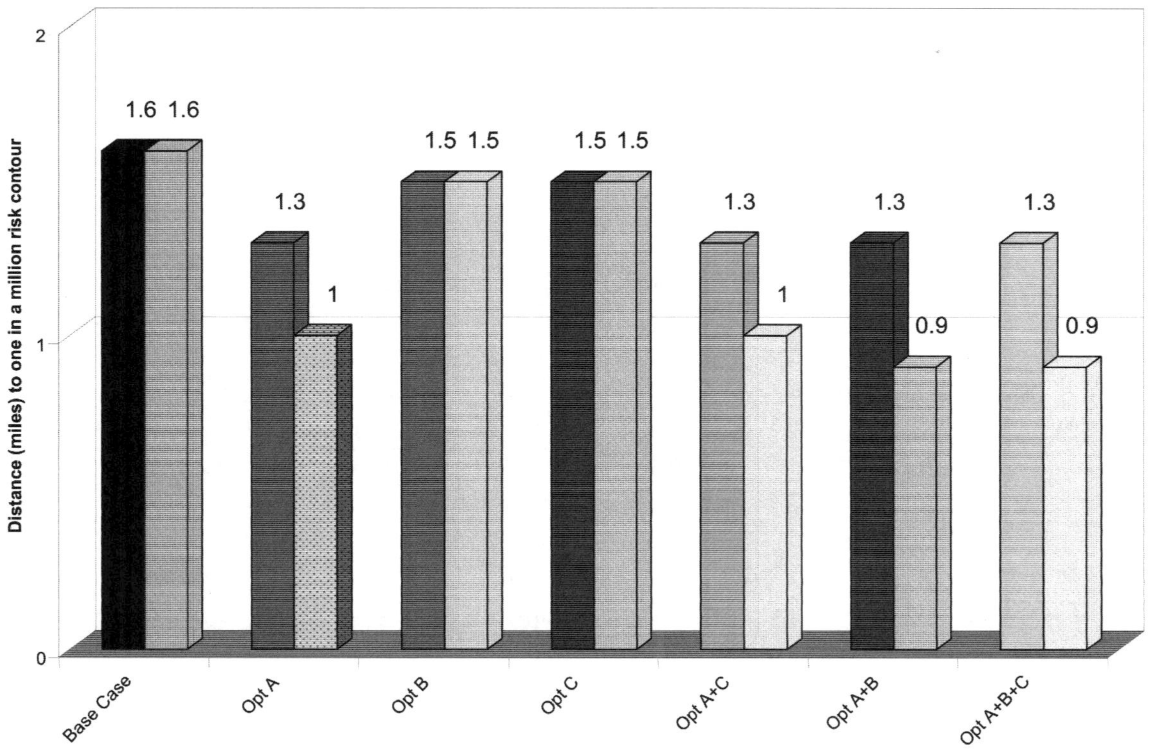

Figure 10. Geographic Risk Comparison of Potential Mitigation Options

The comparison of Societal Risks associated with the base case and proposed mitigation options are presented in Figure 11.

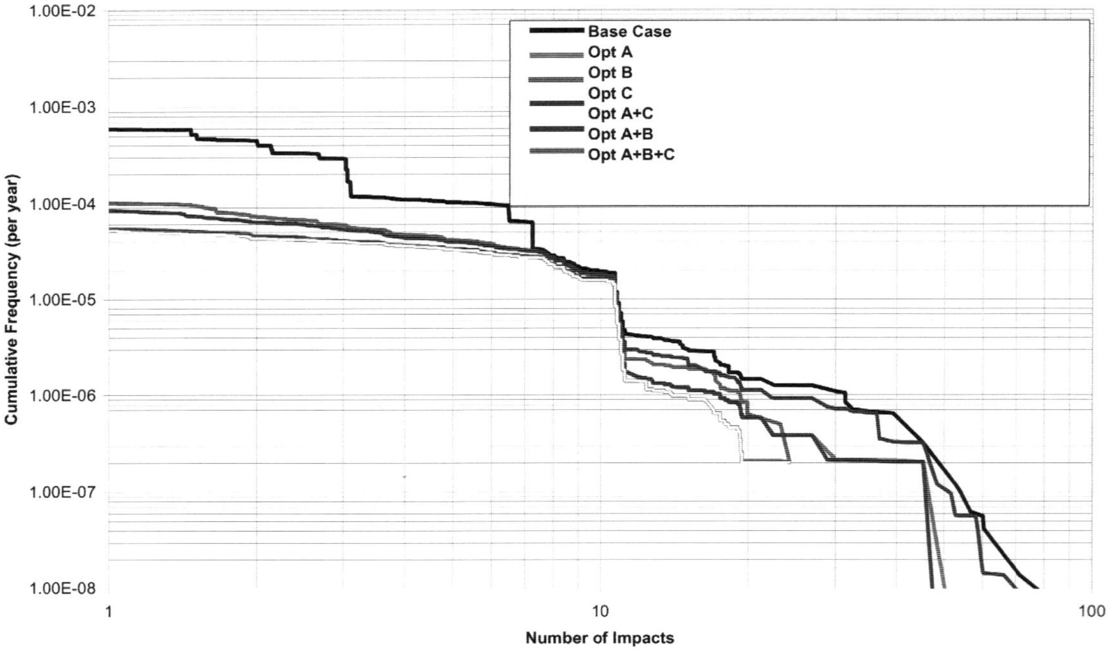

Figure 11. Societal Risk Comparisons

When compared to the As-Is unit risk profile, each proposed mitigation option showed a reduction to the public risk exposure. Although the analysis showed how some of the risk reduction options were better than others at reducing risks, management still had the following questions:

- In the absence of risk guidelines, when should/can we stop?
- Can we prioritize the order of mitigation options?
- What is the incremental risk reduction for remaining mitigation options once we install the most effective mitigation option?
- When have we reached a diminishing point of return?

Since the comparison of each risk reduction option or combination was always to the As-Is case, it was decided that an incremental risk analysis could provide the insight necessary to address management's questions. Of all the risk mitigation options reviewed in the QRA study, the five most effective risk reduction options as compared to the base case were evaluated as part of the incremental risk analysis.

Since there were no applicable quantitative risk guidelines for this site and internal quantitative risk guidelines had not been developed, there was an issue about the level of risk aversion to utilize in the incremental risk analysis. It was decided to utilize the slope of established governmental risk guidelines to see if that made a difference in the order of the mitigation options that should be implemented.

- AF = 1 No Risk Aversion (normal SRI) consistent with Hong Kong guidelines
- AF = 2 Consistent with Dutch guidelines, a factor of 10 increase in impacts requires a factor of 100 reduction in likelihood
- AF = 1.5 An intermediate level

The investment costs were all compared based on the "cost factor" (CF) of being one for Option 1. The cost factor for the other mitigation systems are shown in the table on the right.

The results of the incremental risk analysis are shown in Figure 12.

Mitigation Option	(CF)
Option 1	1.00
Option 2	3.88
Option 3	7.78
Option 4	9.33
Option 5	85.00

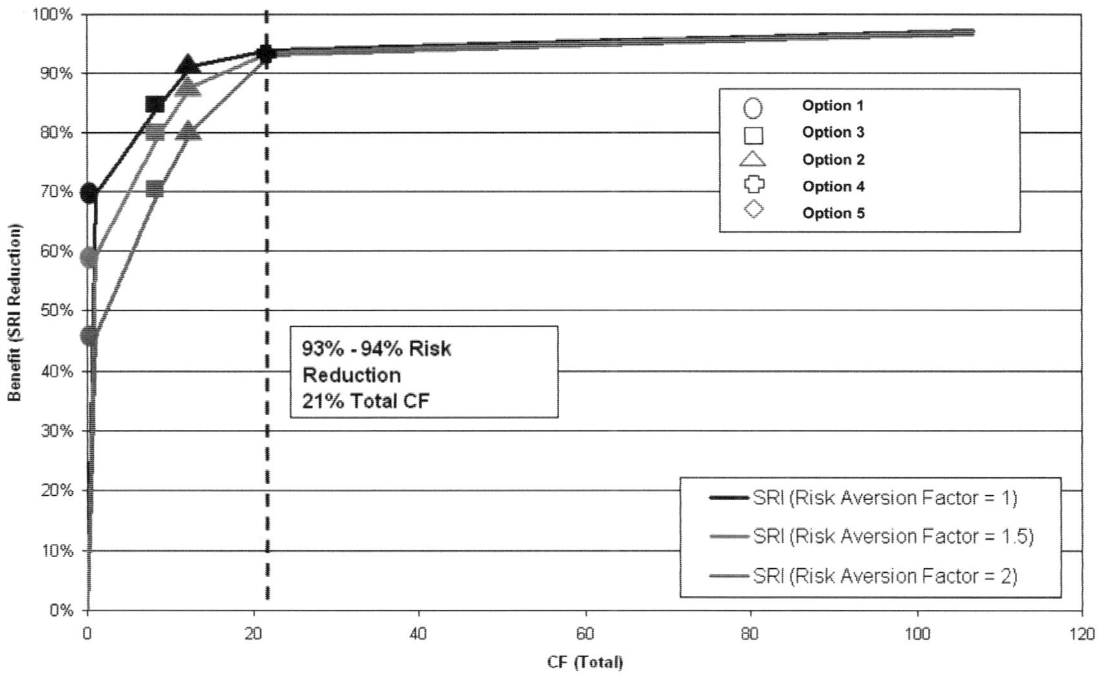

Figure 12. Comparison of Risk Reduction Benefit to Investment (CF)

The results clearly show the order of the mitigation options that should be utilized to achieve the most risk reduction benefit for the investment. Furthermore, this order does not change with increasing risk aversion until an aversion factor of 2.3 is utilized, which is more conservative than what the Dutch government has used for their offsite risk guidelines. Under all situations reviewed, Option 1 was always the most effective and should be implemented first, while Option 5 was always the last and should be implemented last if at all.

The incremental risk analysis provides a very objective way to prioritize the order by which mitigation options should be implemented to give the biggest risk reduction benefit for a given investment. Utilizing a relative cost factor approach allows us to evaluate and economically compare risk reduction strategies. This technique is very useful:

- When risk guidelines are not available or are vague
- When there are multiple options that all provide a benefit from the base case
- When one wants to identify the effective components of a complex system
- When one wants to identify the point of diminishing returns.

References

1. John S. Hammond, Ralph L. Keeney, Howard Raiffa, *Smart Choices – A Practical Guide to Making Better Decisions.* Harvard Business School Press, Boston MA. 1999
2. Center for Chemical Process Safety, *Guidelines Hazard Evaluation Procedures.* Second Edition. American Institute of Chemical Engineers, New York. 1992
3. Center for Chemical Process Safety, *Layer of Protection Analysis, Simplified Process Risk Assessment.* American Institute of Chemical Engineers, New York. 2001
4. Center for Chemical Process Safety, *Guidelines for Chemical Process Quantitative Risk Analysis.* Second Edition. American Institute of Chemical Engineers, New York. 2000
5. Krishna S. Mudan, Jatin N. Shah, and Phil M. Myers, *Financial Risk Assessment – A Uniform Approach to Manage Liabilities*, Proceedings of the 29th Loss Prevention Symposium, Boston MA 1995

Abstract for paper for the 2005 CCPS conference.

Title: Use of Process Risk Analysis and System Assessments to drive PSM Improvement

Author & Presenter – John Herber, Manager, Process Hazard Resource Center - 3M Company, St. Paul Minnesota

3M is a $18 billion diversified technology company and the manufacturer of over 50,000 products ranging from Scotch® tapes and Post-it ® Notes to flexible electronic circuits and industrial chemicals and specialty films. Headquartered in St. Paul, Minnesota, the company has operations in more than 60 countries and serves customers in nearly 200 countries. This diversity in processes, products, business units, countries and cultures creates a challenge to consistently manage process risk.

This paper will present the methods used to develop a process risk management process. It starts by screening locations based on hazards. It sorts Low hazard locations that do not need PSM from locations with higher hazards where PSM should be employed by 3M's procedures. The process has helped cull approximately 150 world-wide manufacturing locations to about 70 who are actively using PSM to manage their process hazards.

The second step uses a semi-quantitative risk screening tool to assess the processes operated at the location. This has allowed 3M to review their global operations using a consistent risk criteria. The tool also includes a consequence modeling section that allows the determination of on-site and off-site effects. The highest risk level measured for a process at a location determines the risk level for that location. Each risk level has a criteria for acceptable process safety systems

The third step is spreadsheet-driven Self Assessment tool that covers 9 key elements of PSM. Each question requires a response for both the program quality and the degree of implementation. Data from this self-assessment is being used to determine the areas of PSM needing the most improvement and facilitates locations to share PSM practices.

Goals have been set for locations through 2008. The goals require some locations to develop extra safeguards to address their risk level.

John W. Herber
3M Company
Corporate Safety
PHMRC - 42-2W-09
Tel: 651.778.4507
Fax: 651.778.6745

Use of Process Risk Analysis and System Assessments to Drive PSM Improvement

Proposed presentation Outline:

I. 3M Company who we are, what we make, where we operate.
How safety is staffed

II. The history of PSM in 3M
a. Initial implementation
b. PHM policy
c. EHS Management System
d. Plant reviews
e.

III Screening tool – breadth of 3M processes
Which one we wanted
How we got them

IV QPRST
Internally developed tool
Based on Dow F&E index
Sorts processes into 5 risk classes
Risk class determines the management system employed

V MSSA – how to measure across 70 locations in 20 countries
Adapted self assessment tool from Rhodia
Needed to be consistent with audits, and global regulations & had to be simple

VI Goal setting – raising the bar
Fit score to risk level,
Choose sequence for emphasis
Guidance to users

VII Annual renewal cycle
Re-screen to find new locations or to determine change
Renew/expand QPRST's
Re-assess PSM performance vs. expectations

20th Annual CCPS International Conference
RISK MANAGEMENT: THE PATH FORWARD
April 11 – 13, 2005
Atlanta, Georgia

Risk Analysis and Management for Critical Asset Protection (RAMCAP)

David A. Moore[1], Michael Hazzan, Brad Fuller, David Heller
AcuTech Consulting Group
Chemetica, Inc.
88 Kearny Street
San Francisco, CA 94108
www.acutech-consulting.com

Lawrence M. Stanton, CPP
Protective Security Division
U.S. Department of Homeland Security
Washington, DC
lawrence.stanton@dhs.gov

Abstract:

The Department of Homeland Security (DHS) has initiated an ambitious effort in 2004 by contracting the American Society of Mechanical Engineers (ASME) to undertake Phase 2 of the project 'Risk Analysis and Management for Critical Asset Protection' (RAMCAP). The purpose of the project is to provide an overall strategy and methodology to allow for a more consistent and systematic analysis of the terrorist threat against the U.S. infrastructure using a risk-based framework. Phase 1 developed the overall risk framework while Phase 2 is the further refinement and development of the methodologies at the sector level.

The RAMCAP project involves three key tasks:

1) The development of sector-specific screening tools and collection of industry specific information to supplement the information in the National Asset Database (NADB), in order to develop an improved understanding of the U.S. assets that are most important to protect against terrorist attack, and to rationalize the priority of those assets.

2) The development of a sector-specific Security Vulnerability Analysis (SVA) methodology, together with supporting tools, metrics, and terminology for the analysis of threats and vulnerabilities, that allow cross-sector comparisons in line with the overall RAMCAP approach, and;

---

[1] Mr. Moore serves as the Coordinator for the Chemical, Petroleum, and Liquefied Natural Gas (LNG) sectors for the RAMCAP project as a subcontractor to ASME Innovative Technologies, LLC.

3) The documentation of a compendium of industry security practices as a supplement to the analysis guidance.

This project involves numerous leading industry trade associations, industrial companies, professional societies (including AIChE/CCPS in the Chemical Sector), and security consultants to ensure wide representation of technical knowledge and sector experience. This wide representation ensures a review process that covers the spectrum of possible stakeholders. This is an important aspect of this program because it is a voluntary effort by DHS to enlist industry assistance in fulfilling its purpose.

The RAMCAP project will likely have a profound impact on the chemical sector as well as all other sectors designated by Homeland Security Presidential Directive-7 (HSPD-7). It will help to identify the facilities and operations that deserve national and regional interest with respect to the threat of terrorism, define methods and assumptions for analyzing threats and vulnerabilities, and describe best practices of the industry.

This paper will describe the details of the program including the mission at hand, the results of the process to date, and the expected future of the chemical sector under this initiative, and will also cover the current status of work of the RAMCAP Chemical Sector Team as of February, 2005.

Introduction

Recently, significant efforts by government and private industry have been underway to assess the vulnerability of critical assets and systems, and to develop ways to mitigate the risks. In addition to these efforts, DHS is undertaking a national program to develop a risk-based approach to the analysis of infrastructure vulnerability and to develop guidance for security vulnerability analysis for homeland security decisions. A national effort will be necessary to effectively integrate all of the public and private sector approaches in accordance with national plans and strategies. The project 'Guidance on Risk Analysis and Management for Critical Asset Protection' (RAMCAP) is managed by ASME Innovative Technologies, LLC, as the prime contractor.

Objectives of RAMCAP Project

The primary objective of the RAMCAP project is to develop guidance for security vulnerability analysis for homeland security decision-making by government and private industry. This guidance provides common terminology and a common basis for reporting of results. It is intended to have broad applicability and use by governmental agencies and private industry in coordinating decisions regarding protection of critical infrastructure and key resources (CI/KR).

A fundamental premise is that the allocation of available resources must be based on the results of a risk-informed process, so that the United States can most effectively apply these resources to combat terrorism. A common framework will allow the risk in one infrastructure to be appropriately compared to the risk in another to better support resource allocation decisions.

The guidance document provides critical factors to be used for screening of CI/KR to determine whether implementation is justified. The methodology builds on established security vulnerability analysis methods that have been used over the past several years in industry.

The first phase of the RAMCAP project in 2003-2004 developed a set of terminology and an overall risk framework. The second phase of RAMCAP in 2005 will involve the development of sector-specific guidance and pilot testing of the methods. Paramount to the program's goals is the voluntary inclusion of sector specific industry experts to develop the guidance.

National Strategies and Plans for Homeland Security and Vulnerability Assessment

Several national objectives and plans support the need for this process. The RAMCAP project is entirely consistent with the goals and approach of these national strategies and plans, as described below. Industry groups are working with DHS in a partnership to develop all aspects of the program. For example, over 30 volunteers are working on the Chemical Sector Team representing chemical industry company security experts, trade associations, professional societies, and homeland security consultants.

National Strategy for Homeland Security of the White House (2002)

The National Strategy for Homeland Security of the White House contains three strategic objectives that are provided in order of priority as follows:

- Prevent terrorist attacks within the United States;
- Reduce America's vulnerability to terrorism; and;
- Minimize the damage and recover from attacks that do occur.

Achieving these fundamental objectives requires a vulnerability assessment process that provides uniform, repeatable, consistent results that allow cross-sector comparisons.

National Strategy for the Physical Protection of Critical Infrastructures and Key Assets (2003)

The White House's *National Strategy for the Physical Protection of Critical Infrastructures and Key Assets* states that the Federal Government will take the lead role in implementation of the strategy, including several responsibilities and assignments relevant to RAMCAP:

- Taking stock of our most critical facilities, systems, and functions and monitoring their preparedness across economic sectors and governmental jurisdictions.

- Assuring that federal, state, local, and private entities work together to protect critical facilities, systems, and functions that face an imminent threat and/or whose loss could have significant national consequences.
- Providing and coordinating national-level threat information, assessments, and warnings that are timely, actionable, and relevant to state, local, and private sector partners.
- Creating and implementing comprehensive, multi-tiered protection policies and programs.
- Exploring potential options for enablers and incentives to encourage stakeholders to devise solutions to their unique protection impediments.
- Developing cross-sector and cross-jurisdictional protection standards, guidelines, criteria, and protocols.
- Facilitating the sharing of critical infrastructure and key asset protection best practices and processes and vulnerability assessment methodologies.
- Conducting demonstration projects and pilot programs.
- Seeding the development and transfer of advanced technologies while taking advantage of private sector expertise and competencies.

Homeland Security Presidential Directive (HSPD)-7 (2003)

On December 17, 2003, HSPD-7 was issued for the 'Critical Infrastructure Identification, Prioritization, and Protection'. This directive establishes a national policy for Federal departments and agencies to identify and prioritize United States critical infrastructure and key resources and to protect them from terrorist attacks. In particular, HSPD-7 calls for coordination between Government and the private sector:

> *"(25) In accordance with applicable laws or regulations, the Department and the Sector-Specific Agencies will collaborate with appropriate private sector entities and continue to encourage the development of information sharing and analysis mechanisms."*

Additionally, the Department and Sector-Specific Agencies shall collaborate with the private sector and continue to support sector-coordinating mechanisms:

> *(a) To identify, prioritize, and coordinate the protection of critical infrastructure and key resources; and*
> *(b) to facilitate sharing of information about physical and cyber threats, vulnerabilities, incidents, potential protective measures, and best practices."*

HSPD-7 assigns responsibility for implementing the directive's objectives for the chemical sector to DHS. Figure 1 is an excerpt from HSPD-7 which gives an underlying basis of the Homeland Security strategy.

Figure 1 – Excerpt from Homeland Security Presidential Directive (HSPD)-7

"Subject: Critical Infrastructure Identification, Prioritization, and Protection

Purpose

(1) This directive establishes a national policy for Federal departments and agencies to identify and prioritize United States critical infrastructure and key resources and to protect them from terrorist attacks.

Background

(2) Terrorists seek to destroy, incapacitate, or exploit critical infrastructure and key resources (CI/KR) across the United States to threaten national security, cause mass casualties, weaken our economy, and damage public morale and confidence.

(3) America's open and technologically complex society includes a wide array of critical infrastructure and key resources that are potential terrorist targets. The majority of these are owned and operated by the private sector and State or local governments. These critical infrastructures and key resources are both physical and cyber-based and span all sectors of the economy.

(4) Critical infrastructure and key resources provide the essential services that underpin American society. The Nation possesses numerous key resources, whose exploitation or destruction by terrorists could cause catastrophic health effects or mass casualties comparable to those from the use of a weapon of mass destruction, or could profoundly affect our national prestige and morale. In addition, there is critical infrastructure so vital that its incapacitation, exploitation, or destruction, through terrorist attack, could have a debilitating effect on security and economic well-being.

(5) While it is not possible to protect or eliminate the vulnerability of all critical infrastructure and key resources throughout the country, strategic improvements in security can make it more difficult for attacks to succeed and can lessen the impact of attacks that may occur. In addition to strategic security enhancements, tactical security improvements can be rapidly implemented to deter, mitigate, or neutralize potential attacks."

National Infrastructure Protection Plan (2005)

In February, 2005, an Interim National Infrastructure Protection Plan (NIPP) was issued. DHS is developing the NIPP to provide a consistent, unifying structure for integrating critical infrastructure protection (CIP) efforts into a national program. Development of the NIPP is an evolving process that requires the participation of all stakeholders from the private sector, states local, and tribal entities, and the federal government. The NIPP outlines how DHS and its stakeholders will develop and implement the national effort to protect infrastructures across all sectors.

The national CIP program is based on a risk management framework, continuously influenced by an ever-changing threat environment, with the goal of reducing the vulnerability of our nation's assets to terrorist attack. As shown in Figure 2, this framework involves five key steps, which are carried out both within sectors and nationally across sectors to protect CI/KR. These steps are part of an ongoing process, with each step carried out as needed to narrow down the overall set of assets to those most critical at a national level. The specific processes will be further refined and developed as the NIPP is implemented.

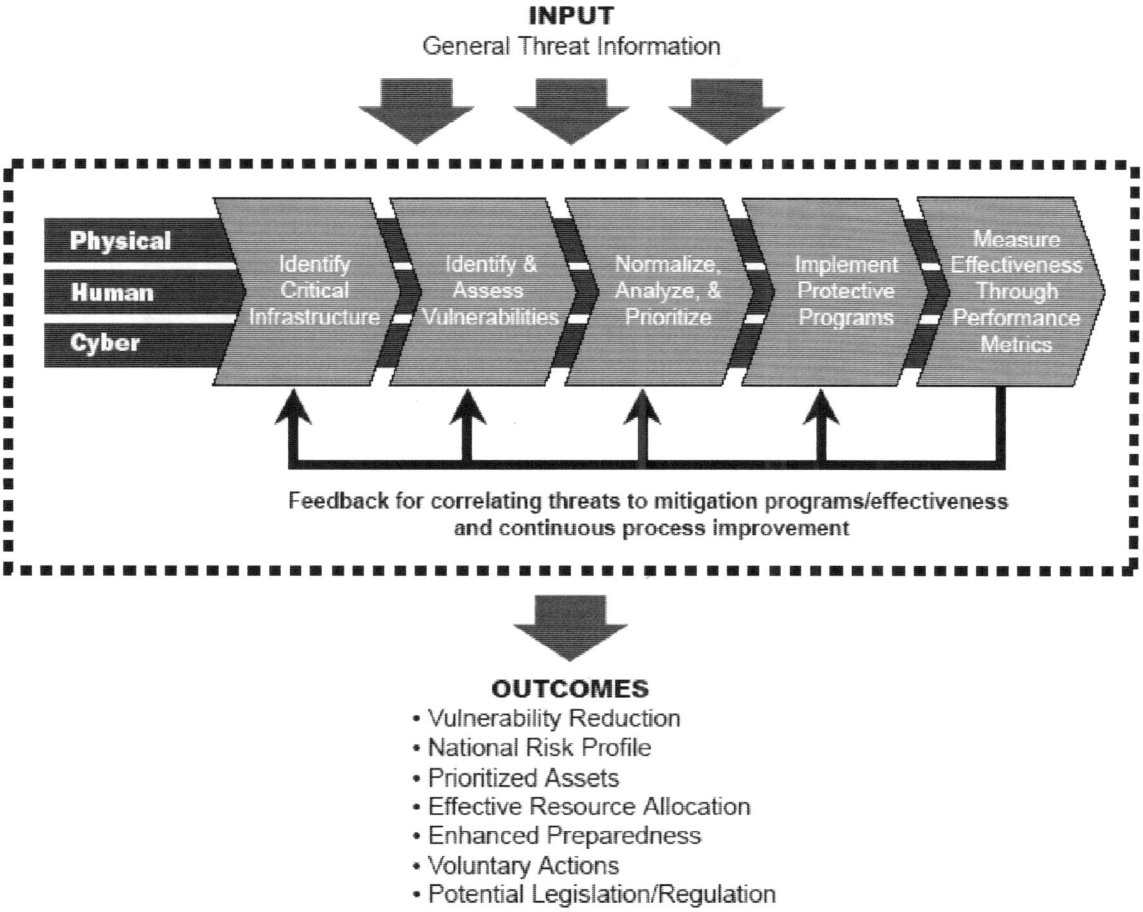

Figure 2 – Interim NIPP Risk Management Framework

RAMCAP Tasks

The RAMCAP project involves three key tasks:

4) The development of sector-specific screening tools and collection of industry specific information to supplement the information in the National Asset Database (NADB), in order to develop an improved understanding of the U.S. assets that are most important to protect against terrorist attack, and to rationalize the priority of those assets.

5) The development of a sector-specific Security Vulnerability Analysis (SVA) methodology, together with supporting tools, metrics, and terminology for the analysis of threats and vulnerabilities, that allow cross-sector comparisons in line with the overall RAMCAP approach, and;

6) The documentation of a compendium of industry security practices as a supplement to the analysis guidance.

This project involves numerous leading industry trade associations, industrial companies, professional societies (including AIChE/CCPS in the Chemical Sector), and security consultants to ensure for appropriate input and then voluntary adoption of the program.

Each of these tasks will be further explained in turn.

Infrastructure Included in RAMCAP

Five categories of assets (sub-sectors of sectors designated by HSPD-7) were identified for the initial RAMCAP process as follows:

- Commercial Nuclear Power Plants
- Commercial Spent Nuclear Fuel Facilities
- Chemical Manufacturing Plants
- Petroleum Refineries
- Liquefied Natural Gas Storage Facilities

These were some of the subs-sectors included in the DHS ‘Liberty Shield’ in 2003. Additionally, the power generation sector was also added to the program for a total of nine infrastructure sectors.

Deliverables

The following items will be delivered to DHS:

- Sector-specific screening tools for commercial nuclear power and chemical manufacturing plants, based on the RAMCAP screening method, which has the general consensus of the sector organizations.

- Sector-specific Security Vulnerability Assessment tools for chemical manufacturing and nuclear power plants, based on RAMCAP, which has the general consensus of the sector organizations.
- Sector-specific recommended guidelines for security best practices and countermeasures, for all named asset categories.

Chemical and Nuclear Sectors Pilot Studies

The chemical and nuclear sectors will lead the Phase 2 sector specific effort by being the first infrastructures to develop guidance and pilot test the methods. These sectors will begin in February, 2005, and continue for up to a year on the project.

Technical Approach

To provide appropriate input into the RAMCAP process for the chemical sector efforts, a Chemical Sector Team was organized. The team is a balanced group of industry professionals representing chemical/processing companies, the trade organizations that represent these industries, and security consultants and experts in SVA working in chemical security providing guidance to the ASME RAMCAP Project Team on industry security practices and assisting in the development and review of project deliverables. Beginning in September 2004, the Chemical Sector Team has held monthly meetings in Washington, DC to review progress and discuss and resolve issues arising during the development of the Phase 2 work products.

1. Screening Methodology

The team was tasked with the development of a screening tool to allow DHS to prioritize the performance of SVAs, and also to prioritize its other programs . This screening tool has taken the form of a cross-sector database. It is designed such that a majority of sites in each sector would answer a relatively short series of key questions about their operations, and then based on the answers, certain sites would then populate the database.

The screening questions and database will collect information to allow for identification of higher priority sites and other DHS uses. RAMCAP SVA methodologies are mostly intended for these higher priority sites.

The Information Analysis and Infrastructure Protection (IAIP) Directorate and other parts of DHS can then use this asset list to:

- Analyze asset information against threat stream data to assist industry in operational security measures.
- Analyze the data potential national impacts using scenario analysis of the data.;
- Analyze the data to determine necessary thresholds for reviewing SVAs or site visits or to conduct other site-specific follow up.;
- Analyze the data to determine the necessity for Buffer Zone Protection Programs (BZPPs) and resource application.;

- Consider the data for use in explaining to regulators and the media the extent of chemical sector security issues with increased confidence than at present.;
- Compare the chemical sector to other lists of assets in other sectors;
- Develop a better understanding of interdependencies between sectors.; and,
- Focus resources on specific unique assets that deserve it based on a common, objective measure of the risk.

Key tasks in the development of the database were:
- Determination of the impacts that were critical for DHS to collect information;
- Determination of screening questions for each of three, progressively detailed, data submittal forms.; and,
- Determination of a list of chemicals that would be used to screen in chemical manufacturers.
- Determination of a correlation factor for estimation of the consequences of a release from the EPA RMP Worst Case Scenario database to reduce the analysis and paperwork burden while providing an improved estimate over the Worst Case Scenario RMP data. As is well known, the Offsite Consequence Analysis report and results were not intended for planning for terrorist events, but were release data for accidental events that fit a specific, common, and very conservative set of criteria and analysis assumptions. They were not even intended by EPA to be used for planning of emergency response activities in response to accidental releases (the Alternative Release Scenarios were intended for this purpose, not the Worst Case Scenarios). These WCS data have been inappropriately used to characterize security-related risk.

- Determine where current and available industry information could be directly used to auto-populate the database.

The RAMCAP Chemical Sector Team worked to develop a process for the collection of information on sector assets to allow for screening of higher priority sites and for other DHS uses. Specific objectives included:

Creation of a national asset list for the chemical sector with the following qualities:

- Consistent and rational with asset lists for other sectors to allow for cross-sector comparisons.
- Fairly granular (of sufficient detail) to allow for definition of assets across the spectrum of assets included.
- Include the following specific datum points:
  - Estimated fatalities
  - Estimated replacement cost impacts
  - Psychological, symbolic or iconic value of asset or proximity to iconic targets
  - Basic interdependency (particularly for critical Government services – military mission supply and continuity)
- Amenable to data verification.;

- Make use of a prescriptive set of assumptions and methodology to produce consistent results.; and,
- User-friendly and minimize burden on the sector.

DHS Asset Database Concept

The DHS Asset Database is conceived as a collection of common information that will allow DHS to better analyze and manage the risks associated with the security of the chemical sector as well as other key infrastructure sectors. The database concept is depicted in Figure 3 and the RAMCAP data collection effort is illustrated in Figure 4. The NADB will be influenced by several initiatives, including RAMCAP, an expert panel elicitation, the original nominations from government agencies and state/territory homeland security advisors, and other Federal databases. RAMCAP's data collection effort will consist of three forms that would be completed by industry as shown in Figure 4.

Figure 3
RAMCAP Influence on NADB

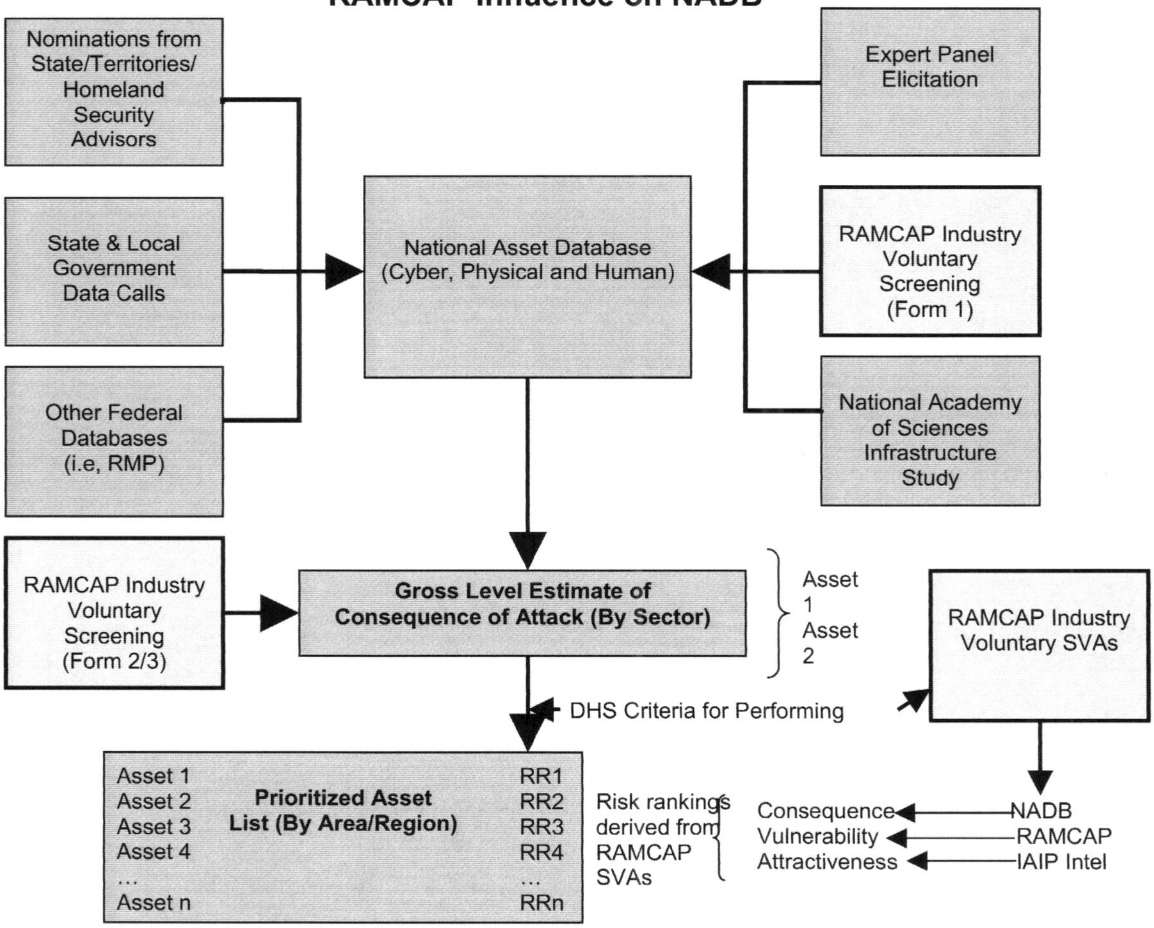

RAMCAP Cross Sectional Database Concept Diagram

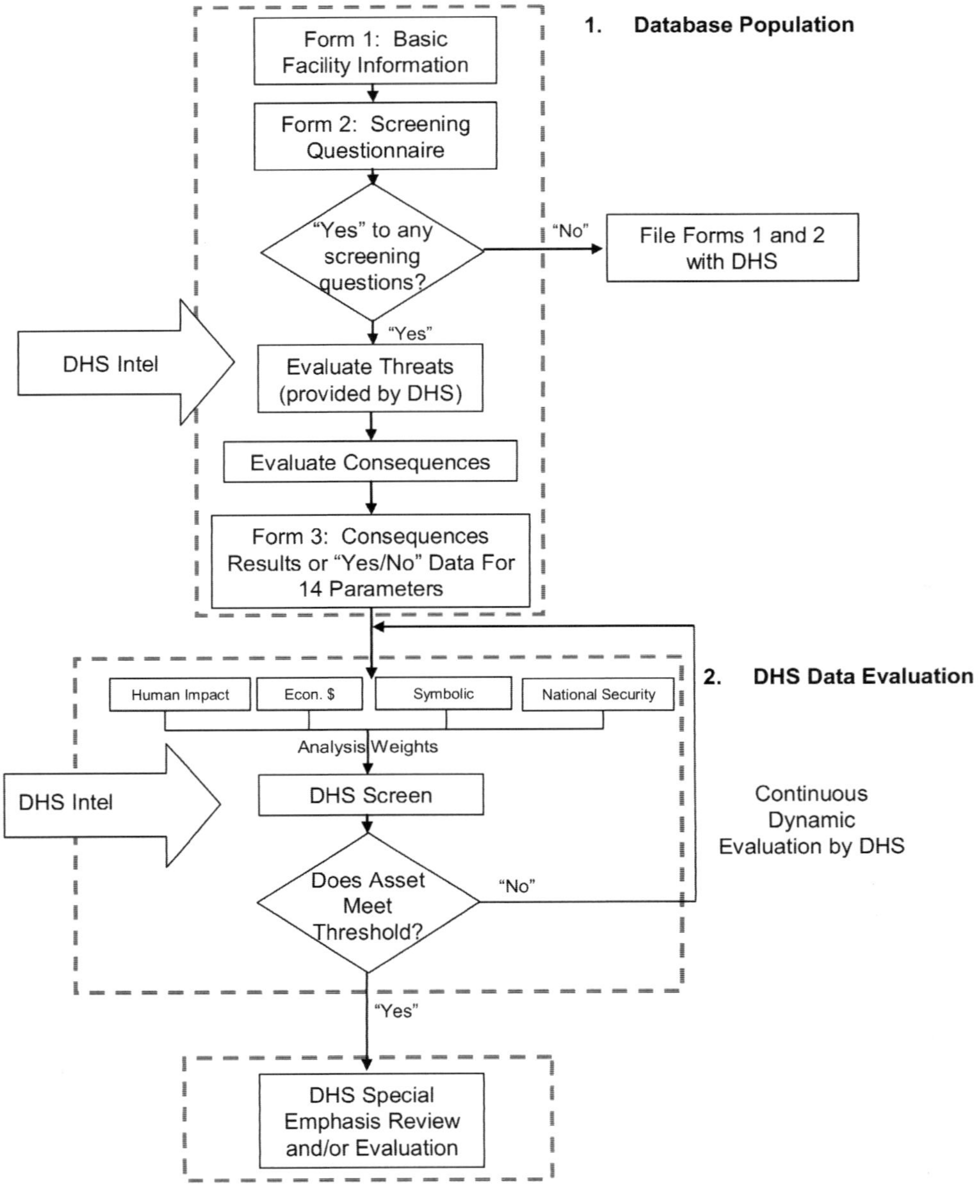

Form 1: A majority of relevant chemical sector facilities would complete Form 1. This form consists of basic facility information that will allow DHS to identify the facility, supporting administrative information regarding the facility (e.g., ownership, EPA identifier numbers, etc.), location, and the chemicals held at the site and their quantities.

Form 2: The same chemical sector facilities would then also complete Form 2. This form consists of a set of screening questions that are intended to filter out those facilities that, by the nature of their business, location/proximity to significant population groups, or other parts of the infrastructure of the nation, or importance to the national economy, or military capability represent a low national security risk and will therefore not be required to populate the database with detailed information, nor envisioned to perform a RAMCAP SVA.

Those facilities that are screened-out (i.e., those facilities that did answered any of the screening questions with a positive answer) would not, at this time, be expected to perform any further RAMCAP activities. Forms 1 and 2 would provide DHS with the necessary documentation to identify and categorize these facilities as low risk and exclude from any further prioritization activities.

Form 3: Only those facilities that are not screened-out after completing Form 2 (i.e., those facilities that answered positive responses to one or more of the screening questions) would complete Form 3. This form consists of the actual parameters that will populate the database and provides data to describe a facility's security-related risks in the following four general areas:

- Human health and safety impacts
- Economic impacts
- National security & government functionality impacts
- Symbolic impacts

Correlation Factor

DHS is requiring an estimate of potential fatalities and injuries resulting from a postulated threat as part of the screening analysis. Due to issues surrounding industry providing this type of human impact estimate it has been proposed that a Correlation Factor(s) or "C" Factor(s) be developed.

The goal of the development of a C Factor is to minimize the impact on industry in performing human impact calculations, while providing a better estimate of potential fatalities and injuries than is provided by current government and industry data related to the human health impacts of chemical releases asset database. The C Factor is based on existing industry information, specifically the EPA RMP Worst-Case Scenario (WCS) calculations for endpoint distance and exposed population. The factors convert the submitted RMP exposure estimates to provide an improved estimate of security-related consequences as opposed to estimates for accidental releases.

To develop the C Factor, the ALOHA dispersion model was used to evaluate downwind distance and impact areas and probit equations were used to develop toxic endpoint concentrations consistent with onset of fatality/injury. As a quality check, the C Factor results have been benchmarked against actual events and reviewed using more sophisticated consequence modeling and impact methods.

The next steps in the database effort are:

- Work with an IT consultant to translate the three forms into a web-based application for submittal of information.
- Development of a User's Guide on data submission and use of the web-based product.

Protected Critical Infrastructure Information (PCII)

The RAMCAP voluntary survey for screening assets will make use of the Protected Critical Infrastructure Information (PCII) program rules. The team is working with the PCII Program Office of DHS to ensure that this is done completely.

Protected Critical Infrastructure Information (PCII) is covered under Section 214 of the Critical Infrastructure Information Act of 2002, Title II, Subtitle B, of the Homeland Security Act of 2002. Information can be protected under the CII Act only if it is voluntarily submitted to DHS. It is protected in the same manner as Classified (Secret) data, but it is not Classified. The PCII Program Office provides training to government employees/contractors who need to process or have access to PCII data. Individuals and their companies must sign legal documents in order to handle PCII data. Breaching these agreements carry significant penalties.

2. SVA Methodology

The RAMCAP project objective is to create a set of sector-specific SVA tools that are:

- Consistent across sectors;
- Appropriate to sector capabilities;
- Reflective of asset owner/operator concerns, strengths and weaknesses; and,
- Able to capture those datum points which support DHS information needs.

These SVA tools should:

- Be based upon specific metrics, the use of which is repeated sector to sector;
- Employ specific, defined consequence generators (threat scenarios);
- Evaluate:
    - Consequences (impact produced by the defined consequence generator define this term);
    - Vulnerabilities (potential point targets and/or attack vectors define this term);
    - Attractiveness of the target(s) (a broadly accepted surrogate value for probability/frequency);
    - Countermeasures (including factors in mitigation, deterrent factors, detection factors, delay factors, response capability, and inherent robustness);
    - Actions/countermeasures at different threat levels; and,
    - Residual security vulnerability (gap analysis).

The purpose of the SVA tool is to advance sector organization efforts to:

- Integrate key features of RAMCAP that cover Security Vulnerability Assessment (including threat and consequence analysis) into existing sector-specific methods, metrics and documentation; or,
- Assist sector organizations in developing new sector-specific Security Vulnerability Assessment methods, metrics and documentation as appropriate.
- Comparison of cross sector security risks with common metrics.

The RAMCAP team will develop the Security Vulnerability Assessment methodology and guidance on the use of the SVA method. This is being modeled on chemical/process industry SVA methods that have been successfully, used, vetted by regulators, and withstood the test of time. Work on the SVA methodology and guidance is on-going.

RAMCAP Path Forward

RAMCAP is an example of a public/private sector cooperative effort for improving homeland security consistent with national goals and plans. It is an ambitious project, and relies on the input of security experts from industry. The challenge is to find solutions that effectively reduce the nation's vulnerabilities without creating unnecessary new burdens for industry and government.

Process Control Systems in the Chemical Industry: Safety vs. Security

Jeffrey Hahn, Donna Post Guillen, Thomas Anderson

Idaho National Laboratory, Control Systems Security and Test Center
Idaho Falls, Idaho
Jeffrey.Hahn@inl.gov, Donna.Guillen@inl.gov, Thomas.Anderson@inl.gov

ABSTRACT

Traditionally, the primary focus of the chemical industry has been safety and productivity. However, recent threats to our nation's critical infrastructure have prompted a tightening of security measures across many different industry sectors. Reducing control system vulnerabilities against physical and cyber attack is necessary to ensure the safety, reliability, integrity and availability of these systems. The U.S. Department of Homeland Security has developed a strategy to secure these vulnerabilities. Crucial to this strategy is the Control Systems Security and Test Center (CSSTC) established to test and analyze control systems and their components. In addition, the CSSTC promotes a proactive, collaborative approach to increase industry's awareness of standards, products and processes that can enhance the security of control systems. This paper outlines measures that can be taken to enhance the cybersecurity of process control systems in the chemical sector.

1. INTRODUCTION

Evidences of safe working environments and practices and secure working places are found in almost every chemical plant. The evidence for cybersecurity is not as readily apparent. However, keeping computers and networks free from viruses and hackers (and available for productive use) is just as important as keeping thieves and terrorists out of the plant. The purpose of this paper is to establish the need for cybersecurity for the chemical industry and to identify the common vulnerabilities of control systems.

2. DISCUSSION

The Safety Mind-Frame

Security, like safety, must be an integral part of any chemical process. Operating procedures are developed with the safety of workers, the public, and our environment in mind. Equipment is designed with safety in mind. The proper protective equipment is worn with safety in mind. Physical security surrounds the chemical plant. "Guards, gates and guns" are used to ensure the environment is safe for the workers. It protects the documents and processes that are business sensitive and classified. The right people are let through the guard gate and the wrong people are kept out.

Both physical security and cybersecurity are put in place with safety in mind. cybersecurity protects control systems to keep the chemical processes working safely and efficiently. It ensures the data are not compromised. It keeps the computer viruses, worms, Trojans, etc. from infecting the computers on the network and from affecting the control systems. It lets the right people access the controls and information, and keeps the wrong people off the controls denying them access to sensitive and proprietary information and out of the network.

Cybersecurity – Is the Threat Real?

Cybersecurity threats are real and they happen every day to people in all walks of life. With the modernization of control system equipment more systems are interconnected and more importantly, more systems are linked at some level to the internet. With each additional connection comes one more doorway by which a hacker, curious or malicious, can enter. Additionally the control system networks are becoming more public. Operations within a chemical or electrical have traditionally been closed and little was known about them to the outsider. Last year at one of the principle cyber forums in this country, a hacker gave a presentation on control systems. Interest and curiosity is rising along with the visibility.

Cyber attacks can slow the computer's response and the networks speed or bring everything to a complete halt, causing a denial of service. Cyber attacks can also cause unwanted and unexpected results. In the world of chemical processing where efficiency is critical to making a profit, an undetected cyber attack can slow the process and reduce the efficiency of the plant. An article from the Rand Corporation stated, "Attacks in cyberspace blur traditional boundaries between nations and private interests, cannot be foreseen or tracked via classical intelligence methods, and are all but indistinguishable from accidents, system failures or even hacker pranks."[1] The critical infrastructure of the United States provides many lucrative targets.

A well known example of a cyber attack that had disastrous results occurred in Australia. A contractor who had been instrumental in installing the network for a waste management company "caused millions of litres of raw sewage to spill out into local parks, rivers and even the grounds of a Hyatt Regency hotel. . . Marine life died, the creek water turned black and the stench was unbearable for residents."[2] A series of cyber attacks had been occurring for at least 2 months before the waste management company learned what was happening and identified the source of the trouble.

In January 2003, the slammer worm caused the Internet to slow down to a crawl. The Bank of America reported "that customers at a majority of its 13,000 automatic teller machines were unable to process customer transactions."[3] It is not hard to imagine how a

Chemical process could become paralyzed if a similar worm were to infect their network system.

The CERT Coordination Center has kept statistics on the number of incidents that have been reported to them between 1988 and 2003. The number of cybersecurity incidents reported each year has been rising rapidly. In 1999, there were about 10,000 incidents reported. In 2003, there were almost 140,000 incidents reported. This exponential increase in incidents demonstrates another interesting and important fact. There are more and more people who are proliferating cyber attacks. As shown in Figure 1, the cyber attacks are becoming more and more sophisticated yet the average person's knowledge required to spawn an attack is actually being reduced. One of the major reasons for this is because the internet has enabled hackers to share their experiences and techniques and obtain tools. For example, a New Jersey man was charged with computer criminal activity and attempted criminal activity for placing spyware on his landlord's computer to see how she was going to proceed collecting a $10,000 judgment from him. A service provider verified that the spyware came from his email address. He is currently out on $30,000 bail.[4]

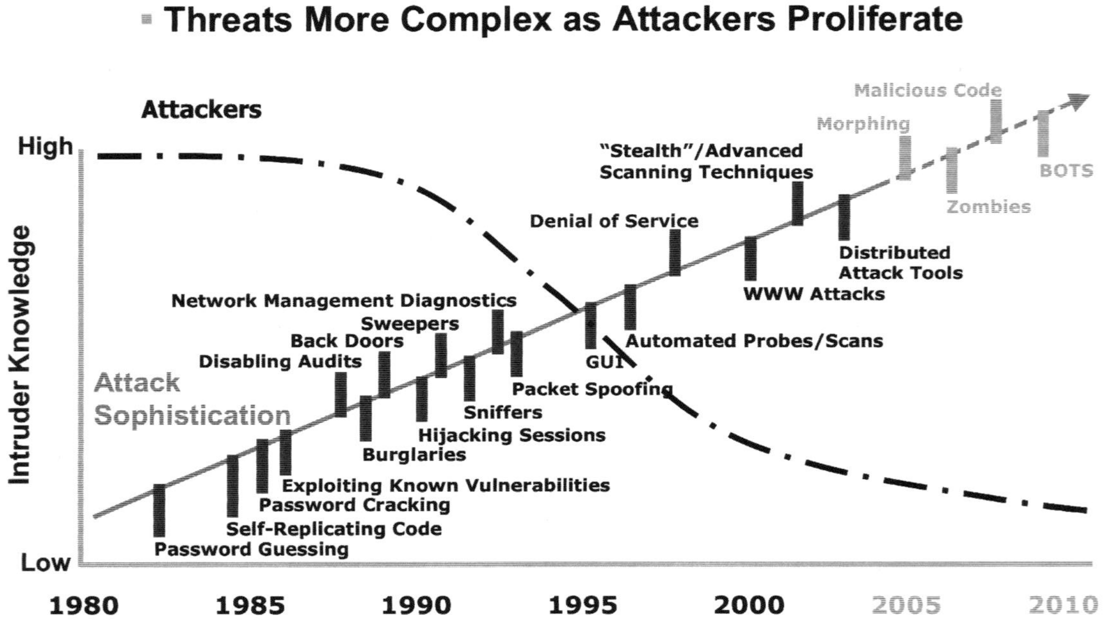

Figure 1 – Intruder Knowledge vs. Attack Sophistication

Cyber threats are real. Cyber attacks come from a wide variety of sources including, viruses/worms/trojans/etc, an insider or a disgruntled employee, neighborhood hackers, industrial espionage, terrorists. Reconnaissance is typically a pre-cursor to an attack. All of the above mentioned sources can cause problems with control systems. However, some of these sources have the intent to create problems on a very large scale. Two examples are note worthy. First, the Washington Times reported that "China is actively developing options to create chaos on the island, to compromise components of Taiwan's critical infrastructure."[5] Then Media Corp News reported that "North Korea has trained more than 500 computer hackers capable of launching cyber warfare against the United States. . . The Military hackers had been put through a five-year university course training them to penetrate the computer systems of South Korea, the United States and Japan."[6]

The U.S. Department of Homeland Security (DHS) has an objective to "create a national-level capability to coordinate between government and industry to reduce vulnerabilities and respond to the threats associated with the control systems that comprise our National Critical Infrastructure."[7] The mission of the DHS National Cyber Security Division is to be the focal point for addressing cyber security issues in the United States."[8] In order to accomplish that mission, the DHS National Cyber Security Division has established the Control System Security and Test Center (CSSTC). The CSSTC is working to identify, analyze and eliminate vulnerabilities associated with the control systems in critical infrastructure applications. A major effort is being made to increase awareness of the need for cybersecurity and what to do to reduce vulnerabilities in control systems.

Taking action to reduce or eliminate a vulnerability will reduce the probability of a successful cyber attack. There are common things that any company can do to reduce the vulnerabilities in their control system. Below are the top 10 vulnerabilities commonly found within control systems.

Top 10 Vulnerabilities in Control Systems

10. **Insufficient network separation (protection) between corporate and real-time control systems.** The corporate and process control networks should be separated by a properly configured firewall. The architecture of the network is essential to good cybersecurity. Figure 2 shows the recommended network architecture.

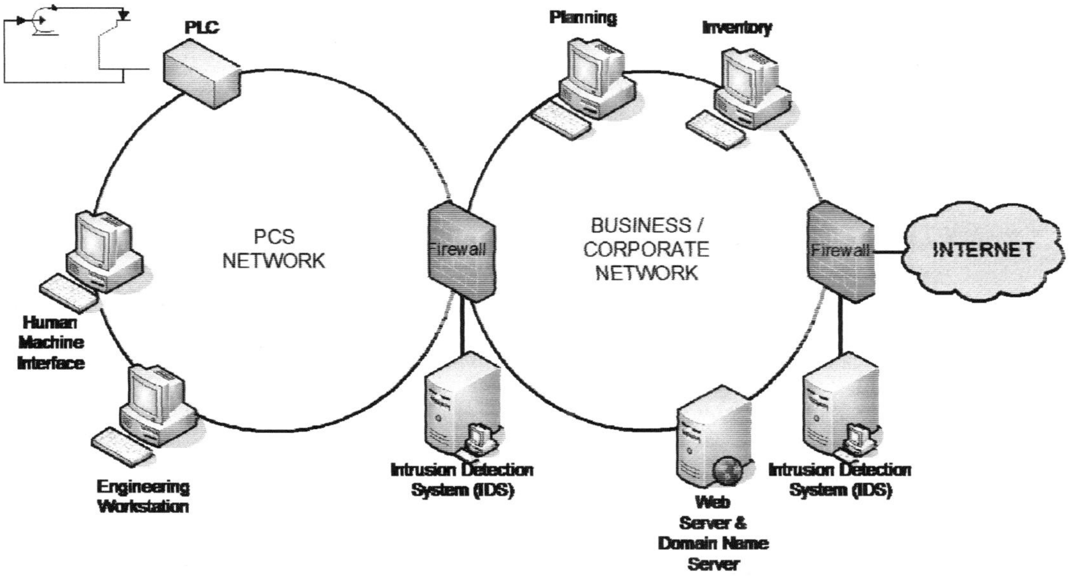

Figure 2 – Recommended Network Architecture

9. **Security efforts focused solely on corporate internet interface point.** When the IT organization assumes that if the corporate network is protected then the rest of the control system is protected it leaves the process control

network open for attack. Some of the best and most secure networks have been compromised by an employee bringing in a laptop w/an embedded Trojan or Worm, connecting to the production network and introducing the malicious code. Therefore, practicing defense in depth is an essential cybersecurity practice.

8. **Remote access to the process control LAN not properly secured.** Many control systems have modems or wireless access. These communication points need to be secured with, for example, encryption or passwords. Equipment vendors frequently leave open ports for remote maintenance of equipment.

7. **Dual-Network cards installed and in use, bridging the two networks**. If a single computer has direct access to both the corporate and process control networks through separate network cards, a hacker can bypass the firewall. This can include remote communications when the remote user connects through a commercial ISP service.

6. **Null-Session Authentication, shared folders, and "everyone" permissions defeat any internal IT controls**. Everyone needs to be trained on good cybersecurity practices. These commonly found security "no-no's" are primarily used for exploiting a system once access is gained. The fewer internal access controls are in place the easier it is for a successful penetration to turn into a successful exploitation.

5. **Process LAN and associated systems easy to find by obvious PC name**. If a hacker can make it into the process control network, it only makes life easier for them if they can easily recognize the names of the equipment, instrumentation and controllers. There are many instances of people making life easier by naming their systems with critical information, i.e., Firewall1, Primary DNS, Historian or History1.

4. **No outbound filtering of data.** Data needs to be filtered as it exits the process control network as well as filtering the data stream as it enters. The only way to know what is being transferred into and out of your network is to look at the data. Know which systems should be talking to which systems and look for anomalies.

3. **Workstations / Servers running process control applications not properly patched.** Nearly every critical exploit for the past several years has had an existing patch. Patches are created by the vendor to make their system work better and operate more securely. Keep your anti-virus software updated, too. Patching in control systems is not easy. We must test, retest and then carefully examine both the need and the net gain. If equal protection or mitigation can be gained another way for the short term then use it while you plan and patch.

2. **Policies either not in place, not followed, inadequate or not enforced (password, backup/restore, etc…).** Event the best training program and the best network architecture will be defeated if the company policies are inadequate or not enforced. Passwords have been found taped to the system, under the keyboard, in a file on the system named "Passwords" or "psswd" and even shared with coworkers.

1. **Social engineering (i.e., human factors) and physical security extremely weak.** Your friendly industrial espionage agent will tell you that a few secrets learned through social engineering or by gaining access to the network system will make the hackers' job infinitely easier. An effective counterintelligence or Operations Security program and strict adherence to policies and procedures are essential.

3. CONCLUSION

Cybersecurity is critical to success in today's business world. It helps defend against cyber attack on control systems for both the casual and the malicious intruder as well as industrial espionage. Cyber attacks are a reality in today's world, and critical infrastructures are increasingly the target. The number of cyber incidences and the sophistication of the attacks have dramatically increased. Companies can defend themselves against these attacks by reducing their process control network vulnerabilities and following good security practices and procedures. Critical infrastructures, such as the chemical industry, need to be proactive and protect their assets from potential cyber attacks and to ensure the safety of workers, the public and our environment.

4. ACKNOWLEDGEMENTS

This work was performed for the Department of Homeland Security under a Work For Others Agreement under DOE-NE Idaho Operations Office Contract DE-AC07-05ID14517.

---

5. REFERENCES

1 *Strategic Warfare*, by Roger C. Molander, Andrew S. Riddle, and Peter A. Wilson, MR-661-OSD. http://www.rand.org/news/Press.95.96/cyberwar.html

2 The Register, Hacker jailed for revenge sewage attacks; authored by Tony Smith, October 31, 2001, http://www.theregister.co.uk/2001/10/31/hacker_jailed_for_revenge_sewage/

3 MSNBC, Virus like attack slows web traffic, authored by Bob Sullivan, January 26, 2003; http://www.netserv.ch/docs/msnbc260103.pdf

4 The Star Leger, Father accused of spying with software, January 12, 2005. http://www.nj.com/news/ledger/morris/index.ssf?/base/news-4/1105511417114630.xml

5 Washington Times, Chinese Information Warfare Threatens Taiwan, authored by Bill Gertz, October 13, 2004.

6 MediaCorp News – Channel News Asia International, North Korea ready to launch cyber war: report, October 4, 2004. http://www.channelnewsasia.com/stories/afp_asiapacific/view/109911/1/.html

7 U.S. Executive Branch, White House Staff, Homeland Security Presidential Directive HSPD-7, White House Press, December 17, 2003. http://www.whitehouse.gov/news/releases/2003/12/20031217-5.html

8 Executive Branch, White House Staff, The National Strategy to Secure Cyberspace, White House Press, February 2003. http://www.whitehouse.gov/pcipb/cyberspace_strategy.pdf

9 Jonathan Pollett, Risk Mitigation – Top Ten Security Issues, UTC SCADA Security Conference, September 22, 2004

CDC's Role in Chemical Weapons Elimination Oversight

Terry Tincher, CSP, Safety Engineer
Centers for Disease Control and Prevention
Atlanta, Georgia
ttincher@cdc.gov

Abstract

The Centers for Disease Control and Prevention (CDC) is required by Congress to oversee Department of Defense lethal chemical warfare agent management and destruction activities to protect public health. Oversight activities include handling, transporting, air monitoring, and destroying chemical munitions containing the nerve agents GA (tabun), GB (sarin), VX, and the blister agent HD (sulfur mustard); and medical oversight of workers.

Public law requires CDC to review "particulars and plans" associated with transportation and disposal of lethal chemical warfare materials to ensure that public health and safety is adequately protected. This law specifies that the plan must include a description of:

- *The methods and facilities to be used*
- *The destruction schedule*
- *The management organization*

The destruction of chemical warfare agents has moved from incineration (baseline) to chemical processing with caustic products (alternate technology) to neutralize and destroy the agent. As a result of this shift, the hazard analysis needs have shifted from the primary use of Quantitative Risk Analysis and Job Safety Analysis to What Ifs, Hazard and Operability Studies (HAZOPS), Failure Mode and Effect Analysis, Human Risk Assessment and other similar qualitative and quantitative studies.

CDC does not create or dictate the process design, but rather reviews the overall plan for destruction, with a focus on hazard identification and hazard assessments. CDC's focus includes the methodology and results of hazard identification, the basis for selecting the hazard assessment methodology, and the results of the hazard assessments/safety studies.

While the Army states "A hazard analysis will be conducted for each chemical operation to include recovered chemical warfare material (RCWM), research chemical agents, and toxic chemical agents and munitions.", they do not specify what type of analysis is required. It is required that the hazard analyses be performed on the total system not just the chemical agents themselves. Furthermore, each hazardous condition is to be assigned a Risk Assessment Code (RAC) as defined in Army Regulation (AR) 385–10 and AR 385–61. Hazard analysis documentation must show

- *The RAC before a control is placed on the hazard,*

- *The use of a control to eliminate or mitigate a hazard*
- *The RAC after application of the control.*

As each phase of the facility designs develops, CDC requires documentation sufficiently comprehensive to fully understand the health and safety risks and protections for the facility involved in start-up, operations, and maintenance. This documentation includes:

- *Technological approach and methodology*
- *Design requirements, standards and regulations*
- *Health/risk assessment (including operational risk) criteria, approach/methodology (including basis for methodology and personnel), and schedule*
- *Design completion schedule*

The final step in analyzing the systems is the establishment by the facilities of Limiting Conditions of Operations (LCOs). LCOs are identified for system configurations that result in a reduction in safety margin. LCOs define the plant configuration for which the plant has been designed to safely operate and the plant condition which must be maintained. At no time is a plant allowed to operate outside the limits established by the LCO and technical baseline

This presentation will focus on the role CDC has in oversight of the destruction of the chemical weapons, the use of the hazard analysis in design and operation and how CDC works with the contractors and the Army to ensure the hazard analysis utilized are adequate.

Background

Recent news reports have often focused on the unsuccessful searches for weapons of mass destruction in locations overseas. What many people do not realize is that weapons of mass destruction, in the form of several hundred tons of chemical agents, currently reside within the borders of the United States. From World War I to 1968, the United States produced chemical weapons as a deterrent against use of similar weapons by other countries. These weapons are now obsolete and deteriorating with age.

Since the September 11th attacks, there have been concerns raised that terrorist groups, rogue states, and foreign governments might gain control of these weapons and threaten homeland security in the United States and other countries. Weapons of mass destruction (WMD) typically refer to three main types of weapons, those that are nuclear, chemical, or biological in nature. Though categorized together as a result of their potential destructive power, the nature of the threat for each of these weapons is very different, and the defensive measures one needs to take to respond to each varies greatly. This article focuses on the destruction of chemical warfare agents and deals primarily with the approach the United States is taking to protect both worker and public safety.

The U.S. Congress mandated that outdated chemical weapons be destroyed in the early 1970s. Public laws 91-121 (1970), 91-441 (1971), and 99-145 (1986) required that the Department of Health and Human Services (DHHS) provide public health review and oversight of the Department of Defense's plans and activities to test, transport, and dispose of chemical and biological weapons. On November 8, 1985, as part of the Fiscal Year National Defense Authorization Act (Public Law 103-160), the U.S. Congress mandated that the Department of Defense be responsible for establishing a Chemical and Biological Defense (CBD) Program. U.S. Code Title 50, Sections 1521 through 1523 provided the legal foundation for weapons disposal activities.[1]

In 1997, the United States ratified the United Nations International Chemical Weapons Convention treaty. By participating in the treaty, the United States agreed to destroy its stockpile of aging chemical weapons -- principally mustard agent and nerve agents -- by April 29, 2007. A 5-year extension has been recently agreed in principle by the CWC.

The U.S. national stockpile of lethal chemical warfare agents primarily involves six toxic chemicals:

Agent	Common and Chemical Name	CAS Number
GA	Tabun or ethyl N,N-dimethyl phosphoroamidocyanidate	CAS 77-81-6
GB	Sarin or isopropyl methylphosphonofluoridate	CAS 107-44-8
H, HD	Sulfur mustard or di-2-chloroethyl sulfide	(HD) CAS 505-60-2
L	Lewisite or dichloro 2-chlorovinylarsine	CAS 541-25-3
T	Bis(2-chloroethylthioethyl) ether	CAS 63918-89-8
VX	O-ethyl-S-(2-diisopropylaminoethyl)-methyl phosphonothiolate	CAS 50782-69-9

Table 1- United States Chemical Stockpile Agent Types

Non-stockpile chemical warfare materiel includes former chemical weapons production facilities; recovered chemical weapons, chemical samples, and binary chemical weapons; and miscellaneous equipment, such as empty aerial spray tanks.

Approach to destruction of the Chemical Weapons Stockpile

As stated earlier, Public Law 99-145 directed the Secretary of Defense to destroy the U. S. stockpile of lethal unitary chemical agents and munitions stored at eight storage sites throughout the continental United States and Johnson Atoll in the Pacific. The sites are shown in Figure 1.

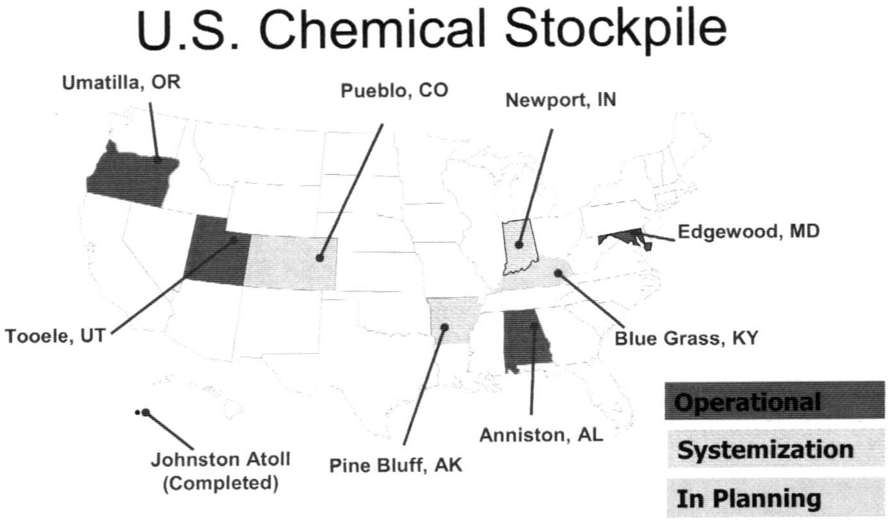

8 Stockpile Storage Sites

Figure 1

Location of 8 Continental United States Storage Sites

The law directed that the destruction be accomplished in such a manner as to provide for:

1) The Maximum Protection of the Environment, General Public and Workers involved in the destruction
2) The Adequate and Safe Facilities singularly designed to destroy the chemical stockpile
3) The Closure including cleanup, dismantling and waste disposal, of the facilities when the stockpile destruction is complete.

During the design of these facilities, their processes and the equipment utilized, the Program Manager for the elimination of Chemical Weapons oversees the system safety program as defined in MIL STD 882D. The objective of this approach is to "achieve acceptable mishap risk through a systematic approach of hazard analysis, risk assessment and risk management". System safety requirements consist of the following:[2]

1. Documentation of the system safety approach
2. Identification of Hazards
3. Assessment of Mishap Risk
4. Identification of Mishap Risk Mitigation Measures
    a. Elimination of Hazards through Design Selection
    b. Incorporation of Safety Devices
    c. Provision of Warning Devices
    d. Development of Training and Procedures
5. Reduction of Mishap Risk to an Acceptable Level
6. Verification of Mishap Risk Reduction
7. Review of Hazards and acceptance of residual Mishap Risk by the Appropriate Authority
8. Tracking of Hazards, their Closures and Residual Mishap Risk

CDC Roles and Responsibilities During Design, Construction and Operation Of Chemical Agent Disposal Facilities

The mission of the Chemical Weapons Elimination Team is "to protect public health and safety by reviewing, advising, and making recommendations on the safe disposal and transportation of stockpile and non-stockpile chemical warfare agents". The focus is prevention rather than response and preparation.

In addition, CDC reviews and provides comments on the design and implementation of the air-monitoring program. CDC will also review and comment on the laboratory management organization, laboratory practices, and quality control/quality assurance parameters. CDC Staff reviews and provides comments on the adequacy of medical clinic facilities, capabilities, and staffing. Finally, CDC will provide a representative to participate in major operational readiness review (ORR) demonstrations or preoperational surveys as an observer that could provide observations or concerns directly to the ORR/ pre-operational survey chairperson.

The destruction of chemical warfare agents has moved from incineration (baseline technology) to chemical processing with caustic products (alternate technology) to neutralize and destroy the agent. The incineration facilities were primarily evaluated for hazards using Failure Mode and Effects Analysis (FMEA), Fault tree analysis and Quantitative Risk Analysis (QRA) for external event studies during the development of the design and the occasional Hazard and Operability Analysis (HAZOP) and Job Hazard Analysis (JHA) at the plant level. After events of 9-11, there was a significant impact and acceleration of the activities of chemical demilitarization destruction and, as such, the projects required expedited hazard analysis. The need shifted from the traditional FMEA and fault tree approaches, which are time consuming, to combinations of What If, HAZOP, FMEA, Human risk analysis and Fault tree methodologies.

To accomplish the review of the studies, CDC has established the following minimum documentation requirements:

CDC requires documentation sufficiently comprehensive to fully understand the health and safety risks and protections for the facility involved in start-up, operations, and maintenance. This documentation includes:

- Technological approach and methodology
- Design safety basis, including requirements, standards and regulations
- Health/risk assessment (including operational risk) criteria, approach/methodology (including basis for methodology and personnel), and schedule
- Design completion schedule

The final step in analyzing the systems is the establishment by the facilities of Limiting Conditions of Operations (LCOs). LCOs are identified for system configurations that result in a reduction in safety margin. LCOs define the plant configuration for which the plant has been designed to safely operate and the plant condition which must be maintained. At no time is a plant allowed to operate outside the limits established by the LCO and technical baseline.

As part of the oversight role delegated to CDC, several partnerships have been formed with state and local agencies such as the State Health Departments, local medical facilities and State Environmental Departments. CDC regularly provides guidance and expertise to address specific issues and concerns of the public and state officials. CDC also works with public interest groups, such as the local Citizens Advisory Committees and federal agencies such as the Department of Defense, the Environmental Protection Agency, and the National Institute for Occupational Safety and Health to address issues for workers and the local areas.

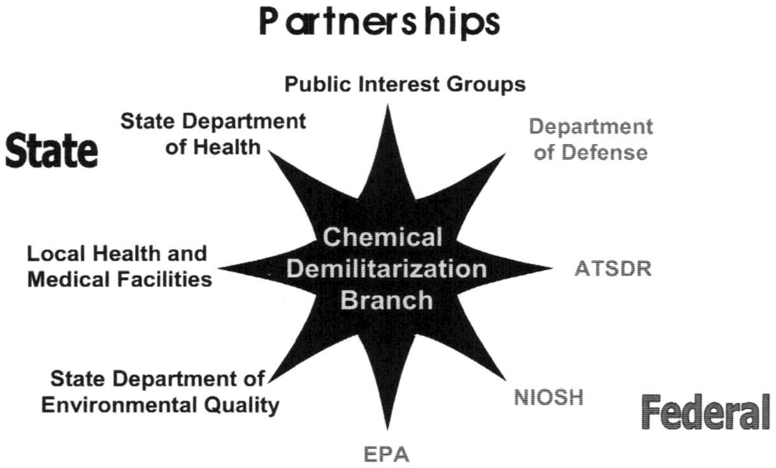

Figure 2
CDC Partnerships in Chemical Weapons Disposal

Considerations for Hazard Review Methodologies for Chemical Weapons Disposal Facilities

In the course of a project to design and later, to operate a Chemical Disposal facility, it is necessary to employ a series of hazard reviews to ensure that an adequate review is performed to identify, assess and either eliminate or control hazards. A series of reviews are necessary in increasing depth as more detailed information is developed during the course of the project. The current requirements include a series of Hazard Reviews which are: [3]

- Preliminary Hazard List (PHL)
- Preliminary Hazard Analysis (PHA)
- Subsystem Hazard Analysis (SSHA)
- Operations and Support Hazard Analysis (O&SHA)
- Systemization Safety Review (SSR)
- Operating Plant Hazard Analysis

In addition special reviews need to be carried out for design and plant modifications. Currently, there is no single hazard identification technique which will ensure all hazards are identified. Each methodology has its own strengths and weaknesses. It is usually necessary to employ a number of different methodologies, some deductive and some inductive, to ensure as broad a coverage as possible.

Typical Hazard Analysis Methodologies include the following: [4]

- What If - brainstorming alone
- Checklist - experience and standards
- Hazard and Operability Study (HAZOP) - brainstorming with guide words
- Failure and Effects Mode Analysis (FEMA) - component failure possibilities
- Layer of Protection Analysis (LOPA) - structured system/ component failure probabilities
- Fault Tree - structured system/ component failures
- Quantitative Risk Analysis (QRA) - structure system/ component failure probabilities based on engineering evaluation and mathematical technique.

The design contractors and the Army, working with risk analysis specialists, developed their own database for failure rates in the chemical demilitarization incinerator facilities for use in Quantitative Risk Analysis. This database was initially used in the Anniston Chemical Disposal Facility QRA and subsequent incineration sites. As the design focus shifted from the traditional incineration process to alternative technologies, the database was utilized along with published failure rates from AIChE, OREDA, IEEE etc for traditional qualitative hazard analysis like FMEA, What If and HAZOP to determine the failure rate of the equipment.

The chemical demilitarization contractors face many unique challenges in the analysis of hazards for the equipment used in the disposal of chemical weapons and chemical agents. The events of 9-11 accelerated the design efforts, requiring quick turnaround of hazard analysis and quantitative information. The teams developed and utilized a combination of What If, HAZOP, FMEA, Human risk analysis and Fault tree methodologies. Human risk analysis became more critical as many of the new designs incorporated a variety of man-machine interfaces to resolve complex operations and procedures.

Each site has a different primary contractor, which can create issues with communication of hazards and issues that were faced. The teams and contractors have developed a database of lessons learned and have worked together to address common safety issues. These contractors and Army personnel have forged a working relationship that recently played a key role in a Human Reliability Analysis (HRA) performed at CDC's request on equipment in a newly designed site. The HRA team consisted of a third party risk assessor, the design contractor and experts from the operating sites. The ensuing study first combined Job Hazard Analysis on existing equipment, lessons learned from an operating site, and focus on operating procedures involving both procedural review and observation. These items were primarily utilized to screen the processes and identify those areas where the HRA was required. The HRA utilized fault trees for quantitative assessment of the hazards and the team suggested possible solutions. Another team developed a methodology called the Semi Quantitative Analysis- Failure Mode and Effects Analysis (SQA-FMEA). This methodology is similar to Layer of Protection Analysis (LOPA). In this methodology, multiple failures are evaluated on selected events by using fault trees and the frequencies of the top event are then used determine the risk assessment level. There will be more information about these methods in a future paper.

CDC is recommending the sites consider a hazard analysis revalidation after one year of operation. The advantage to this is that plant changes, incidents and accidents (along with the changes that were incorporated) and any new information from other sites would be examined in total on a regular basis. The review would allow all the changes that occurred since operations began to be integrated into the hazard review documentation. Incorporation and integration of these reviews allows the hazard review to become "evergreen".

CDC focuses on the management systems and looks for "Characteristics of a good Management System" which include a clear definition of process change and replacement-in-kind, a process change risk ranking or risk screening system, control of changes initiated by the work authorization system, control of new chemicals and spare parts through purchasing procedures and consideration of equipment decommissioning. [6]

Another key area is how the site deals with Process Change both from a design and operations standpoint. The Chemical Center for Process Safety (CCPS) defines a Process Change as "A temporary or permanent substitution, alteration, replacement (not in kind), modification by addition or deletion of Critical process equipment, applicable codes, process control, catalysts or chemicals, feedstocks, mechanical procedures, electrical procedures, safety procedures, emergency response equipment from the present configuration of the Critical process equipment, procedures, or operating limits". Further, British Standard 6488 [1] defines a process change as "The discipline of identifying the components of a continuously evolving system (taking into account relevant system interfaces) for the purposes of controlling changes to these components and maintaining integrity and traceability throughout the system life cycle".

Finally, the U.S. Chemical Safety and Hazard Investigation Board stated the following in a 2001 Safety Bulletin on MANAGEMENT OF CHANGE to emphasize why process change is considered so important (and why CDC focuses on it): [7]

"The Occupational Safety and Health Administration's (OSHA) Process Safety Management standard and the U.S. Environmental Protection Agency's (EPA) Risk Management Plan require covered facilities to manage changes systematically. It is good practice to do so, irrespective of the specific regulatory requirements."

"If your organization has an MOC policy, review it to be sure that it covers operational variances in addition to physical alterations. If you do not have a systematic method for handling changes, develop and implement one."

Prior to the start of chemical agent destruction facility operations or after major changes requiring reconfiguration of the process equipment, a comprehensive Operational Readiness Review and Readiness Assessment (ORR) is conducted. The ORR is based on a similar program that is utilized by the Department of Energy and is a disciplined, systematic, documented, performance based examination of the facilities, personnel, procedures, management control systems and Safety Management programs. The purpose of the ORR is to ensure the readiness of the facility and personnel to start the facility within the boundary of acceptable risk and to ensure the facility has adequate Safety management programs and controls in place to operate within those boundaries. DOE defines acceptable risk as that within the safety basis of the authorization agreement and in compliance with applicable environmental, safety and health regulations. What the ORR is not is a tool of line management to achieve readiness, instead the reviews provide an independent confirmation of readiness to start or restart operations.

Summary

CDC's role must be consistent with the public law that requires CDC to review "particulars and plans" associated with transportation and disposal of lethal chemical warfare materials to ensure that public health and safety is adequately protected. This law specifies that the plan must include a description of the methods and facilities to be used, the destruction schedule and the management organization.

CDC does not create nor dictate the design itself, but rather reviews the plan, with an emphasis on hazard identification and hazard assessments. CDC comments and recommendations address findings regarding the methodology and results of hazard identification, the basis for selecting the hazard assessment methodology, and the results of the hazard assessments/safety studies. Hazard and risk analysis are used throughout the life cycle of development of a chemical destruction facility to identify and evaluate

Today, the U.S. Army is recognized as a world leader in international chemical weapons elimination efforts. Over a decade of experience has demonstrated that these weapons can be destroyed safely, without harm to employees, to the community, or to the environment. The use of hazards analysis and the meticulous approach to risk assessment have in no small part, contributed to this record.

REFERENCES

[1] Centers for Disease Control and Prevention. "Overview of U.S. Chemical Weapons Elimination." <http://www.cdc.gov/nech/demil/overview.htm>

[2] Department of Army AR-385-16, "System Safety Engineering and Management." May 1990, Washington, DC.
[3] Guidelines for Technical Managers of Chemical Process Safety, 1989, Centers for Chemical Process Safety, New York, NY, pp 45-57.
[4] Guidelines for Hazard Evaluation Procedures, 2nd Edition, 1992, Centers for Chemical Process Safety, New York, NY, pp 51-71.
[5] Guidelines for Auditing Process Safety Management Systems, 1983, Centers for Chemical Process Safety, New York, NY, pp 5-6.
[6] Guidelines for Integrating Process Safety Management, Environment, Safety, Health and Quality, 1996, Centers for Chemical Process Safety, New York, NY, pp 1-10.
[7] Management of Change, Safety Bulletin No. 201-04-SB, August 2001, U. S. Chemical Hazards and Safety Investigation Board, Washington DC. http://www.csb.gov/safety_publications/docs/moc082801.pdf

20th Annual Center for Chemical Process Safety (CCPS) International Conference - Risk Management: The Path Forward
April 11 – April 14 2004, Atlanta, Georgia
Plenary Session 4 Monday April 11 – Risk Analysis

Integrity of Piping Systems: Nature and Scope of the Problem Viewed from a Regulators Perspective

A paper by **M Skellett** of the Health and Safety Executive, Newcastle upon Tyne, UK

ABSTRACT

The integrity of piping systems poses significant challenges for Operators and Dutyholders. Historical failure data suggests that piping failures are among the most common on process chemical sites and there have been a number of recent serious incidents in the UK. These together with issues such as extending life and aging plant, especially in the petrochemical and process industries, continue to be a source of concern to the regulators.

Piping systems are not individual discrete items of plant like machines or vessels. They do not have readily identifiable boundaries and can be extensive, fragmented and often located in the most in accessible places. UK legislation is complex and can serve to confuse Dutyholders in the measures required to deliver integrity assurance.

Incidents involving piping already evident to the UK Health and Safety Executive (HSE) became even more obvious with a number of serious failures in 2000/2001. A revitalising target for Hazardous Installations Directorate (HID) was to secure progress towards a reduction in the numbers of loss of containment incidents by 20% over a five year period. Plant Integrity interventions were to form one of four core elements in the strategic plan. As a pilot a project study into pipework integrity management at the UK refineries was carried out in 2002/2003.

This paper seeks to outline the nature and scope of the problem viewed from a regulators perspective. It gives a summary of the project findings and discusses aspects of vulnerabilities of piping systems, failure modes, relevant legislation and adoption of best practice for in-service integrity.

Integrity of Piping Systems: Nature and Scope of the Problem Viewed from a Regulators Perspective

1. Introduction and Background

Maintenance of the integrity of pressure systems and avoidance of loss of containment is the essence of loss prevention. Catastrophic failure of properly designed and constructed pressure vessels is rare. Most failures occur not in the vessel but in the rest of the system, including pipework, valves and other fittings.

Incidents involving piping already evident to the UK Health and Safety Executive (HSE) became even more obvious with a number of serious failures in 2000/2001. A revitalising target for Hazardous Installations Directorate (HID) was to secure progress towards a reduction in the numbers of loss of containment incidents by 20% over a five year period. As part of the operating plan, a project study into pipework integrity management at the UK refineries was carried out in 2002/2003.

This paper gives a summary of the project findings and seeks to discuss aspects of vulnerabilities of piping systems, failure modes, relevant legislation and adoption of best practice for in-service integrity.

2. Failure data

There is a vast amount of data available on pipework related failures, but the range of values is wide and can be confusing. In general, failure rates tend to be higher for small than for large diameter pipes. Some statistics drawn from a number of sources are given in the table below.

Failure Data

Source	Period	No of incidents	Cause
MHIDS	1973 - 2003	221	Various
OSD Hydrocarbon Releases	1992 - 2002	2312 total 241 in 2001/2002 (62% pipework related	corrosion/erosion - 28% small-bore fatigue - 25% flange failure - 17%
HSL Loss of Containment Analysis	1991 - 2002	2500 investigations 718 studied	flange - 7.0% weld - 1.3% body - 6.4%
EU Reportable Incidents	1999 – 2003	21 reported (8 pipework related)	bellows leak - 1 fatigue failure - 3 open ends - 1 erosion/corrosion - 3

The MHIDAS[1] database identifies 221 events between 1973 and 2000 with many direct and indirect causes. Typical causes however, are unqualified engineers and thieves cutting through ammonia or fuel oil pipelines with no persons killed, injured, or evacuated, to worst case events such as the fracture of a bypass with 28 killed, 89 injured and 3000 evacuated. (The 1974 Flixborough disaster).

Over the past eleven years HSE's Off-Shore Division (OSD) has conducted a number of studies of hydrocarbon leak incidents on oil and gas platforms in the North Sea. In summary, the latest information - 2002 release statistics[2] - concluded that the distribution as a percentage, according to equipment, revealed that the majority were related to piping. Current data suggests that 62% of all releases were associated with some form of piping failure, 28% from corrosion, 17% were from flanged joints and 25% from small-bore piping. When analysing underlying causes the largest contribution was from inadequate inspection/condition monitoring.

Usefully these studies also point to where the likely problems are found e.g. flanges, corrosion, small-bore fatigue failure etc. The data also aligns closely to a survey carried out on behalf of HSE by Engineering Consultants ABB Engineering Services[3]. That study revealed that the causes of failure ranked in order were: -

- leakage at flanged joints
- leakage at corroded pipe (especially underlagging)
- leakage at small-bore pipework (e.g. due to fatigue)
- failure of supports
- leakage at bellows
- leakage at instruments
- failure of steam trapping
- modifications
- wrong materials
- overpressure

Research undertaken by HSE's Health and Safety Laboratory on Loss of Containment[4] using the Reporting of Injuries, Diseases and Dangerous Occurrences Regulations (RIDDOR) database revealed that pipework (including pipe flange, weld, body and open end) was the site of 23% of all releases. Of these 51% occurred during normal operation and 32% during maintenance.

As part of the Competent Authority HSE has a duty under COMAH legislation to report back to the Commission of the European Union those accidents deemed as 'Major' on COMAH regulated sites.[5] The overall sample size is small, however, in the four reporting years since the legislation came into force 21 have been recorded. Of these 8 are associated with piping failures. By comparison the next nearest in numbers for equipment types are valves, which account for 3 failures, and 3 more involving Road Tankers.

3. Piping Systems Complications

Assuring integrity of pipework systems, especially on large, major hazard sites, is a significant challenge for Dutyholders. It can represent a far more complex process than that for other plant such as vessels and machines.

Some of the issues include: -

- The pipework of interest may involve many systems and vast number of lines.
- Access can be difficult, dangerous and expensive. Much process pipework is in racks at high level, buried, or laid in trenches. Geometries can be complex. A large proportion is lagged, some within asbestos insulation.
- Boundaries of pipework systems can be hard to identify compared to a vessel for example. They often cross many areas of ownership and this can lead to confusion and risks of unmanaged sections.
- Identification of 'at risk' pipework and decisions on formal registration is a complicated process and no definitive rules apply. Size and complexity can vary widely. Within a single system there may be significant differences in duty and/or materials of construction. Larger and more complex systems may have to be broken down into manageable sub-sections.
- Inspection processes for pipework are not nearly as well developed as those for other plant.
- There can be complications and difficulties with legislation.
- The amount of data from inspection is considerable with ample possibility for errors and misjudgements.
- Aging plant, particularly with carbon steel the material of choice

In many ways maintaining pipework integrity may be considered more a logistical, economic, and management challenge, than a technical one.

4. UK Legislation, Codes and Standards

In the United Kingdom Operators and Dutyholders chemical process sites are governed by many statutory regulations and often align their own integrity management systems to differing codes and standards. For piping systems these can encompass: -

4.1 Legislation and Regulation

- Health and Safety at Work Act 1974 (HSW 1974)
  An enabling act which requires employers to maintain safe systems of work to protect their employees and third parties.

- Management of Health and Safety Regulations (MHSWR 1992)
  One of a suite of European Regulations a core part of which requires employers to undertake risk assessments of the activities undertaken on their premises.

- Provision and Use of Work Equipment Regulations (PUER 1998)
  Regulations which set out important health and safety requirements for the provision and safe use of work equipment.

- Control of Major Accident Hazards (COMAH 1999)
  In particular Regulation 4 of the Regulations which require that Operators and Dutyholders demonstrate that all necessary measures to prevent a major accident have been taken.

- Pressure Systems Safety Regulations (PSSR 2000)
  This legislation focuses on the hazards relating to 'stored energy' of 'relevant fluids' in pressure equipment and prescribes integrity assurance through specifying duty limits and formal Written Schemes of Examination.

- Pressure Equipment Regulations 1999 (PER 2002)
  Derived from a European Directive to promote trade free barriers the regulations affect the manufacture of pressure equipment within the EU.

- Pipeline Safety Regulations 1996 (PSR 1996)
  These Regulations apply only to those systems which extend beyond the boundaries of premises. They are particularly relevant to the UK national gas transmission pipelines.

4.2 Codes of Practice

- Piping Inspection Code: API 570 2nd Edition 1998[6]

- Institute of Pressure Piping Systems Examinations: Model Code of Practice IP 13 2nd Edition[7]

- Safety of Pressure Systems – Approved Code of Practice L122[8]

The primary legislation in respect of piping system integrity for on-shore chemical sites is COMAH and PSSR (2000). There are important limitations however. COMAH legislation is only relevant where COMAH substances exceeding threshold quantities apply and there are many hundreds of chemical Petroleum process sites, which by virtue of lower inventory, do not fall into this regime.
Note: there are approximately 400 COMAH Top Tier and 1000 Lower Tier establishments within the UK

Likewise with PSSR 2000 it is key for Operators to recognise that these Regulations do not address safety or environmental issues that may result from the nature and properties of the conveyed fluid. In many ways the Regulations are not phrased to be pipework integrity friendly. The wording of the Regulations, and its associated Approved Code of Practice L122 in respect of pipework, has long been a matter of debate.

Any enforcement action considered by the regulator where COMAH and PSSR (2000) Regulations do not directly apply may then have to be taken under one or more of the others above.

5. UK Refineries Project (2002/2003) – Piping System Integrity

In the 2002/2003 work year, Inspectors in HID's Chemical Industries Division carried out a Plant Integrity Initiative on pipework integrity management at UK Refineries. Nine sites were visited over a period of just less than a year.

It is worth stating here that inspection and maintenance on its own is recognised as only representing a part in affording overall integrity assurance. Other elements such as designing to code, selection of appropriate materials of construction, good operational controls, avoidance of abnormal conditions, corrosion monitoring, use of treatments and inhibitors, change control procedures etc. play an equally important role. Nevertheless, this study was limited to, and focussed on, in-service inspection and maintenance issues of piping systems contained within the site boundary.

Refineries were first chosen to inspect because:

- High standards of integrity management were expected.
- They include the most complex systems
- Degradation mechanisms are numerous and occasionally unpredictable
- They employ a wide variety of inspection techniques
- The age of some systems is becoming significant
- There was a background of recent serious incidents

The primary objectives were to:

- Raise the overall profile of pipework integrity assurance and increase awareness of the problems of releases
- Benchmark good practices at sites, and promote their use across the whole chemical sector

The inspections were focussed on a number of key areas: -

- The corporate and site procedures used to drive the integrity management process
- Examination of the rationale for the selection and classification of piping systems
- Methodologies and standards of in-service integrity through inspection and maintenance
- Resources and competencies
- Physical inspection of piping systems

A key part of the interventions was the physical examination of a range of piping systems. This allowed judgements on the actual condition of the plant, afforded a check of the methodology, effectiveness and exactness of the process. It also provided the root of enquiry into registration and classification practices.

Identifying representative systems for verification was not straightforward. Selection was made to cover those with aggressive deterioration mechanisms and a high probability of failure, and those with a low failure probability but where potential consequence was high. Both on and off-plot systems were selected. Where available, information and intelligence was drawn from site COMAH Safety Reports.

Although not considered in a major accident context, utilities were also examined during the inspection.

As part of the exercise, checks were also made in regard to topic specific areas including:

- Corrosion under insulation
- Small-bore vibration fatigue related failure
- Deadlegs and injection points
- Buried piping
- Bolted flange joints
- On-line sealing

A report of the findings of the project is available on the HSE web site www.hse.gov.uk/chemicals/spctechgen33.htm [9] and a summary of the main conclusions is given below.

6. Project Findings

6.1 Policy and Procedures

HSE recognises that good policy and procedural documents do not necessarily translate to first-rate integrity management and controls. Nevertheless there was an expectation that the companies visited, all of national and international repute, would adopt and maintain high quality policies and procedures. There was also an expectation that references to industry specific codes described in this paper would be widely used. However, this was not always the case.

Considerable variation of both policy and technical procedures was seen and common shortcomings were found. Almost without exception documentation was not up to date e.g. referencing outdated legislation or inspection procedures that were no longer employed. Navigating some documentation was a challenge in itself. Inclusion of corporate with local site procedures complicated the understanding of the processes in some cases and in others inconsistencies and duplication of information was common.

At most sites integrity management procedures were migrating from conventional time-based schemes to fully or partial risk-based techniques. Substantial investment has been made by some in order to facilitate these changes. The changes were often linked to the introduction or updating of computer based maintenance/inspection software systems. Only in a handful of cases was this transition complete. A degree of catch up was taking place regarding implementation, training and administrative elements.

HSE supports the move to more robust and systematic regimes which RBI can deliver. Though resource intensive in the first instance, the analysis and rigour of the assessment process carried out by a multi-disciplinary team is arguably the most powerful means to delivering a successful integrity management system. From a regulators point of view we found it disappointing that so many providers of RBI architecture emphasised cost saving when describing the benefits. However, for most sites, the move to RBI for their piping systems was driven by a desire to advance the integrity process.

The majority of procedures did provide some basic information on failure modes and inspection techniques. However, most did not take account of site or unit specific issues. With the exception of one site where very detailed guidance had been produced, the value of providing specific information on those factors impacting on integrity at a unit level had generally not been recognised.

There appeared to be little detailed technical information and guidance to assist plant inspectors in targeting inspection effort. The link between anticipated degradation, the likely places affected, and the most appropriate techniques to identify it were not well documented. Considerable reliance was placed upon inspector judgement to correctly identify the vulnerable areas and scope out the degree of inspection. This in turn relied heavily on the competence of inspection engineers. Although occasionally captured by audit, almost no peer review and oversight of inspector judgements was apparent.

6.2 Registration Practices

One of the main objectives of the intervention was to quantify the relative proportions of registered and unregistered systems or lines and determine the rationale behind selection. The question put to Operators was, "are there any pipeline systems at this site, inside or outside of battery limits, containing hazardous fluids, which are unregistered, not inspected, and, should failure occur, have the potential to cause an unacceptable event?" Many found this question difficult to answer. The following discussion offers some possible reasons for this.

In the UK, registration of pipework systems i.e. the process of identification and inclusion within formal pro-active periodic inspection and maintenance regime, only really began in the early to mid 1990s. A principal motivational factor was arguably recognition of the requirements in 'Pressure Systems and Transportable Gas Containers 1989' (the fore-runner to PSSR 2000) for 'Written Schemes of Examination' of pressurised plant. Previously, inspection of pipework was generally more ad-hoc. Consequently, the processes for pipework inspection, integrity management and data recording were not nearly as mature and well developed as those for pressure vessels.

The process of registration varied across the sites visited. But almost all began by identifying those lines which fell within legislative requirements (e.g. PSSR 2000) and often described as Class A systems. Those that did not (Class B) were left for a second pass assessment. Estimates varied considerably, but the former may account for as little as 30% of the total on most chemical process sites.

Operators demanded unambiguous evidence of <u>both</u> serious consequences and deterioration mechanisms before registering Class B systems. Often, it appeared that methods were very much biased towards confirming that systems could be left 'unregistered' rather than capturing those for which formal registration was appropriate. Rightly or wrongly, advantage, it seems, has been taken of a section in L122 the ACoP to the Pressure Systems Safety Regulations, which states '*the effect of Regulation 8 is to enable the exclusion of most pipework from Written Schemes of Examination where appropriate*'.

A majority, if not universal position appeared to be that where no deterioration mechanisms were identified then there was no need to register or include in the inspection programme. This is at odds with most other equipment at high hazard plants, especially vessels. There was wide variation in the extent of inspection coverage and abundant evidence of systems with no inspection history.

As a result the likelihood of failure from material degradation was generally decisive in the decision making process. Systems where the probability was assessed as low or negligible were often not registered even though the potential consequences of failure were high. The obvious weakness is that if the degradation assessment is flawed then the system may be open to threat. HSE has found it common that failures have occurred because an activity or condition has been overlooked at the evaluation stage.

During the interventions we found almost total agreement among the Operators that present schemes were not as thorough and comprehensive as they should be. Of particular concern were those areas where the predicted consequences of failure are significant. Invariably these had been excluded on the basis of assumptions of low degradation and thus low failure probability.

However, it is HSE's judgement that this argument cannot be sustained without inspection evidence to confirm these assumptions.

It is anticipated that revised programmes will increase the proportion of registered and formally inspected pipework – in many cases from 20% to 80%.

6.3 Corrosion under Insulation

Corrosion under Insulation (CUI - sometimes referred to as underlagging corrosion or ULC) is still a serious deterioration mechanism affecting pipework on most sites. Many Operators had little in the way of evidence of specific policies and procedures to deal with this problem. In some cases operators were using guidance that was over 20 years old, though this was recognised.

Some progress was being made however. Two Operators had up to date information on identification issues and protective measures. One of these was of particularly high quality – their recent internal guidance on prevention and detection of corrosion under insulation identified that many of the worst examples of CUI were either self inflicted or linked to the age of the plant. Pipework 30+ years old is common in refineries.

A leading UK Engineering Consultants had also just produced a guide with recommended approaches for the detection of CUI and its management, especially in regard of equipment with high susceptibility.

Very few sites had carried out proactive insulation requirement surveys. Our inspections revealed many examples where insulation was no longer required. At one site, where CUI was identified as the most credible failure mechanism, insulation standards were well below what was acceptable. Design of cladding, particularly around supports and joints was poor. Reinstatement following maintenance and break-ins and water tightness was not well controlled.

6.4 Small-Bore Piping

Small-bore pipework failures through vibration related fatigue was identified as a significant problem. However, this topic appeared to have been given a lower priority than CUI. Some sites had taken measures such as removal of unnecessary instrumentation and reinforcing supports. Many were also conducting detailed surveys of small-bore pipework in an attempt to identify those branches at highest risk. A number of engineering consultancies have recently developed effective methodologies to identify and prioritise the highest risk areas and some had been contracted to carry out these surveys. There was increasing knowledge of recommendations and methodologies of approach given in a guidance document produced by the Marine Technology Directorate (MTD). [10]

6.5 Deadlegs and Injection Points

All Operators were aware of the problems associated to these vulnerable areas. There is no doubt that recent failures have raised their profile. Deadlegs can be created by the way a plant is operated, or when modifications to the original design are carried out. Management of Change procedures do not always capture where deadlegs have been created. Some had completed identification surveys or had them in progress. One Operator produced a unit-by unit register which included an individual hazard assessment for each deadleg. The usual hierarchy of control measures was removal (where possible), followed by others such as flushing procedures, material changes and increased inspection.

6.6 Buried Piping

Most areas of buried piping or sections through bunds, across roads etc. are found in off-plot areas (i.e. those that fall outside of battery limits). Although these are generally considered to be of lower risk, failures can have severe environmental and safety consequences. Few sites had procedures detailing how in-service integrity of theses pipelines was assured; and in particular guidance where lines might require unearthing. There was ample evidence of systems between production units not being registered.

Cathodic Protection was not the normal measure of choice to prevent corrosion related failure here. Many lines (normally bitumen coated on installation) were simply buried and forgotten and there was evidence of poor installation standards. Maintenance and remedial actions normally followed breakdown/defect identification and was often conducted only on a capital project basis.

There was general agreement that this approach was no longer tenable. Responses included bringing pipework above ground, use of improved wrapping materials, registration and periodic inspection using more modern techniques such as long-range ultrasonics. In general sites were using consequence estimates to prioritise the remedial work much more than they were doing in the past.

6.7 Flanged Joints and On-Line Sealing

Most Operators were adopting radically improved procedures and techniques for making flanged joints and bolted connections. Leaks from flanges and associated failures are well documented. Many were now following recommendations given in a new guidance document produced by the United Kingdom Off-Shore Operators (UKOOA).[11]

Not all were adopting controlled bolt loading techniques and flange tagging, especially for the smaller diameter and lower pressure systems. But there was a clear recognition that improvements could be made. Some Operators were attempting to compile registers of known troublesome joints that required superior controls. At one site, where significant benefits were seen following a turnaround where flange bolting procedures had been radically improved, convinced the Operator to extend the process to all their sites.

There was generally good evidence that strict procedures were being employed when using on-line sealing techniques. However it was apparent that some felt under corporate pressure to regard these as more permanent.

HSE's position is that temporary repairs are only approved once all other options have been considered, that they are treated as a modification and that assessments are undertaken before use, including structural analysis where necessary. Once installed repairs should be registered, made subject to formal inspection and are given a finite service life. They should be returned to the original specification at the earliest opportunity.

Increased use was being made of composite wraps to effect piping repairs. The technology, developed for military and off-shore use is now being employed for on-shore repairs.

6.8 Fitness for Purpose, Defect Assessment, Retirement Criteria and Inspection Scheduling

Only a limited number of operators had formally documented defect assessment procedures and fitness for purpose rules. In these guidance was given on the evaluation process e.g. defect acceptable, defect acceptable but not to next inspection, defect not acceptable, along with action strategies and levels of authorisation/responsibilities

Some differences of approach were observed in relation to predicting the remaining life/retirement dates of piping systems and inspection scheduling. For example, wall-thinning data was used to calculate corrosion rates and retirement limits. Graphs were then extrapolated to predict a retirement date. If the next inspection was due before this date no action was taken. At face value this approach appears reasonable. However degradation is rarely uniform and so inspection programmes should err on the side of caution. This is reflected in the corrosion rate and half-life inspection principles given in API 570.

Methodologies for defect assessment and predicting piping retirement limits were sometimes in place, but often were not referenced and incorporated into site procedures.

6.9 Inspection Processes

Most inspection and integrity departments were adequately resourced and in some cases numbers had been increased. Inspection departments generally provided a strong and competent service and were well led. All had accreditation from the United Kingdom Accreditation Service (UKAS) as authorised inspection bodies. It was gratifying to find, even where there were downsizing pressures, that Engineering Integrity Departments did not appear directly affected.

An estimated 80% of all pipework examinations were carried out on stream. Influencing factors included aspects such as access difficulty, line temperature, availability of plant etc. Unlike most other process equipment they are not bound by plant turnarounds and shutdowns. However this demands careful scheduling (an additional complication). It is in this area that dedicated software can offer significant benefits. Invasive inspection of pipework was not the norm and was usually restricted to circumstances where failures have occurred.

The volume of information arising from pipework inspections is vast, and opportunities for errors and misjudgements here is obvious. Accurate interpretation of the raw data and making informed judgements on fitness for service is a crucial element in the process. The inspection revealed that many operators were now using or considering use of proprietary software. Asset management tools such as CORTRAN and CREDO can help with the manipulation of the data. These packages are able to provide scheduling routines, isometric sketches, data tables, corrosion limits spreadsheets, and retirement dates. A key feature of these systems is that they can also identify safety critical factors by generating alarm flags for certain criteria.

Use of ‘state of the art’ NDT technology was more widespread on some sites than others. Rapid scan techniques were now being more widely used. Examples included long-range ultrasonics, where tens of metres can be checked from a single test point; corrosion under insulation detection without the need to remove insulation, and methods that give thickness data under supports. There is an increasing use of digital radiography. Full evaluation of the strengths and weaknesses of these recent developments is still taking place; however there appears to be far greater levels of confidence now than in the past.

In general, good use was made of piping isometric drawings for targeting and recording inspection findings, however the detail varied considerably. Isometrics can also be of great value during the degradation assessment stage. They show configurations such as deadlegs, elbows, branches etc, where geometry can significantly influence the degradation likelihood.

The appropriate level of detail on piping isometrics is a debatable point. Nevertheless, whilst their production is expensive and resource demanding, the value of drawings which accurately represent the piping configurations cannot be overstated.

The availability of deep specialists such as materials/corrosion engineers varied. Overall however they were not well represented. In some cases where problems had arisen specialists were outsourced. The obvious disadvantage is the lack of site-specific knowledge. When fundamental changes such as moves to RBI assessment and corrosion loop studies are in progress their input is even more demand. One operator employed two full time specialists and the value they added to the overall process was obvious. They were also able to provide corrosion training for the plant engineers and inspectors.

All Inspection departments were subject to one form of auditing or another. However, little evidence of the auditing of actual performance was found. No audits were carried out to determine whether regimes were technically compliant with good practice. Time constraints are thought to be the main reason. It is considered that departments would benefit from having clearly measurable key performance indicators.

7. Verification

Even though system verification was, by necessity, a sampling exercise, it is fair to say that many anomalies were found. Some related to data inaccuracies and dubious interpretations when setting next inspection dates. Flaws and misjudgements were also found in the registration and assessment processes. Where significant weaknesses were found Operators were advised to review their procedures.

None of the process piping systems in the sample inspected gave cause for serious concern. Most appeared well maintained although the condition and standards of insulation was poor in certain cases. Process induced vibration was also noticeable, particularly at the front end of crude units where 2-phase flow is common. The condition of utility systems, especially utility steam, was unsatisfactory at a number of sites. Individual shortcomings were reported back at close out meetings and through follow up inspection. No enforcement notices were served during this round of visits, though a number have been issued as the inspections have continued.

8. Good Practice Benchmarks

The following list summarises good practice identified during the inspection initiative together with examples drawn from existing guidance and standards. Collectively these represent a benchmark standard to be sought at all Major Hazard sites in the UK. It is intended that the list will provide site regulatory inspectors with a basis for advice or enforcement where shortcomings are found.

- A method for hazardous line identification
- Procedures for selection, registration and inspection of 'at risk lines'
- Information on measures and procedures to assure mechanical integrity of all site plant including process pipework
- Appreciation and use of relevant codes and standards
- Guidance on the credible site or unit degradation mechanisms that take into account piping material, process fluids, piping configuration and operational factors
- Use of appropriate detection techniques
- Suitable schemes of examination which take into account the likelihood and consequence of failure
- Means by which assumptions on degradation and failure modes can be confirmed and reassessed
- Means to identify specific deterioration modes such as corrosion under insulation, vibration fatigue, flange failure, deadlegs, injection points etc.
- Suitable data recording and evaluation systems
- Defect assessment procedures and fitness for service rules
- Established retirement criteria
- Temporary repair registers
- Resources to target, gather, evaluate and review data
- Use of failure and near miss data as intelligence to target and focus actions.

References

1. Major Hazard Incident Data Service (MHIDAS): AEA Technology plc

2. HSE Off-Shore Division Hydrocarbon Release Statistics 2002

3. Research report RR253 - Piping System Integrity: Management Review 2004: ABB Engineering Services

4. Loss of Containment Incident Analysis: HSL Report HSL/2003/07 Health and Safety Laboratory

5. COMAH Major Accidents; Accident Reports. www.hse.gov.uk/comah/index.htm

6. Piping Inspection Code: API 570 2nd Edition 1998

7. Pressure Piping Systems Examinations: Model Code of Safe Practice
   Institute of Petroleum 13 2nd Edition 1993

8. Safety of Pressure Systems – Approved Code of Practice L122

9. Refinery Pipework Project www.hse.gov.uk/chemicals/spctechgen33.htm

10. Guidelines for the Avoidance of Vibration Induced Fatigue in Process Pipework:
    Marine Technology Directorate 2002

11. Management of Integrity of Bolted flanged Joints: UKOOA Guidelines 2002

Development of an Risk Based Inspection Implementation Manual

John Aller,
Capstone Engineering Services, Inc.
Houston, TX USA

Ricardo Valbuena,
ConocoPhillips Ltd.
Humberside, United Kingdom

Michael Renner,
Bayer Technology Services AG
Leverkusen, Germany

Emory Ford,
Materials Technology Institute
St. Louis, MO USA

ABSTRACT

The Materials Technology Institute (MTI) has recently developed a manual for use in implementing Risk Based Inspection. This presentation will cover the process MTI used to develop and roll out the manual to MTI members and provide an overview of its contents.

It will present the "How to's" on a variety of topics including "How do you Demonstrate the Value of RBI vs. Conventional Inspection Strategies to your Company", "How To Implement Risk Based Inspection", "RBI Considerations in Plant Design", "RBI – An Insurer's Perspective", and "What are the barriers what keep you from applying RBI".

1. INTRODUCTION

The chemical, petrochemical and refinery sectors have been facing increasingly tougher safety, environmental and mechanical integrity regulations as well as challenges associated with the need for to improve competitiveness. Under these circumstances, it is crucial to manage operational risks through the use of effective technology and best practices for inspection and maintenance planning.

MTI is a non-profit Institute that serves its member companies from the chemical, refining and allied industries. It is a unique organization whose members include both producing and servicing companies. Both types of companies brought their unique perspectives to bear on a recent MTI project to develop a manual on Risk Based Inspection. The purpose of the MTI manual is to present a "tutorial" or "how to" on Risk Based Inspection for the Process Industry. This tutorial enables operating companies to understand Risk Based Inspection, both its benefits and limitations. This book goes beyond promoting a methodology or Assessment Tool, which has often been the main purpose of what is found in the prior literature.

2. DEVELOPMENT PROCESS

MTI is functionally organized with a Board of Directors, a Technical Advisory Committee, Project Development Committees, and Project Teams. The concept for the RBI project was brought to a Project Development Committee by a representative of a member company. A Project Team was formed to review the idea, sharpen the focus and create a written Strategic Project Summary, which was presented to the Technical Advisory Council (TAC). At this point, the potential project deviated from a normal MTI project because the proposed work was to be performed by volunteers from within the member companies of MTI, rather than using a contractor to do the work. This approach provided a low-cost solution to MTI, but also created some discussion about whether the project could be completed in a timely manner. The TAC was responsible to approve the technical merit of the project. After TAC approval, the project was presented to the Board for authorization of the funding. Since the work on this project was to be done by MTI members, the Board assigned a Conflict of Interest Committee to oversee the business ethics issues of the project.

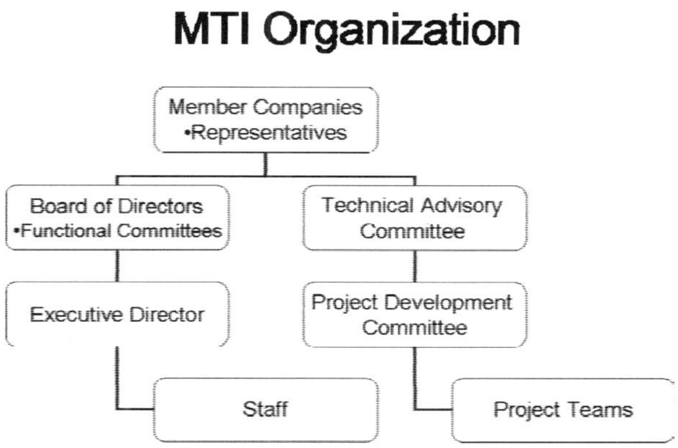

Figure 1

The RBI Project Team stayed in place during the writing and roll-out of the book. The Project Team met during each of the scheduled MTI meetings, as well as by conference call on an as-needed basis.

The book was written by four authors that each contributed one or more chapters as well as commented on the other chapters in the book. Three of the authors were employees of MTI member companies, while one of the chapter authors was not associated with MTI. At least eight other persons reviewed and helped edit the document. It was turned over to a technical editor for final editing, and then published by MTI using their normal publication process. The project duration was slightly more than two years.

3. ROLL OUT

Due to the significant industry interest in the topic of Risk Based Inspection, both within the MTI member companies and non-member producing companies, it was decided to roll-out the book in a one-day workshop. The RBI Project Team was also responsible for planning and delivering the Workshop. The design of the workshop was based on series of presentations on different topics covered in the book by subject matter experts in each of these topics. The agenda included an overview of the book, the value and limitations of RBI, the perspective of the insurers, work going on in Europe, and the future of the technology. An Open Forum following the presentations allowed the attendees to exchange information and get answers to specific questions from recognized industry experts. Attendance included copies of the presentations and a copy of the book.

4. THE PRODUCT

The book is titled **"Implementing and Evergreening RBI in Process Plants"**[1]. RBI is being increasingly used in the Chemical Process and the Petroleum/Refining industry. Vendors are offering a variety of services in connection with RBI. There is a significant amount of confusion and misunderstanding about this work process and technology. The purpose of this book is to present a "tutorial" or "how to" on Risk Based Inspection for the Process Industry. This tutorial will enable operating companies to understand Risk Based Inspection, both its benefits and limitations. This book is intended to go beyond selling a methodology or Assessment Tool, which has been the main purpose of what is found in the literature. This MTI document will present the "How to's" on a variety of topics including "How do you Demonstrate the Value of RBI vs. Conventional Inspection Strategies to your Company", "How To Implement Risk Based Inspection", "RBI Considerations in Plant Design", "RBI – An Insurer's Perspective", "What can be done in states that do not recognize ASME & API Standards to allow RBI", among others. Input from both owner-users and vendors is included to ensure maximum objectivity.

Figure 2

The Table of Contents of the book is shown below.

A brief synopsis of some key elements in the book is presented in the following sections.

4.1 Defining Risk Based Inspection

Risk Based Inspection (RBI) is a systematic approach to the management of inspection programs by prioritizing equipment inspection activities. The approach consists of a detailed inspection, corrosion, material, process, plant operations and consequence analysis to identify and mitigate risk and to make informed decisions on inspection frequencies, level of detail and methods of non-destructive evaluations.

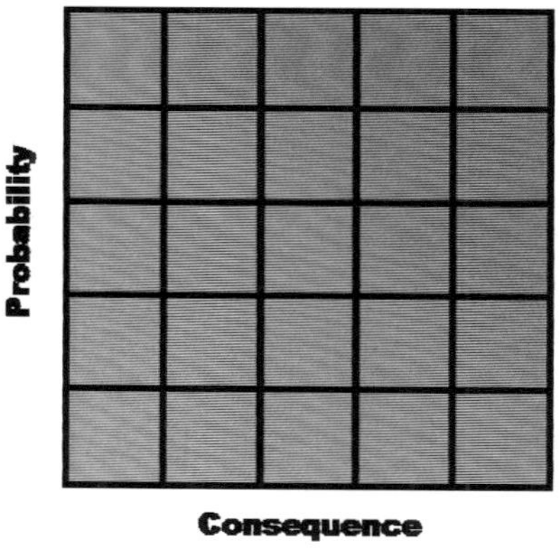

Risk in the context of RBI is the combination of likelihood and consequence of failure. Likelihood or probability of failure is driven by material damage mechanisms, their rate of progression, the tolerance of the equipment to damage, and the amount and type of inspection activities that have been performed in the past. Consequence is the outcome of a failure that may be expressed, for example, in terms of safety to personnel, economic loss, or damage to the environment. Risk can be represented in the form of a matrix displaying the level of likelihood and consequence.

4.2 Benefits of Risk Based Inspection

For any company to take on the technical and cultural changes to begin managing their assets using a risk-based approach there must be some benefits that will provide the impetus for change. One obvious benefit is the potential reduction of inspection costs in a facility, but that should not be considered the only driver for implementing RBI. The book identifies a number of other significant benefits, including the following:

- Those equipment items that are the largest contributors to the risk in the facility are identified so they can be managed to maintain the plant risk at acceptable levels.

- Typically, sites have broadly applied inspection programs with excessive ultrasonic thickness measurements and sometimes buried themselves with measurement data. RBI will focus the inspection program so that the proper amount of inspection is being planned for each asset.

- For some cases, the RBI study may determine that inspection alone will not adequately manage the risk, and additionally will identify material changes, operational procedures, instrumentation upgrades, etc. as the appropriate means to reduce risk.

- In any operating plant, a relatively large percentage of the risk is usually associated with a small percentage of the equipment. A Risk Based Inspection methodology permits the shift of inspection and maintenance resources to provide a higher level of coverage on the high-risk items and an appropriate effort on lower risk equipment.

- RBI will determine the required level of inspection effectiveness, or the ability of the *inspection method* to detect the damage type being inspected for. This will minimize over inspection and under inspection.

- The RBI work process also provides the opportunity to challenge the types of inspections in the current program. The effect of alternative inspections, such as non-intrusive inspections instead of intrusive inspections can be studied.

- RBI uses objective, reproducible methodologies to assess risk and establish inspection plans. Comparisons can be done within a plant or between plants to determine priorities and relative benefits of changing inspection strategies.

- RBI data and tools can be used in a variety of studies or "what-if" analysis of interest to plant management such as:
  - Operational benefits or risks associated with extending time between unit shutdowns.
  - Risk versus increased maintenance and inspection costs associated with potential loss of production
  - Risk versus increased maintenance and inspection costs associated with refining "opportunity" crudes.

- Forecasting of risk into the future to study alternate scenarios of inspection, new metallurgy, new design or other mitigation efforts.

Because RBI Implementations require a large organizational commitment, companies may find it difficult to financially justify the start or continuation of such programs. An appropriate way to deal with the justification is to perform a Cost Benefit Analysis and a Risk Benefit Analysis before and during the project.

The RBI process will document the risk picture of the facility at a point in time, and updates in the risk analysis will be able to show how the risk changes with time. This change in risk can be used to define a Risk Benefit Analysis, as shown in Figure 4.

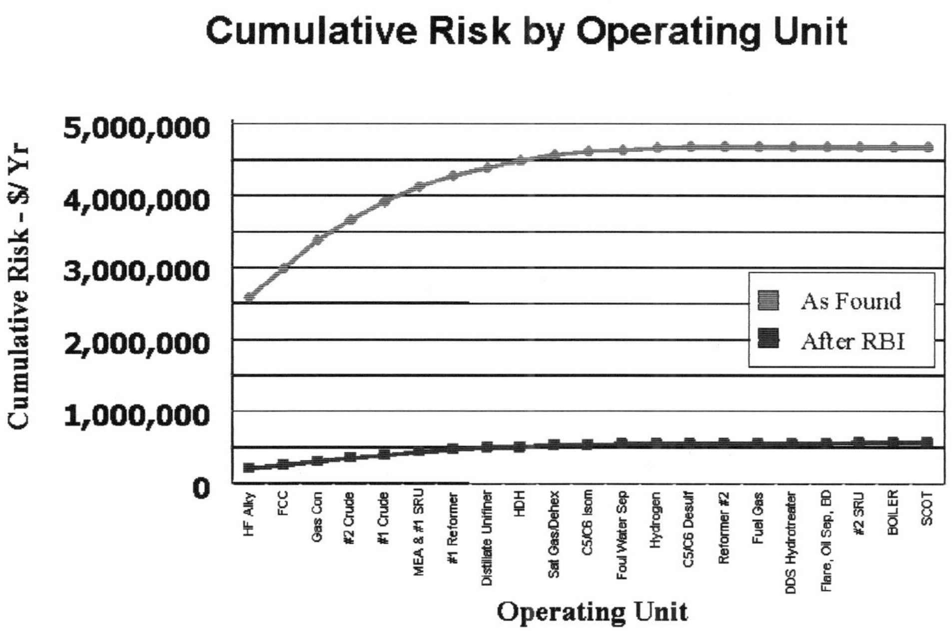

Figure 4

The Cost Benefit Analysis from a RBI study may be calculated as follows:

- Reduction of inspection costs as a result of equipment being removed from upcoming scheduled turnarounds
- Reduction of maintenance costs as a result of equipment being removed from upcoming scheduled turnarounds
- Increased production time as a result of inspecting less equipment (i.e. shorter turnarounds)
- Use of non-intrusive inspections instead of intrusive inspections

The Cost Benefit Analysis may include the Return on Investment (ROI) of the RBI analysis. This number should equal the cost savings as agreed to by the plant divided by the cost of the RBI analysis, as shown in the following example:

Table 1. Example of Cost Benefit Analysis

Cumulative savings from turnaround scope reduction and interval shortening	$1,600,000
Additional non-intrusive inspection and associated maintenance costs	$200,000
1 Unit RBI analysis cost	$55,000
Total costs	$255,000
Net Benefit	$1,345,000
Return on Investment	5.3 to 1

4.3 How to Implement Risk Based Inspection

Implementing Risk Based Inspection starts with a process at a level above just performing risk analysis. The following five steps have been found useful by actual practitioners to start the overall process.

1. Select Acceptance Criteria:

   Acceptable risk limits for safety, health and environment need to be considered as well as acceptable risk limits for business risk. These limits may have already been developed at a corporate level and any RBI program should be in compliance with existing corporate risk limits. If no risk limits are currently defined, they need to be clearly documented by the RBI team and may require approval from corporate staff.

2. Select Plant or unit for assessment

   The basis for selecting the facility that will be the subject of the RBI process will often be tied to understanding the value of performing the work. Selection criteria can be based on turnaround schedules, an implicit understanding of the risk of the facility and a desire to reduce the higher risk, or a facility with staffing most likely to support a successful implementation.

3. Perform the risk assessment to identify the main risks associated with the facility.

   It is important to understand the consequence of failure and the factors that are driving those consequences, especially the high consequences. Likewise, it is important to know the likelihood of failure and the damage mechanisms that are governing the failure probability.

4. Develop an Inspection Plan to reduce risks

   Once the likelihood of failure and the governing damage mechanisms are understood, an Inspection Plan can be developed to properly manage the damage mechanisms and the resultant likelihood of failure.

5. Update Inspection Plans on a defined basis

   The basis for updating the Inspection Plans needs to be clearly defined. This may occur after each turnaround, when new information is available, such as process changes or new inspection data, or on fixed intervals (such as every ten years) as a revalidation process.

The work process is shown in some detail in Figure 5.

Figure 5. The RBI Implementation Steps

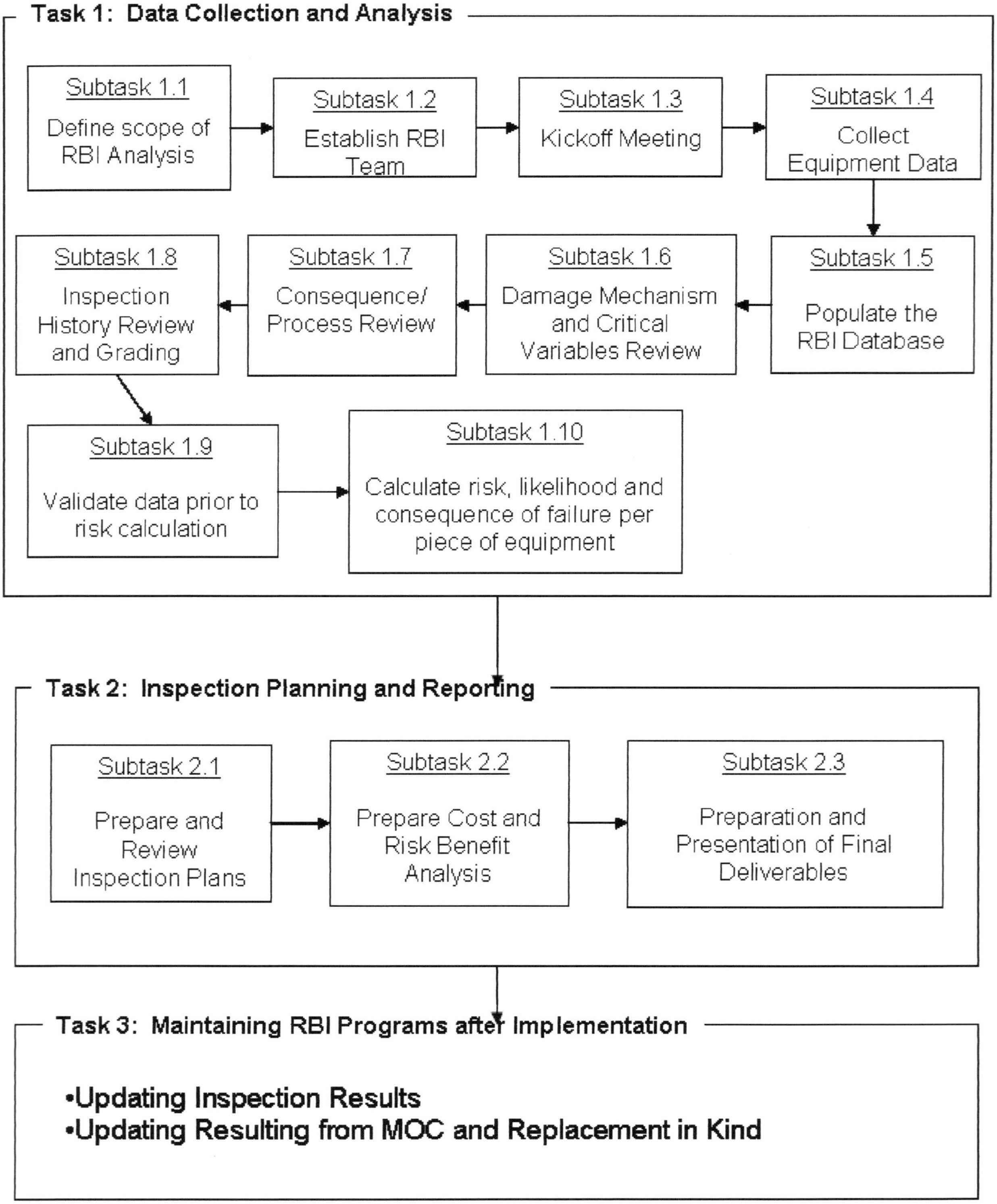

4.4 RBI Considerations in Plant Design

Using RBI during the Design Stages of a plant can provide many of the benefits described above from the start-up of the plant. This process will involve experts from Design, Engineering & Construction disciplines. Their experience in building plants combined with operating, maintenance and inspection experience on in-service plants should enhance the RBI study.

Where necessary, use more conservative criteria to compensate for lack of equipment history and industry experience.

Use RBI as early into the design process as possible. Input from RBI teams in the early stages have resulted in important design, process and materials of construction changes which have prevented in-service damage and unnecessary inspections.

4.5 Training and Competence Requirements

Training of staff in RBI is fundamental for the correct analysis, inspection planning and evergreening of these programs. All staff involved in these programs require a minimum understanding of RBI. In order to help ensure the integrity of the RBI process, training requirements are suggested in Table 2.

4.6 Impact of RBI on existing PSM Mechanical Integrity Concepts

The Risk Based Inspection program should be considered as an integral part of the site Mechanical Integrity Program, which in turn is part of the overall plant Process Safety Program. In the US and many other countries Process Safety Management is an initiative mandated by a regulatory body. In other organizations it is a corporate driven initiative to improve the safety of the operating facilities. In that context, the RBI program must meet all of the appropriate requirements of the PSM program.

This chapter will present the elements of Process Safety (including Mechanical Integrity) as defined in the US OSHA 1910.119 Regulation[2] that pertain to or have a connection with the RBI methodology. This applies to nearly all of the process facilities in the US, and is similar to the Process Safety Requirements imposed on the process industry in most other countries as well.

A person should be assigned the responsibility to integrate the RBI program into the site Mechanical Integrity and Process Safety Management program. Although many of the Process Safety Management (PSM) elements are interrelated, a mechanical integrity program that includes RBI is related to all the other elements and must be carefully integrated to avoid inefficiencies and confusion. The book provides recommendations on how an RBI program should be integrated with each element of a PSM System.

Table 2: Suggested RBI Training Requirements and User Grades

	RBI Expert	RBI Data Supervision	RBI User 1	RBI User 2
Example Functions	• Site or Corporate RBI Champion or Expert	• Information Coordinator • RBI Data Administrator	• Unit Inspector • Equipment Engineer	• Likelihood Expert (Corrosion Engineer) • Consequence Expert (Process Engineer) • Reliability Engineer
Qualifications	• Aware of corporate risk assessment methods • Understanding of the Company's risk management philosophy and targets • Thorough understanding of the RBI concepts and the software or Assessment Tool if one is being used	• Familiar with the chosen RBI Assessment Tool • Good computer skills. Excel desired	• Understanding of RBI concepts and chosen Assessment Tool(s) • Understanding of Inspection methods and their effectiveness • Understanding of the Company's risk management philosophy and targets.	• Likelihood Expert: understanding of potential damage mechanisms and the factors that may increase likelihood • Consequence Expert: knowledge of chemical processes, compositions and operating history • Reliability Engineer: understanding of industry and plant experience regarding failure history • All: knowledge of equipment design and design code requirements
Responsibilities	• Provide on-going quality assurance and consistency between studies	• Maintain the RBI Master database as directed • Along with User 1 is responsible for all data entry.	• Update RBI Database or paper-based reference book • Ensure that RBI data updates are directed to the RBI data administrator • Produce RBI Reports and direct work from Equipment Inspection Plans • Arranges for updating of PFD or other drawings	• Likelihood Expert: provide guidance on assignment of damage mechanism • Consequence Expert: have knowledge of PHA results and likely failure scenarios • Reliability Engineer: keep abreast of industry experience and failure data • All: advise of changes to process that would affect Probability of Failure or Consequence of Failure
Recommended RBI Training Hours	• 80 hours of course work and on-the-job training	24 hours of course work and on-the-job training	• 80 hours of course work and on-the-job training	• 40 hours of course work and on-the-job training
Recommended Training Topics or Related Experience	• RBI Assessment Tool and relationship of data fields • RBI Principles • NDE Methods • Damage Mechanisms • Plant Chemical Process principles • Consequence theory • Preparing RBI Reports	• General RBI Assessment Tool structure and data entry.	• RBI Assessment Tool and relationship of data fields • RBI Principles • NDE Methods • Damage Mechanisms • Plant Chemical Process principles • Consequence theory • Preparing RBI Reports	• All: RBI Assessment Tool introduction and relationship of data fields • All: RBI Principles • Likelihood Expert: Damage Mechanisms • Consequence Expert: Consequence theory • Reliability Engineer: Plant process principles • All: understanding RBI Reporting

4.7 Barriers to the Implementation of RBI Programs

There are a number of barriers that are typically encountered as an RBI program is being implemented. Experience indicates that many of these barriers are common to most organizations, and can generally be classified into a number of broad categories, such as:

- **Regulatory**

Overcoming this barrier will require the operating company to become familiar with the regulations and regulators within each country where they operate. Often there will be options for variances from strict, time-based inspection codes. In addition, in many of the developing countries there are the opportunities to influence the developing or changing regulations by working as an expert resource with the regulatory bodies.

- **Insurers**

The barriers imposed by insurers is similar to the regulatory barriers discussed in the previous section, and will normally be overcome by including the Insurer's Authorized Inspector in the early discussions and planning for the implementation of RBI.

- **Organizational / Cultural**

A typical barrier might be the syndrome, "We've never done it that way before", or "If it ain't broke, don't fix it." By definition, there is normal conservatism within an inspection department, and it can intensify if the department personnel feel threatened by the proposed changes. When RBI is perceived as non-essential or extra work, it can be neglected. This is a typical cultural resistance to change, and is often a more significant barrier with staff that is highly experienced or longer in position. Change management processes, training and staff movements can help with this barrier. When the inspection staff starts to see the value they are achieving with the RBI process, there will generally be a change in the culture. The end goal of implementing RBI is to have the inspectors believe in it and use it efficiently. This will result in the process being driven from the bottom-up.

The cost of implementation is often perceived as a barrier to an RBI program. It is normal for the implementation costs to significantly exceed the on-going maintenance costs of the existing program. A careful value analysis that looks at both the costs and the long-term value will generally overcome this barrier.

- **Resources**

One resource barrier is the lack of qualified and available staff. Some companies do not have adequate Subject Matter Experts (SME's), such as corrosion engineers, to handle the day-to-day operational and project needs as well as be deeply involved in RBI studies. In other companies, the required SME's are at a corporate or remote site location, and are not readily available.

This barrier can often be overcome by assigning the appropriate persons to the RBI Implementation Team as a full time position. For critical resource unavailability, a contracted resource can become a part of the implementation team. Another way this barrier can be overcome is to provide the classroom and on the job training to bring additional people of the required level of competency to work on the RBI Implementation Team.

- **Technology / Software**

The RBI technology can also create some barriers to the implementation of RBI. Any technology selected has some limitations, as discussed in Chapter 2. The limitations are the result of the technology choosing to use simpler (lower cost) analysis models, which compromise their ability to discriminate between the risks of different assets. A more quantitative analysis generally provides better discrimination but it comes at a higher cost, and may require a higher level of expertise of the analyst. Once the technology is selected, it carries its limitations into the implementation process. These limitations become barriers when they are not properly understood and communicated to the stakeholders of the results of the analysis. In some cases these inherit limitations, such as consequence models that do not match the fluids being analyzed and deliver results that are questionable. When the stakeholders do not understand the reason for the results, they may begin to question the technology.

The RBI practitioner can overcome this barrier by clearly understanding and communicating the technology and its limitations, and how to deal with the situations where the limitations do not provide adequate results.

- **Data**

The most challenging part of implementing a Risk Based Inspection Program is the collection of high quality data used in the analysis of risk and inspection planning. Since this is the most significant effort in the process, it also creates significant barriers to the implementation. One of the most prevalent data issues is lack of good data when the RBI effort starts. Many facilities are older and have incomplete records. Files are incomplete, inspection results have not been well documented, and management of change has not always been performed. The net result is that available data has to be validated and critical gaps have to be filled in around missing data.

Inconsistent data gathering can also lead to inconsistent results in the risk analysis or inspection planning. The inconsistent results can become a barrier to the process with the stakeholders.

5. SUMMARY

MTI has taken an innovative approach to develop a timely manual on a topic of interest to its member companies and the broader Chemical Process Industry. The book, **"Implementing and Evergreening RBI in Process Plants**", is in the publishing process and will be available first to MTI member companies then to the general public. It was presented to the industry through a successful roll-out workshop. The next focus will be on how to "evergreen" the book itself.

6. REFERENCES

1. Valbuena, Ricardo, Aller, John, Renner, Michael, and Feigel, Gene, "*Implementing and Evergreening RBI in Process Plants*, St. Louis, MO, Materials Technology Institute, 2005

2. OSHA 29 CFR 1910.119, Process Safety Management of Highly Hazardous Chemicals, U.S. Department of Labor, Occupational Safety and Health Administration, Washington, DC, February 24, 1992

145c A National Reactive Chemicals and Flammability Testing Database: Progress and Plans

Eric Clark and Dave Frurip

The concept of a database of reactive chemical testing data is currently being discussed as part of a broader effort to reduce risks associated with reactive chemical incidents. A public database, including flammability, would enable information-sharing as a means to increase awareness of potential hazards and mitigation of such hazards. The private and public benefits of such a database include:

· the ability of all companies (small and large) to access shared and evaluated test data to improve the safety design aspects of chemical processes associated with reactive hazards

· a potential reduction in incidents due to use of the information in the database to identify hazards.

· the database might stimulate research activities in industry and academia to develop new, more accurate testing methods and prediction techniques

· the accessibility of reactive chemical information from the database might stimulate the development of additional industry accepted and industry driven best-practices for the mitigation of reactive chemicals hazards

· the voluntary use of reactive hazard test data to reduce chemical reactivity hazards by industry could reduce external pressures for increased regulatory requirements associated with reactive chemical testing.

The original concept of this idea has been discussed by industry and the National Institute of Standards and Technology for several years, but progress was slowed due to legal/liability issues. Momentum returned when the U.S. Chemical Safety and Investigation Board (CSB) published a report in 2002 making a formal recommendation to NIST and to industry groups (ACC and SOCMA) to work together to establish such a database.

This paper seeks to outline the critical issues for the establishment of a national database. Since this is the first time an in-depth discussion of the topic appears in print, we attempt to make the paper quite general, identifying important topic areas and issues without necessarily coming up with all the answers. For example, we include a discussion of what our vision is for the database's structure, our proposals on how it could be managed, solutions to the liability concerns regarding data donation, the types of data which should be included, how to insure data quality, how to ensure the data are used properly, and the need for economic incentives to encourage data donation. We are hoping that this paper will stimulate further interest in the topic and foster broad stakeholder involvement in the process. Clearly, there is a need for the development of a coherent strategy for this idea based on a collaborative effort between CSB, NIST and ACC / SOCMA membership.

Beyond Compliance – The Future Role of Risk Tools?

Dr Nic Cavanagh
Global Product Manager
DNV Software, London, UK
Nic.cavanagh@dnv.com

Dr Jeremy Linn
Regional Manager, UK, BeNeLux and Americas
DNV Software, London, UK
Jeremy.linn@dnv.com

ABSTRACT

The management of risks associated with major accident hazard facilities has historically focussed on safe operation rather than areas such as improved financial performance or increased productivity. With major advances in Information and Communication Technology (ICT) and the maturing of technology in other domains, the emphasis has moved beyond straight-forward compliance with safety legislation.

With greater global competition and much more challenging margins, there is now less opportunity to invest in activities that do not contribute directly to the bottom line, or are perceived as such. ICT advances, along with growth in Internet technology, has brought a highly connected environment where corporations can share information and collaborate globally. Combined with this is a need for corporations to be more transparent in their corporate governance and demonstrate their performance in relation to the environment and society as well as financially.

Technological advances mean that the move from pure compliance based systems to performance based risk management or Risk Based Operations is now possible and desirable. This paper reviews the history of Quantitative Risk Assessment (QRA) software in this context, assesses how much of the technology developed over the last 20 years for safety management can be integrated with mainstream information systems and applied to improving and managing operations. It goes on to look at the state-of-the-art with regard to risk based operational management and to discuss how latest technology in terms of risk modelling, information management software and ICT can be used to improve operational performance.

1. INTRODUCTION

Over the last 30 years, the management of risks to life associated with the operation of major accident hazard facilities has evolved steadily. This has been driven by a number of major accidents including Flixborough (1974), Seveso (1976), Bhopal (1984), Piper-Alpha (1988) and, more recently, Toulouse (2001) and Fluxys (2004), and subsequent legislation such as the Seveso directives in Europe and the EPA Risk Management Plan regulations in the US[1]. The management of these risks has generally been focussed on safe operation rather than other areas such as improved financial performance or increased productivity. With major advances in Information and Communication Technology (ICT) over this period and the maturing technology used for software tools and related models, the emphasis has moved beyond straightforward compliance with safety legislation.

During the 1980's, oil and chemical companies were cash rich and highly proactive over safety issues with large budgets and often setting internal standards beyond the needs of legislative compliance. However, the business environment has progressed with greater global competition and much more challenging margins. There is now less cash available for activities that do not contribute directly to the bottom line, or are perceived as such. ICT advances, along with growth in Internet technology, have brought a highly connected environment where corporations can share information and collaborate globally. Combined with this is a need for corporations to be more transparent in their corporate governance and demonstrate their performance in relation to the environment and society as well as financially.

The combination of these ICT advances with the maturity of models for QRA and related technologies means that the move from pure compliance based systems to performance based risk management or Risk Based Operations is now possible and desirable. The "Q" of QRA need not only be fatalities, but effects on the environment, downtime or dollars. Also, these technologies can be linked with other enterprise wide applications such as Enterprise Resource Planning (ERP) and Enterprises Asset Management (EAM) systems to provide a company wide risk based performance management system, providing real-time information on the operational and financial impacts of decisions on safety and risk. By quantifying impacts on people, operations or assets analysts are better able to estimate the likely costs of an incident or incidents in terms of down-time, asset damage, personal injury and loss of life, brand damage, environmental clean-up, litigation and compensation, and so on. With the use of real-time information on the status of operations, mangers can assess the likely impacts of decisions on safety and business risks.

2. HISTORICAL EVOLUTION OF QRA TOOLS

QRA in the context of process plant safety provides a methodology for quantifying the risks associated with the activities involved in the production and processing of chemicals and petrochemicals. In order to quantify risks it is necessary to first identify all possible risk situations, quantify them in terms of event consequence and likelihood and compare them with acceptance criteria. The main questions to be answered by a

QRA are what can go wrong, what are the potential effects if it does go wrong, how often will it go wrong and is it important. Or, in QRA terms, identify the hazards, analyse the consequence, estimate the likelihood, combine consequence and likelihood to quantify the risks and put measures in place to mitigate and manage those risks. The key objectives of any QRA are to identify the major hazards, quantify the overall risk, optimise the risk reduction measures to be implemented and to help the decision making process with regard to acceptable risk criteria.

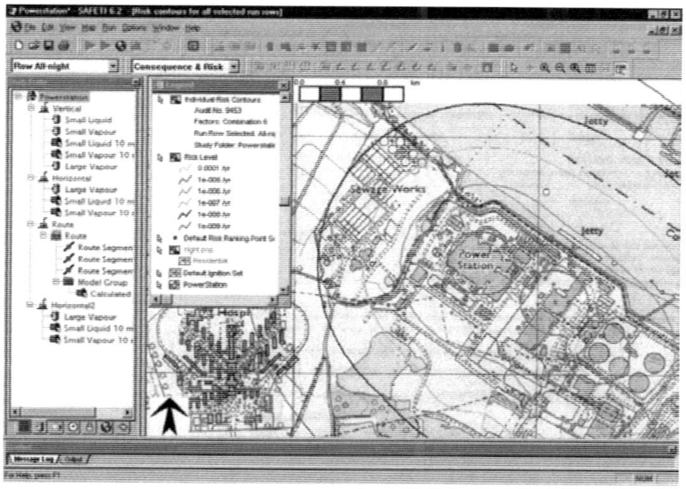

Figure 1 – Typical Individual Risk Contours displayed using SAFETI

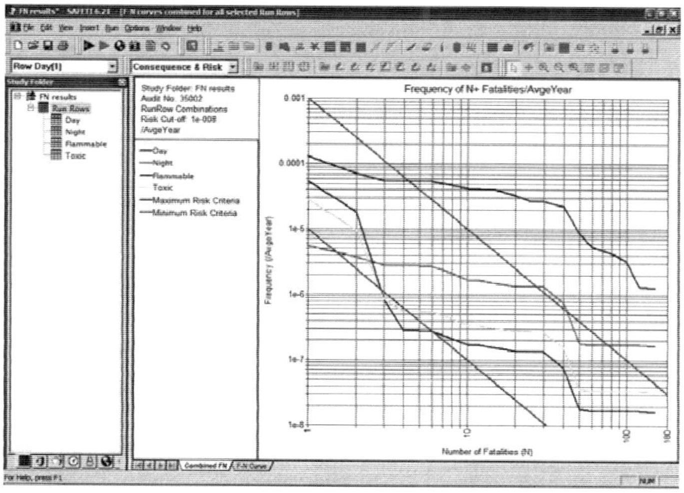

Figure 2 – Typical FN Curves for Societal Risk displayed using SAFETI

Typical outputs of a QRA study are individual risk contours as illustrated in Figure 1 and the FN curve for representation of societal risk as illustrated in Figure 2. Individual risk can be defined as "the frequency at which an individual may be expected to sustain a level of harm from the realisation of specified hazards" and is usually taken to be the risk of death expressed as a risk per year. Societal Risk is defined as "the relationship between the frequency and the number of people suffering a given level of harm from the realisation of specified hazards". It is normally taken to refer to the risk of death expressed as a risk per year and displayed as FN curves.

The first commercially available software tools for Process QRA[2] were developed in the early 1980's. The key drivers at this time were a number of major accidents, like Flixborough and Seveso, and later Bhopal. These resulted in legislation being implemented in a number of countries, particularly the Netherlands, which made QRA obligatory to ensure risks of further major accidents were minimised applying acceptable risk criteria and principles of ALARP[3].

At that time, the techniques and methodologies developed to enable large scale QRA to be performed challenged the computing power available. This limited the possibilities and meant that the software architecture had to be carefully designed to enable the necessary calculations to be made within the IT limitations of that time.

Since then ICT has developed rapidly and continuously[4] and the last 20 years has seen massive advances in the technology available. However, the methodologies used in performing QRA have generally remained relatively static. Although individual components of the QRA have improved in-terms of both modeling accuracy and speed of operation, the underlying architecture still largely supports the classical risk analysis methodology illustrated in Figure 3. These analyses are usually performed with single-user PC applications, which are not connected to the larger ERP or EAM systems, by an experienced risk analyst, with updates being infrequent and driven by legislation or changes to the plant or process.

This is the case for a number of reasons, not least the fact that QRA studies in tools like SAFETI[3] have been created over many years and have taken many man years of effort which their owners are loath to "throw-away". But there are limitations to this approach. For example, domino effects or escalation is difficult to account for using the classical approach, which is largely a series of sequential calculations or summations and assumes that hazardous events are independent of each other. Also, this approach where individual failure cases are assumed to be unrelated does not require a spatially based model to be developed, However, such a model containing plant geometry is a pre-requisite when extending these models to business or financial risk[5,6].

The data collection requirements when performing QRA are extremely labour intensive. Historically data has been collected manually and is often stored in databases which are essentially standalone and non-generic. Although technologies like Computer Aided Design (CAD), Enterprise Asset Management (EAM) and Computerised Maintenance Management Systems (CMMS) contain much of the data required for a QRA study links, where they exist, are rudimentary and largely unintelligent. Because of this data re-use has been difficult and thus limited. Plant data may have been acquired in a number of separate places for process design, process simulation, maintenance management, inspection planning and QRA. If this were available within a data warehouse, like Intergraph's SmartPlant Foundation[7] for example, providing facilities and procedures for change management could be comparatively straight forward. Much of this data is just as applicable during the operational phase of a plant life-cycle as it is to earlier phases like design and construction. The first step to using risk information in making operational decisions is to create this kind of data re-use capability throughout the asset life-cycle.

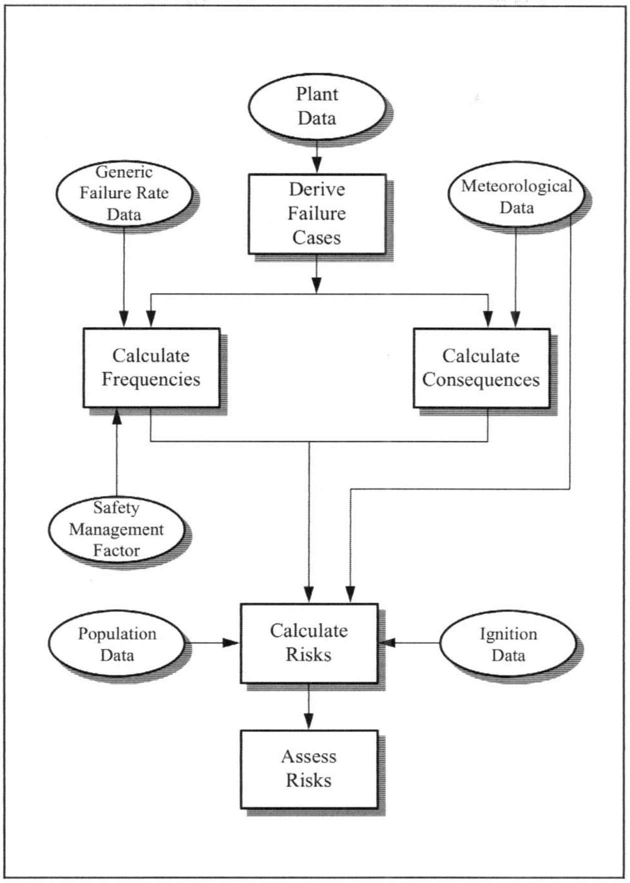

Figure 3 – Classical risk analysis methodology

The advent of the PC and other developments in ICT have revolutionised business work processes. When QRA software was first developed it was typically installed on a central computer. Now it is most commonly installed and run on local PCs. Calculations that might have taken days to run on a mainframe 20 years ago can now be run quickly on a PC. Software now uses advanced graphical user interfaces enabling input data and results to be manipulated and presented in a far more visual way as illustrated in Figures 1 and 2.

Handling data using ICT has developed substantially over the last 20 years. State-of-the-art approaches to managing data within an organization now involve enterprise-wide systems with integrated financial, personnel and production software applications. They either handle the data within an application from one vendor or, more commonly, make use of a number of applications that are interfaced so that they can share data. These developments have had a dramatic effect on the way data is used and shared. The data itself has a value to the organisation in proportion to the extent it can be distributed and shared.

The synergy with advances in network and internet technology makes enterprise-wide data sharing possible, adding enormous value to knowledge based organisations. From the perspective of risk information, these developments make it possible to re-use this in other parts of the organisation and with potential for direct relevance, and benefits, to operational phases of the plant life-cycle. The combination of maturity in risk modeling, large quantities of related data and advances in ICT begin to make Risk-Based Operations (RBO)[8] a real possibility.

If it can be shared, re-used and kept up-to-date, this wealth of information is of enormous value. The concept of the 'information asset' is illustrated schematically in Figure 4 in the context of the process industry. As a plant (or any other asset in fact) progresses through its lifecycle, the information asset grows as information is added and shared from other processes and applications.

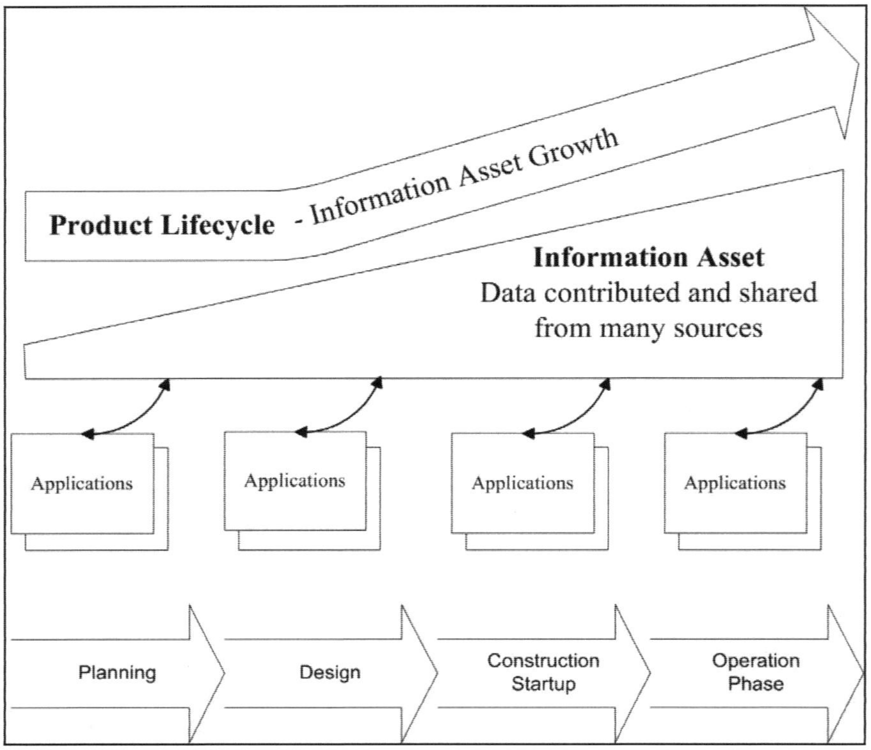

Figure 4 – Information Asset Growth through Asset Lifecycle

Much of the information required before a classical QRA can be performed will exist in datasets belonging to different departments within an organisation. The software applications used to manage the data may come from different vendors and it is likely to be structured in different ways. This requires a common data modelling system so that the applications can communicate and data can be published to and retrieved from the system. Such systems are now emerging, like Intergraph's SmartPlant Foundation[7], for example, as mentioned earlier.

Even though the data may already exist in other applications within the organisation, the entry and manipulation of this data for QRA remains a largely manual process. One of the first steps to extending the role of risk tools beyond compliance to meet the broader business needs of industry is to take advantage of latest ICT to facilitate the kind of data management described above. As an example, in the latest version of SAFETI we have taken a first step towards enabling data integration with other applications by incorporating Intergraph's GeoMedia GIS system. We are now working on a prototype to extend this to "Business Risk" which, as well as looking at the potential for loss of life, also includes financial losses associated with other drivers like business interruption, equipment damage and so on. We now go on to discuss our vision for the future role of risk tools in the two key areas of RBO and Business Risk.

3. BUSINESS NEEDS

At the heart of the Business Risk approach is the need to manage risks and to enhance safety within operations. Pitblado[8] has shown that although the industry has greatly reduced injuries there seems to be less evidence that there has been a significant reduction in major accidents. Analyses of US and EU data have failed to identify any noticeable downwards trend in major accidents.

The one exception to this is within the UK offshore oil sector which is managed by a safety case regime[9] which has taken recommendations from Lord Cullen's report from the Piper Alpha Inquiry. The focus has been on addressing major accidents through a number of mechanisms, some of which are not present in other regulations.

In assessing performance of the UK legislation the incidence of leaks has been evaluated with the belief that major leaks are a good indicator of the potential for a major accident, although there are insufficient data to identify a correlation between the two. The UK data indicate a significant drop in the number of major leaks reported. Interestingly medium leaks reduced less and minor leaks increased with this approach.

The UK Offshore safety case regime requirements have the following elements

- Safety management system
- Risk Assessment
- Quantitative risk assessment demonstrating meeting a defined risk target
- Identification of safety critical elements (critical barriers)
- Performance standard for all safety critical elements
- Lifecycle programme to maintain critical barriers (written schemes of examination)

The implications of this are that safety is no longer assessed and addressed through a snap-shot assessment made every few years but assessment of the elements controlling risk and thereby safety are monitored and maintained as part of the on-going operational processes.

The industry has made many initiatives in the application of risk based methodologies to the control of safety critical elements. Examples include Risk Based Inspection (RBI), Risk Based Maintenance (RBM) and more lately RBO. These initiatives continue with innovations such as Bow-Tie diagrams and Matrix of Permitted Operations. These innovations will continue to improve safety and drive down the numbers of accidents by pulling safety more into the mainstream operations.

When accidents happen not only can there be loss of life but many other losses can occur that can have a significant impact on the environment and business often costing many millions of dollars. Fewtrell and Hirst[1] examined the costs of accidents post Flixborough and from UK accidents reported quoted losses at the time of the accident up to £100M (1996 values). When looking at accidents internationally the maximum reported loss from a single accident was nearly $1.5 billion when including business interruption costs. Kersten and Mak[10] reported that in the explosion at the Atofina plant in Toulouse in September 2001, 29 people were killed, 2500 injured (30 severely) and material losses were assessed at €2.3 billion.

As well as costing lives these kinds of losses can obviously have a huge impact on a business, the environment, the local population, on surrounding businesses and those that depend on them. There is a clear opportunity for businesses to apply the knowledge and expertise of the safety practitioners to the broader business risk environment. Innovations that have been used to improve safety can be extended to other identifiable types of loss and used to manage a broad range of risks thus minimising many of the potential major losses to a business.

Companies have suffered massive financial losses and entire economies have been disrupted by incidents where asset loss has been minimal and there have been few fatalities. Two well documented examples are the Longford gas explosion in Australia[11] in 1998 and the Seveso Dioxin release[12] in July 1976. In the case of Longford there were only 2 fatalities but the cost to industry was estimated to be in excess of $1.3 billion. So, from the health and safety standpoint both incidents were relatively minor compared to, say, the Toulouse Ammonium Nitrate release in France in 2001 or the Fluxys Natural Gas release at the Ghislengien Industrial Estate in Belgium in 2004. But from the Process Business Risk standpoint they have all resulted in significant preventable losses.

4. THE PROCESS BUSINESS RISK CONCEPT

The Process Business Risk concept provides an opportunity for extending conventional safety focused QRA to manage these additional risks. Typical outcomes that may result in financial losses are:-

- Impacts on people in terms of both injuries and fatalities
- Property damage in terms of both capital costs to replace damaged equipment and damage to other property
- Business interruption
- Inventory loss
- Environmental damage including clean-up costs, fines and impact on flora and fauna
- Plus many others (legal costs, loss of reputation, brand image, compensation, etc.)

The first of these items above has traditionally been considered when performing a QRA for a hazardous facility. Typical output produced by extending this to cover the additional items above through a Process Business Risk analyses may be total loss rate or F/$ curves, analogous with the F/N curves from a QRA and as illustrated in Figure 5 (reproduced from reference 5).

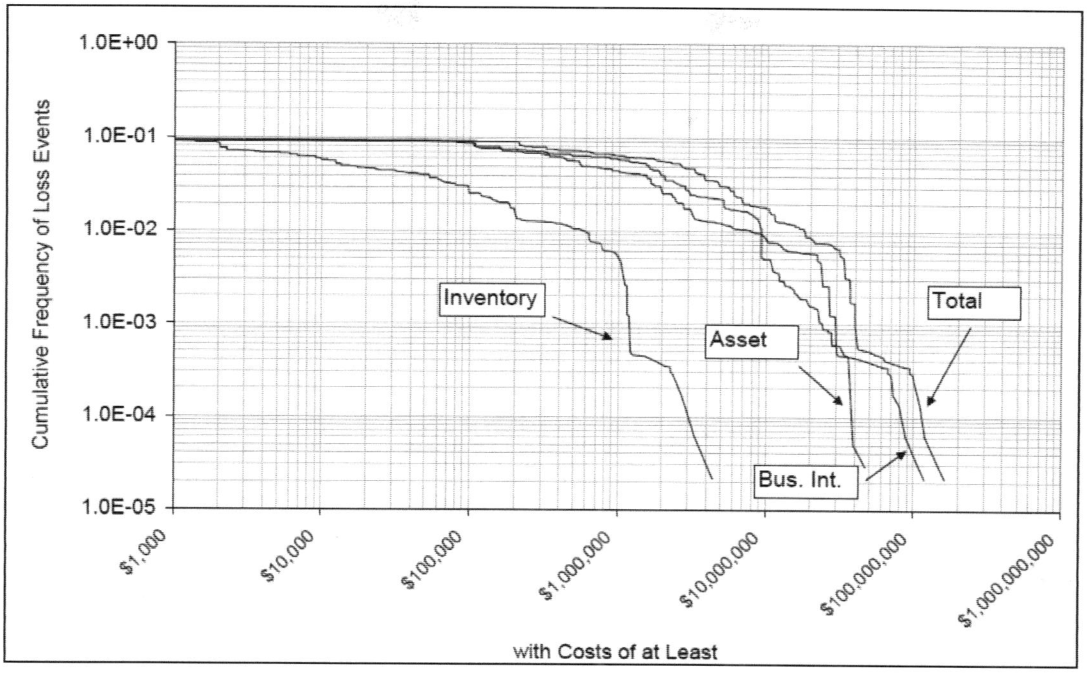

Figure 5 – Typical F/$ Curves for different Components of a Business Risk Model[5]

Similarly, typical uses of this approach may include some or all of the following:-

- Benchmarking operational risk levels
- Trending risk over time
- Comparison of risks for different operational conditions
- Demonstrating a culture of CSR (Corporate Social Responsibility) and adherence to the principles of triple bottom line (3BL) reporting
- Assessment of exposure to financial risks due to major hazards on process plants
- Evaluation of the relative effectiveness of different risk mitigation strategies for reducing exposure to financial risk
- A decision support tool for implementing cost effective risk reduction measures (e.g. using cost benefit analysis)
- Justification for reduction of insurance premiums and deductibles by demonstrating reduced exposure to financial losses

By developing Process Business Risk technology which focuses on more than compliance with safety legislation, this is a first step to linking more directly with other business processes. This is a key facilitator for longer term extension to full Risk Based Operational monitoring using the same risk based techniques linked directly with more mainstream operational systems.

5. FROM COMPLIANCE TO OPERATIONS

The concept above could easily be performed as an infrequent analysis by a risk expert with a report to the management. However, the true value of the approach is realised when it is integrated with the operational management and planning systems. Steps have been taken in this direction. For example, today we see the integration of RBI analysis tools with inspection databases which store detailed technical inspection results, to interchange data and optimise inspection strategies. Further integration with ERP systems allows an optimised inspection plan to be implemented. This is a huge step forward but the inspection plans are optimised infrequently and there is no real dynamics with the operations.

Within plant operations there are generally three main systems supporting the operational processes; Enterprise Resource Planning (ERP); Enterprise Production Management (EPM) and Enterprise Asset Management (EAM), all of which overlap to some degree as sown in Figure 6.

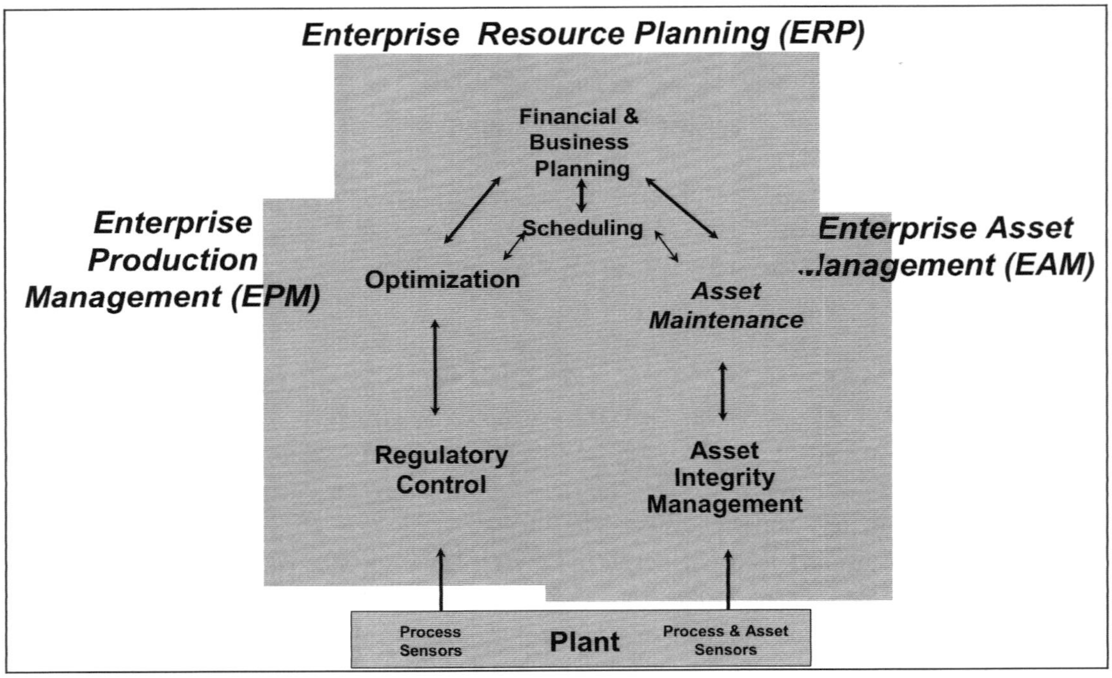

Figure 6 – Relationship between ERP, EPM and EAM systems for a typical plant

ERP systems are an integrated set of information systems designed to tie together more closely a variety of company functions including human resources, inventories and financials while simultaneously linking the company to customers and vendors.

EPM software is designed to manage, monitor, control and optimise the production processes within a process plant.

EAM software is designed to help companies maximise the value from its capital assets. The software is used to track, depreciate and maintain their fixed assets and capital equipment. EAM software is often integrated with ERP software for preventative maintenance planning as well as asset accounting.

Through integration of software for analysing business risks with these systems, real-time information can be used to determine the current level of risk to which a plant is exposed. It can also be used to determine the effects on risk of alternative courses of action and improve decision making whilst managing risk and safety.

Corporate knowledge and experience can be captured within the system and utilised real-time in decision making to help avoid high risk scenarios. As data are real-time the system can also indicate when a high risk situation has developed giving operational managers another source of information early to help mitigate such situations.

Classical risk analyses are performed on plants operating under normal conditions. With appropriate set-up a business risk based operations system could be highly effective at assessing and assisting managers in minimising risks during non-standard operating conditions such as start-up and shut-down and during maintenance activities.

6. CONCLUSIONS

QRA has traditionally been used to ensure safe operation, often from a compliance perspective. But, as demonstrated above, as well as costing lives, major accidents can have other social, environmental and economic consequences. Classical methodologies like QRA can be extended to assess a broader range of business risk and ICT is now at a stage where it can support such extensions.

Risk results can be linked through other techniques, like Bow-Tie diagrams, to risk indicators and other decision support systems. By doing this in real-time, operational decisions can be made based on current risk, not based on a single risk snap-shot taken at some time in the past. Furthermore, "risk" in this context does not just refer to risk to life. By linking with Process Business Risk, the risk from a number of standpoints can be considered simultaneously, like life, property, the environment and so on. Setting priorities in each risk category allows appropriate operational decisions to be made.

Live links from Process Business Risk tools to enterprise systems could be used to update the status of barriers to aid operational decision making, enabling decisions to be made based on actual risk, not perceived risk. Using this kind of system, major contributors to risk can be identified for action and the consequences of these actions will be apparent immediately.

The first step in achieving this kind of integration between risk and operations is to further develop risk tools beyond the domain of compliance with safety legislation. We are in the process of developing a prototype Business Risk software tool which takes this first step.

7. REFERENCES

1. Fewtrell, P. and Hirst, I.L., A review of high cost chemical/petrochemical accidents since Flixborough 1974, IChemE Loss Prevention Bulletin, No. 140, April 1998.
2. Worthington, D.R.E. and Witlox, H., 2002., SAFETI Risk Modelling Documentation – Impact of Toxic and Flammable Effects, DNV Software Risk Management Solutions, October 2002.
3. Cavanagh, N.J., Calculating Risks, Hydrocarbon Engineering, Volume 6, Number 6, Palladian Publications, London, June 2001.

4. Worthington, D.R.E. and Cavanagh, N.J., The development of software tools for chemical process quantitative risk assessment over two decades, ESREL 2003 Conference, Safety and Reliability, Maastricht, The Netherlands, 2003
5. Chippindal, L. and Butts, D., Managing the Financial Risks of Major Accidents, Annual CCPS Conference, Emergency Planning: Preparedness, Prevention and Response, Orlando, Florida, 2004.
6. Evans, J. and Thakorlal, G., Total Loss Prevention – Developing Identification and Assessment Methods for Business Risks, 11th International Loss Prevention Symposium, Prague, 2004.
7. Intergraph, The Engineering Framework, Integrating the Plant Information Asset throughout the Plant Lifecycle, Intergraph Process, Power and Offshore, 2002.
8. Pitblado, R.M., Real-Time Safety Metrics and Risk-Based Operations, 11th International Loss Prevention Symposium, Prague, 2004.
9. HSE Offshore Safety Case Regulations, 1992.
10. Kersten, R.J.A. and Mak W.A., Explosion Hazards of Ammonium Nitrate; How to Assess the Risks, International Symposium on Safety in the Manufacture, Storage, Use, Transport and disposal of Hazardous Materials, Tokyo, 10-12 March 2004.
11. Kenney, G.D., Boult, M. and Pitblado, R.M, Lessons for Seveso II from Longford Australia, SEVESO 2000, Risk Management in the European Union of 2000: The Challenge of Implementing Council Directive 96/82/EC, Athens, November 10-12 1999.
12. Pocchiari, F., V. Silano, and G. Zapponi., The Seveso Accident and its Aftermath – Insuring and Managing Hazardous Risks: From Seveso to Bhopal and Beyond, pp. 60-78, Editors P. Kleindorfer and H. Kunreuther, Berlin: Springer-Verlag, 1987

The DOE Emergency Management System and Mitigation of Chemical Accident Risk

James D. Jamison, CHP
Science Applications International Corporation (SAIC)
Richland, Washington
Jamisonj@saic.com

James T. Powers, Ph.D.
U.S. Department of Energy Office of Emergency Management (NA-41)
Washington, D.C.
jim.powers@nnsa.doe.gov

ABSTRACT

The U.S. Department of Energy (DOE) uses a systematic approach involving hazard identification, quantitative analysis, emergency planning, and response to minimize chemical accident risks resulting from its operations. The DOE Comprehensive Emergency Management System, which also addresses radiological and biological hazards, is founded on three basic principles:

- *Emergency response is the last line of defense against unacceptable health consequences to workers and the public,*
- *Emergency planning and preparedness measures must be specific to (and commensurate with) the hazards, and*
- *Early recognition of emergency conditions is essential if response measures are to be effective in minimizing health risk.*

The DOE facility hazards assessment protocol requires that all hazardous materials in quantities above certain thresholds be identified and the relevant properties (e.g., toxicity, dispersiblity) determined. A spectrum of scenarios by which each material could be accidentally released to the environment is then developed for analysis. Exposure or dose, and the implied health consequences of each scenario are estimated for three population groups: the workers in the facility where the release occurs; workers at other facilities on the same site; and, members of the public. Opportunities for moderating the consequences of potential releases, using measures such as inventory limits, engineered features, and controls on material storage locations are also identified. To aid in rapid recognition of chemical release events, the observable indications that are associated with each scenario are compiled.

Finally, the health consequence estimates are compared to exposure/dose standards to determine the locations where protective measures will be needed to minimize the health risk to each population group. Specific predetermined actions are then taken immediately upon recognition of an actual or impending release, thereby providing the most rapid and effective response to protect the groups at greatest risk.

1. Introduction

The United States Department of Energy (DOE) manages a wide variety of sites and activities throughout the country. These include:

- ❖ National Laboratories
- ❖ The nuclear weapons complex
- ❖ Legacies of the Manhattan project and attendant cleanup sites
- ❖ Power Marketing Offices
- ❖ Petroleum Reserves

The National Laboratories conduct basic and applied research, ranging from high energy particle physics to the human genome, to alternative sources of energy and applications of technology to homeland security. Activities at other sites include development, stockpiling and dismantling of nuclear weapons, production of nuclear power sources for space vehicles, environmental cleanup at legacy sites, transportation of waste materials generated by Departmental operations, and storage and isolation of radioactive waste. The Department also manages the nation's Strategic Petroleum Reserve and is the home of the Power Marketing Administrations, which function both as generators of electrical power and brokers of regional energy supplies. Finally, the Department is involved in energy restoration activities following major disruptions of supply or distribution.

As the diversity of these activities would indicate, DOE sites and facilities store, use, produce and dispose of many different hazardous substances, including radioactive materials, toxic chemicals and biological agents. In managing these hazards, DOE holds the safety of workers and the public to be paramount. By maintaining high standards of facility design, conduct of operations, safety oversight and personnel training, DOE activities consistently achieve accident and injury rates that compare very favorably with those of similar operations in the private sector. In addition to the systems of workplace hazard controls that prevent incidents and ensure worker safety, the DOE Emergency Management System superimposes additional protections over all operations involving dispersible hazardous materials in quantities that could harm people outside the immediate workplace.

The overall mission of DOE Emergency Management is to be ready to respond promptly, efficiently and effectively to any emergency involving or affecting DOE facilities or activities by applying the necessary resources to mitigate the consequences and protect workers, the public, the environment, and national security. The DOE "comprehensive" Emergency Management system provides a framework within which to address *all* hazards, from natural phenomena to terrorist attacks, and *all* of the components of emergency management. As defined in DOE directives, the standard components of a DOE emergency management program are *planning*, *preparedness*, *readiness assurance*, *response*, and *recovery*. "Planning" involves determining, in advance, what will be done in response to specific emergencies. "Preparedness" is putting in place procedures, equipment, and personnel capabilities that will be needed to respond. "Readiness Assurance" is the ongoing process of verifying and demonstrating readiness to respond. "Response" is the actual mobilization of people, equipment and resources to take action during an emergency. "Recovery" is planning for the return to normal operations following an emergency.

The current DOE emergency management requirements are the result of evolution that began after the Three Mile Island reactor accident in 1979. At that time the Department was still operating several reactors and producing materials for nuclear weapons. Accordingly, throughout the early- and mid-1980's, DOE's emergency management requirements focused heavily on nuclear and radiological hazards. After the Bhopal tragedy and some other accidents turned the world's attention to chemical release hazards, the Department expanded the emergency management requirements in 1991 to include toxic chemicals. After the Occupational Safety and Health Administration enhanced its worker protection regulations and included hazardous waste operations, the DOE expanded its emergency management requirements to stipulate a level of basic emergency planning for *all* facilities and activities, with more specific and detailed requirements for activities using radioactive materials and toxic chemicals. At present, there is work ongoing to address biological toxins and infectious agents in the emergency management requirements, because some national laboratories are using these materials in the development of detectors and in supporting other homeland security-related programs.

2. The "Foundation Stones" of DOE Hazardous Materials Emergency Management

The DOE approach to hazardous materials emergency management is constructed around three guiding principles, the "foundation stones" of DOE hazardous materials emergency management.

Principle Number 1: ***Effective response is last line of defense against adverse consequences.*** Regardless of how sound the fundamental safety programs and hazard controls may be, bad things sometimes happen. This principle expresses DOE's position that if hazard controls should fail, the site/facility *must* be prepared to take actions to limit or prevent adverse health and safety impacts to workers and the public.

Principle Number 2: ***Planning and preparedness must be specific to and commensurate with hazard.*** The Department is responsible for a large number of different hazards that could threaten the health and safety of workers or the public if released to the environment. The hazards are very different in the nature of their impacts on people, their behavior in the environment and the distance at which adverse impacts would be experienced. While the basic emergency management framework is the same for all DOE sites and facilities, the specific planning and response measures for each hazard are to be tailored to the hazard, such that they are *specific to* (i.e., technically appropriate) the hazard and *commensurate with* (in size, scope or scale) the magnitude of the hazard and its potential impacts.

Principle Number 3: ***Early recognition is vital to timely, effective response.*** In many cases, mitigating the health impacts of toxic releases to the environment can only be accomplished by warning the potentially affected population and directing them to take actions to prevent or limit their exposure. Early recognition of a release event is essential if warnings are to be delivered in time to be executed effectively. By developing a full understanding of possible release scenarios and the indications that would point to actual or impending release, emergency management will increase the likelihood of successful warning and intervention to limit health impacts.

3. The Role of DOE Emergency Management in Reducing Chemical Accident Risk

Implementing the emergency management program requirements of DOE Order 151.1B, *Comprehensive Emergency Management System,*[1] reduces chemical accident risk at a DOE site and facility in several ways. Although the change in operating risk is not usually quantified, it is easy to demonstrate that the incremental reductions in risk are very real, and in some cases, total (i.e., eliminate a hazard – eliminate the associated risk).

3.1 Understanding the Hazard—the EPHA

DOE Order 151.1B requires that every DOE facility identify and document the *types* of hazards (such as natural phenomena, industrial accidents, fires, toxic material releases, etc.) that that might apply to it and to *qualitatively* assess the potential effects of each type of hazard on the facility, the workers, the public and/or the environment. This process, termed the "Hazards Survey", identifies any facility that holds significant quantities of dispersible radioactive materials or toxic chemicals (i.e., quantities that might pose a threat to the health and safety of people outside the immediate area of storage or use). Such facilities are then required to do a *quantitative* assessment of the hazard to provide a technical basis for emergency planning. The preparation of this quantitative analysis, termed the Emergency Preparedness Hazards Assessment (EPHA), yields a detailed understanding of the hazardous substance, the potential for accidental releases, and the possible health effects that could result from releases.

3.1.1 The Hazards Survey-EPHA Process

The DOE Office of Emergency Management has published guidance[2] on how to conduct the Hazards Survey and EPHA. The guidance is detailed, but the process really quite straightforward, consisting of five basic steps:

- Identify/screen hazardous materials
- Characterize the materials
- Identify potential release scenarios
- Analyze/assess impacts of releases (to workers and public)
- Identify recognition factors

Step 1: Identify and Screen Hazardous Materials. The Emergency Management staff begins by identifying the hazardous materials stored, used or produced in the facility. Using all the available information sources, both documented (material inventories, procurement records) and otherwise (interviews, physical inspections) a complete list is developed of the materials in the facility that

[1]DOE O 151.1B, *Comprehensive Emergency Management System*, U.S. Department of Energy, Washington, D.C., 2003. Available at http://www.directives.doe.gov
[2]DOE G 151.1-1 Volume II, *Hazards Surveys and Hazards Assessments.* U.S. Department of Energy, Washington, D.C., 1997. Available at http://www.directives.doe.gov

may represent a dispersible hazard. Then the list is "screened" using both quantity thresholds and the properties of the materials. Screening excludes from further consideration any substances that are not dispersible, non-toxic, or present only in small quantities. Because DOE facility emergency management programs are required to address hazards to the workers in and near the facility as well as the public, the screening quantity thresholds for most materials are quite small by industrial standards. The screening process yields a "short list" of substances that are candidates for quantitative analysis. If it is not clear that a substance should be eliminated from consideration based on the screening criteria, it is retained for quantitative analysis.

Step 2: Characterize the Materials. Materials remaining on the "short list" are characterized to the extent necessary to quantitatively analyze their potential health impacts. Material properties such as concentration, vapor pressure, boiling point, acute toxicity and particle size are determined from Material Safety Data Sheets, reference books or scientific literature. The location(s) and manner of use or storage are noted, including the barrier(s) to release, container size, system pressure, and the amount available for release under different operational conditions or by different initiating events.

Step 3: Identify Potential Release Scenarios. A systematic examination of toxic chemicals identifies potential release scenarios for each substance in each location where it exists in the facility. A range of different failure modes and initiating events are considered, including accidents; natural phenomena (e.g., earthquakes, tornadoes); external events (e.g., aircraft crashes, offsite transportation accidents involving hazardous materials, pipeline explosions); and terrorism, sabotage, or other malevolent acts. A spectrum of potential events ranging from low-consequence, high-probability events to high-consequence, low-probability events are postulated and realistically analyzed to determine the amount of the substance that would be released, the rate of release, the pathway by which it would reach the environment and operational indicators associated with the release event. By comparing the features of the different release scenarios, a number of specific cases are selected to represent the range of possibilities.

Step 4: Analyze Impacts of Releases. Using source and atmospheric transport models appropriate to the substance and the release conditions, the hazardous material concentration or dose at several locations is calculated. The receptor distances are chosen to represent a nearby worker (30 meters away), workers at other facilities on the same site (100 meters) and the public (at the nearest DOE site boundary). For toxic chemicals, the health impact at each of these receptors is judged by comparing the concentration (calculated as a 15-minute average value) to criteria that are specified in the DOE Order and guidance. These criteria, in order of precedence are: the 60-minute Acute Exposure Guideline Level (AEGL) -2 value published by the Environmental Protection Agency; the Emergency Response Planning Guideline (ERPG) -2 value published by the American Industrial Hygiene Association; and the Temporary Emergency Exposure Level (TEEL) -2 value developed by DOE. In keeping with DOE emergency management doctrine, specific planning and preparedness measures to warn and protect people are considered necessary for any location where the calculated impact exceeds the applicable criterion.

Step 5: Identify Recognition Factors. Rapid, effective response to protect people begins with recognizing the release. In characterizing the material (step 2) and identifying release scenarios (step 3) the hazards analyst is expected to identify the indications that might signal an actual or impending release. These "recognition factors" are recorded by the analyst, along with any additional insights reached in the course of analyzing the release scenarios (such as, "this indication would be seen during scenario A but probably not in scenario B or C because…..."). Recognition factors are not just things that indicate a release is already in progress. From the standpoint of effective response, the most valuable recognition factors are those that indicate a *degrading* safety condition or a *potential* release at some future time (such as increasing pressure in a storage tank). Recognition factors are an exceedingly important product of the EPHA, because the facility staff needs to know *what specific facts and indications* they must be alert for to start the protective action process.

3.1.2 The EPHA Outputs

The EPHA process yields a *detailed technical understanding* of the hazards upon which to base the emergency plans and procedures. Answers to the following questions are among the more important facts and insights that emerge from the EPHA.

- What exactly is the hazard?
- What are its properties?
- Where is it stored/used/produced in the facility?
- What are the barriers/controls that keep it in a safe state?
- What events or failures could cause it to be released?
- What would be the indications of an ongoing or impending release?
- What areas would be affected by each release scenario and in what time frame?
- What would be the effects on people who might be exposed?
- How does the material behave in the environment?
- What can be done (before the fact) to reduce or eliminate the hazard?
- What can be done to mitigate the health/safety impacts once a release occurs?
- What will it take to accomplish that mitigation (rapid warning, public education, preparations for evacuation, shelter enhancement)?

3.2 Advocate for Hazard Reduction

With the quantitative understanding of the hazard that is developed through the EPHA process, the emergency management staff is well prepared to advocate for reduction or elimination of hazards. At some facilities, it is not uncommon for production- or operations-oriented people to obtain and store inventories of hazardous materials well in excess of immediate needs as a hedge against future funding or procurement constraints. Also, when processes and programs change or are terminated, hazardous material can be left stored at end-user facilities with no expectation that it will ever be consumed, but with no plan for disposal. The job of the emergency management people is to conceive of really bad things and to think about the consequences that would result if, for example, a high-integrity tank failed or if someone *deliberately* opened the drain valve. It is perfectly acceptable--in fact, almost a job requirement—for the emergency management staff to ask questions that make facility operations people and managers uncomfortable. Sometimes all it takes to make a hazard go away is to ask the right question, like:

- What do you use this material for?
- How much do you use per day/week/month/year?
- Why do you store it here?
- How would it impact the operation if you substituted another material/form/cylinder size?
- Wouldn't it make sense to put a lock on this door/valve/gate?
- If a leak occurred right here, how (and how soon) would you know it?
- Where does this pressure relief valve discharge?

In most cases, the changes in inventory or the way the facility does business will never be subjected to a cost-benefit study. It will simply become obvious that there is a low-cost or no-cost way of accomplishing the facility mission with less risk to workers and the public. Some real examples of risk-reduction measures at DOE sites that have been accomplished mainly because they were advocated by the emergency management staff are:

- Transition from 1-ton chlorine cylinders to 150-pound cylinders to hypochlorite injection as water treatment requirements declined.
- Separation of storage tanks holding incompatible chemicals (mixing them would form a toxic gas) so they would not drain into the same sump in the event of a common-mode failure of the tanks.
- Dividing the containment berm surrounding two acid tanks to reduce by half the pool size (and evaporation source term) from failure of a single tank.
- Reduction of end-user facility inventories, with bulk supplies stored in a central, more secure location.

3.3 Provide for Early Recognition of Releases

Exposure to airborne toxic hazards can be minimized if people in the affected area are warned early enough to evacuate or seek effective shelter. A warning will only be issued after the release is *recognized* and a *decision* is made. From the in-depth quantitative understanding of the hazard that is developed through the EPHA process, the emergency management staff attempts to identify the earliest possible indications of each release scenario in order to ensure that warnings can be issued in time to be effective. In addition to indications of *actual* releases (equipment failures, alarms, human sensory impact, etc.), signs of an *impending* release (indications that barriers or controls are failing) are identified. Events/conditions that may be *precursors* to barrier failure and release (natural phenomena, loss of utilities, security incidents, etc.) may also be used to initiate actual warnings or raise the state of readiness for making critical decisions.

Using the recognition factors thus identified, the emergency management staff works closely with facility Operations staff to develop clear, unambiguous *decision criteria*. The operators and managers need to be completely familiar and comfortable with the criteria and the decision process in order to make time-critical warning decisions with the greatest possible speed and certainty. It is a DOE-wide requirement that every facility assign the responsibility and authority for making those time-critical decisions to an individual who is:

- on duty or immediately on-call, 24/7,
- knowledgeable of the facility hazards and potential emergencies, and
- trained in the facility emergency procedures and use of the decision criteria

Armed with a quantitative understanding of the hazard and what it takes to mitigate the consequences, the emergency management staff is often able to argue effectively for new or improved means of detecting releases. Preventing death or severe injury in the nearest population groups can sometimes be shown to depend *completely* on early recognition, warning and protective actions. In such cases, it becomes fairly easy to sell the idea of installing a new instrument or alarm. Often the simplest and most effective risk reduction measure is to make sure that the operating staff is trained to *recognize the significance* of certain indications that are already available to them.

3.4 Plan, Prepare and Train for Effective Response

Using the knowledge and insights gained through the EPHA process, a DOE facility emergency management organization will develop detailed plans and procedures and train the staff to carry out response actions to reduce the severity of release events and minimize health impacts. The emergency management program for a facility with toxic chemicals will typically include the following measures to ensure the most effective response to releases.

- Mobilization of response assets. The resources and personnel needed for different types of emergencies are determined during the planning phase. Their activation or mobilization is built into the procedures and checklists that guide the very early response actions for different events or types of emergencies. A number of different specialized response elements (e.g., spill teams, medical response units, offsite support organizations) can be called into action with a *single decision*, reducing the time required to field the most capable and appropriate "package" of assets.
- Rapid warning and notification of affected population. Minutes lost deciding who to warn of a release and how to warn them can be a matter of life and death. Determining in advance just when (in response to what indications) a warning will be issued and exactly how it will be done (what method and message) gives the potentially-affected people the most time to protect themselves.
- Planned (predetermined) protective actions. The most effective action to mitigate the impact of exposure will not be the same for all releases. The properties of the toxic material, the type of release, and the availability and effectiveness of protective equipment or shelter are all considered during the planning phase. The most effective course of action for each population group or area is selected, spelled out in the immediate response procedures and checklists, and practiced during facility drills and exercises.

- ❖ Prepositioned equipment and supplies. If they are to be effective, equipment and supplies for mitigating toxic chemical exposures must be stored where they will be needed and maintained in serviceable condition. Personal protective equipment, communications devices, and materials that can be used to enhance shelter effectiveness (tape, plastic sheeting) are a few examples of supplies that might be determined necessary for the protection of specific groups.
- ❖ Provisions for personnel decontamination and medical intervention. Timely and effective decontamination and/or specialized medical treatment can reduce the health detriment from exposure to many toxics. The planning and preparations (decontamination methods, antidotes, etc.) for these measures are guided by the results of the EPHA. The on-site preparations and arrangements for offsite medical support are coordinated by the emergency management organization to help minimize the health impacts of any exposure.

4.0 Summary

Although the change in operating risk is not usually quantified, it is easy to demonstrate that the incremental reductions in risk are very real. Emergency management makes substantial contributions to the reduction in chemical accident risk at DOE sites and facilities by:

- ❖ Developing the detailed, quantitative understanding of chemical hazards that provides a technical basis for emergency plans and preparedness measures.
- ❖ Serving as advocate for both the reduction/elimination of hazards and for enhancement of means to recognize actual or impending releases.
- ❖ Mitigating human health impacts by providing for early recognition of releases, rapid decisions, timely warnings, and effective response measures to prevent or reduce exposure.

Risked Based Process Safety

Jack McCavit
Manager, Process Safety and Risk Assessment
Celanese
PO Box 937
Pampa, TX 79065
jlmccavit.pampa@celanese.com

For presentation at the
20th Annual CCPS International Conference
Risk Management: The Path Forward
April 11 – 13, 2005
Atlanta, Georgia

Abstract

Chemical process industries face a dilemma. Resources have been cut throughout the industry for cost control, yet expectations for continuous reduction of process safety incidents remain. This paper describes a CCPS book, *Risk Based Process Safety*, that is being written to address the need for better results with fewer resources. Applying the principles in *Risk Based Process Safety* will reduce incidents in the chemical process industries.

Introduction

The Center for Chemical Process Safety, CCPS, provides guidance to the chemical industry by publishing books that address the process safety needs of the industry. Topics for new books come from brainstorming sessions at various CCPS meetings throughout the year. The brainstorming ideas are then consolidated and developed into proposed projects. The CCPS Technical Steering Committee then votes to select the projects that will be funded. Once the project is funded, a committee of volunteers from the Technical Steering Committee is assembled to guide the project. This paper describes a CCPS project that has been funded and is in the early stages of development.

The project is titled *Risk Based Process Safety Management*. It is the next generation process safety management book to be published by CCPS. Four CCPS management books have been published in the past:

- "A Challenge to Commitment" – An overview and an outline of technical management of chemical process safety, 10 page brochure, published in 1989. (1)
- *Guidelines for Technical Management of Chemical Process Safety* – Expanded on the brochure. Offered alternatives for implementation of each of the CCPS elements. Guideline book, published in 1989. (2)
- *Plant Guidelines for Technical Management of Chemical Process Safety* – Highly detailed instruction and example materials suitable for direct implementation at plants. Guideline book, published in 1992. (3)
- *Guidelines for Implementing Process Safety Management Systems* – Designed to assist those who are called upon to design, develop, and install process safety management systems with their plant or company. Guideline book, published in 1994. (4)

These books were well received by the industry, but do not address the vast changes that have occurred in the chemical industry in the last 11 years. A universal theme across the industry is reduction of costs and increase in efficiency while maintaining or improving performance. To remain competitive, essentially every company has been forced to reduce headcount both in corporate central resources and at plants. Tools such as six sigma are being applied to achieve efficiencies.

One issue has remained constant however; societal and management demands for continual improvement in process safety results. Process Safety incidents are not acceptable. Deaths or serious long-term injuries from fires and explosions in the chemical processing industry have been few, but they are newsworthy, sensational, and any deaths from an industrial fire/explosion are too many.

Therein lies an anomaly. The industry has fewer resources to implement and maintain process safety systems at the same time there are expectations of ever improving results. Using a sports analogy, it is like demanding our defense allow fewer scores, and at the same time asking them to use fewer players.

It is easy to get caught in denial that the expectations are realistic. Process safety professionals must get past those initial perceptions and recognize that the dilemma is real. In order to remain competitive, industry has to maximize the efficiency with which it spends resources, spending less and at the same time improving performance. Process safety professionals must recognize the change and accept the challenge to find ways to get better results with fewer resources.

That challenge is why there is so much passion around this project. It is exciting because the book will provide solutions to the dilemma described above. Resources are spread thin, maybe too thin if process safety management systems are not efficient. If the limited resources are not applied appropriately and holes develop in the management systems, the frequency of incidents will increase, not decrease. This book will allow us to cope with the low resources and actually improve process safety performance. Applying the principles in *Risk Based Process Safety* will reduce incidents.

The Book

As the book developed, several titles were considered. *Risk Based Process Safety* was selected because it clearly places the emphasis on risk as the determining factor for designing process safety systems. When resources are at a premium, risk needs to be the determining factor to apply resources appropriately.

Risk Based Process Safety will be a step change improvement from previous CCPS process safety management books. It will include

- new elements
- more of the life cycle of an operating unit

  Most elements currently are focused on a normally operating unit. For some elements *Risk Based Process Safety* will include suggestions for applying the principles to more of the life cycle such as commissioning a new unit, maintenance, and in some cases, decommissioning.
- Risk based implementation for each element

  Focus of the new guidance will be using the understanding of risk to apply resources appropriately. This will also include examples of activities to avoid that are unnecessarily resource consuming.
- Management review

  Since resources are stretched thin, meaningful management review is critical to be certain systems are working as expected and to identify ways to improve performance.

When OSHA 1910.119 was first promulgated in 1992 most companies had to modify their process safety systems to assure compliance. Today compliance should be a given. The driver today is preventing catastrophic incidents. The book is intended to help companies achieve better process safety results whether their compliance standard is OSHA 1910.119 or SEVESO II or any other international process safety standard.

Several of the elements covered in *Risk Based Process Safety* have been addressed in CCPS books. The focus of previous books has generally been complete documentation about an element. The focus of *Risk Based Process Safety* is efficient systems requiring fewer resources and being sure the efficient systems are actually working as designed. The principles should apply no matter where a company is in the evolution of its process safety management systems. The guidance will be valuable if the company is still trying to come into compliance, if the company has been in compliance and is seeking efficiency improvement, or if the company is simply trying to maintain efficient systems

Elements

The elements of the book are below. Note that they are divided into four groups.

- **Commit to Process Safety** – This includes elements that require management commitment. Without these fundamental building blocks, the rest of the elements will not be able to be done well.
- **Understand Risk** – Once there is genuine demonstrated commitment to process safety, the next step is to understand the risk involved in specific process. This too is critical. Understanding risk is an obvious prerequisite to managing risk.
- **Manage Risk -** Once the risk is well understood, the systems to manage risk can be effective.
- **Learn from Experience** – Finally, when the risk is well understood and managed, there is a need to learn from the empirical observations from operating the process.

Commit to Process Safety

- **Process Safety Leadership and Culture**
  In order to have a successful process safety management system there must be appropriate leadership at all levels in an organization. Executives need to understand the business case for process safety and provide the human and financial resources. Managers must assure systems are in place and functioning as designed. Engineers must be diligent in the implementation of the process safety management systems. Combined, all levels in an organization contribute to a culture of high expectations.

- **Compliance with Standards, Codes, Regulations, and Laws**
  This section will provide practical guidance for compliance with regulations as well as codes and standards. The OSHA regulation requires adherence to RAGAGEP, recognized and generally accepted good engineering practices. Adherence to RAGAGEP is required by the OSHA regulation, but just as importantly, adhering to standards and codes results in well designed mechanical systems. Understanding RAGAGEP is challenging, but can yield much improved process safety results and in many cases reduce costs. Unfortunately, strictly following codes and standards that are not applicable or necessary can be costly with no added benefit. This section will help companies sort out compliance with standards, codes, regulations, and laws.

- **Workforce Involvement and Information Sharing**
  This section will give practical guidance for involving employees in process safety management systems so the employees will be committed to the systems and so the knowledge of people close to the operating process will be used to reduce risk on an every day basis.

- **Community Outreach and External Sharing of Information**
  There is general lack of understanding in the public about chemical processing plants. The more the public understands about our process safety systems the more comfortable they will feel. This section will help define how to develop a relationship with the community.

- **Maintenance and Enhancement of Process Safety Competence and Technology**
  This element will address methods to develop and maintain competent people and unit specific technical knowledge. It is an element that truly requires management commitment because with the frequent changes required by today's business demands, there is the danger, if change is not managed properly, that the appropriate number of resources and the technical knowledge of a unit will be lost.

Understand Risk

- **Process Safety Information**
  Process safety information is a basic building block in understanding the risk involved in operating a process. This section will define expectations for various types of process safety information as well as give practical guidance for management of PSI to keep it current. This will be an important element both because of the critical need to have good information for a safe process and because poorly managed process safety information can easily become out of date and is costly to update.

- **Hazard Identification: Process Hazard Analysis**
  Process Hazard Analysis, PHA, is viewed by some as the backbone of process safety management, but countless human resources have been wasted doing PHA's poorly. Human resources have been wasted spending excessive time on parts to of the plant that had very low potential consequences and therefore low risk. *Risk Based Process Safety* will identify wasteful practices and will describe methods of identifying hazards using fewer resources.

- **Risk Assessment**
  Assessing risk requires understanding potential consequences and understanding the frequency at which the potential consequences may occur. There are several ways to approach risk assessment, some rigorous and resource intensive, some not rigorous and not resource intensive. This section will discuss alternative methods to approach risk assessment and clarify the advantages and disadvantages of each.

Manage Risk

- **Training, Competence, and Performance Assurance**
  Risk based process safety management philosophy can certainly have a major role in the resources required for training, competence, and performance assurance. The book will help define appropriate training systems depending on the risk of the process.

- **Operating Procedures**
  Operating procedures vary tremendously from company to company and sometimes from unit to unit. This is an obvious area in which one size does not fit all. The operating procedures for a jet fighter should be written more rigorously than the operating procedures for a tricycle. This section will provide guidance for evaluating the need for rigorous procedures. It will also provide examples of different formats from which to choose.

- **Mechanical Integrity**
  Mechanical integrity is an element that is fraught with opportunity to waste resources. It is also an element which if not done well, can lead to poor reliability and incidents. Yet risk based management is critical to avoid unproductive spending. Guidance will be given about how to balance spending and reliability.

- **Safe Work Practices and Control Procedures**
  Safe work practices were in place for decades before process safety was recognized independently. The key to good safe work practices is the rigor in which practices are implemented. The most important guidance in this element may be in the metrics or audit areas.

- **Management of Change**
  Management of change is one of the most important elements for safe operation. There are several classic examples (Flixborough is probably the most well known) in which safe

units became unsafe due to a change that was not managed properly. Unfortunately, MOC systems can also be resource intensive if not designed properly. Poor MOC systems can also cause costly problems in process safety information. This element will include ways to manage change safely and ways to be sure your MOC system is working well.

- **Pre-Startup Safety Review**
  This element ties closely to MOC and is not normally a major resource drain. The book will emphasize critical activities that will result in the safe start-up of a change.

- **Emergency Response and Crisis Management**
  This element will present alternatives for emergency response and crisis management when resources are limited.

- **Contractor Management**
  While contractor management does not require the expenditure of a great deal of resources, it does require a good system for knowing that the risk associated with the use of contractors is being properly managed. This element will describe contractor management systems for varying degrees of risk.

Learn from Experience

- **Incident Reporting and Investigation**
  If the industry successfully reduces incidents, reporting of minor incidents and near misses will become even more important. This element will provide information about systems to report incidents in addition to incident investigation techniques. Of course a perfect incident investigation adds no value until the recommendations have been implemented, so incident investigation recommendation tracking methods will also be discussed.

- **Performance Measurement, Monitoring, and Continuous Improvement**
  Each element will include metrics specific to that element. This element will include cross cutting metrics for the overall management system. It will likely be inappropriate for any company to use all of the metrics presented, but there will be several from which companies or sites can choose.

- **Auditing and Management Review**
  Auditing is required for OSHA covered processes, but management review is not. Management review is the periodic internal review of process safety systems to be sure they are functioning as intended. When resources are tight, it is easy for process safety management elements to slip into disrepair. Even if a perfect PSM system is implemented, it is critical to monitor the system to be certain it continues to be robust. In addition to auditing practices, this element will give details of implementation of a management review system.

Element Outline

Each element will have a similar outline. The outline is below.

0 Element Title

1 **Element Intent**
This portion of each section is an introduction to the element which describes the importance of the element and what the element is intended to achieve.

2 **Relationship to Other Elements**
The elements are not stand-alone. To have a complete process safety management system, it is important to understand the relationship of the elements to each other.

3 **How Risk, Culture, and Resources Affect Element Design and Implementation**
Risk based system means that one size does not fit all. This part of each chapter may be one of the most valuable to readers. If process safety management is truly to be risk based, the same degree of rigor cannot be used for plants with widely different potential consequences. Every element will include methods and logic to optimize the management system and resources based on risk, culture, and available resources.

4 **Essential Features**
Essential Features describes the critical activities to assure safe operation. These are the activities that must be done well to achieve better results.

5 **Element Work Activities**
This portion of the element describes fundamental activities that assure each element achieves it's intended purpose. It defines who does what. It is the basics.

6 **Examples of Ways to Improve Effectiveness**
Ways to Improve Effectiveness is one of the most important parts of each element. There are a number of ways to develop process safety management systems that will minimize incidents. However, the effort required varies significantly. This portion of each element will include examples of innovative methods of achieving good results with minimal resources. It will also include examples of activities that have been done earlier that are now recognized as inefficient or unnecessary.

7 **Element Metrics**
Metrics are an important part of any good management system. Each element will have specific metrics for that element.

There is also an element devoted only to metrics, but that element is intended be cover broad process safety management topics so those metrics will not normally be element specific.

8 **Element Audit Methodology**
. There is a demand for greater production volumes and more rapid response to customer needs at the same time headcount is reduced. With reduced resources and greater demands, process safety systems can gradually deteriorate if the system is not monitored when faced with these challenges. This portion of each element will include methods to audit process safety management systems to be sure the systems are functioning as intended. Implementation of a management review system including these audits is one of the most critical actions required to be confident of safe operation.

9 **Implementation/Practice Examples**
From Pots and Pans Chemical Company to Giant Conglomerate Chemical Company, units have different levels of potential consequences and different levels of resources available. This portion of each element will contain specific examples for companies with widely varying needs and capabilities.

Risk Based Process Safety Project Committee Members

The CCPS committee members for *Risk Based Process Safety* have worked long and hard to develop the book. The members are:

Jack McCavit, Celanese, Chairman
Bob Perry, CCPS, Staff Consultant
Don Abrahamson, OxyChem
Steve Arendt, ABS Consulting
Scott Berger, CCPS
Susan Brate, GE Advanced Materials
John Bresland, US Chemical Safety and Hazard Investigation Board
Bruce Hartman, ChevronTexaco Corporation
John Herber, 3M
Dan Isaacson, Lubrizol
Shakeel Kadri, Air Products
Greg Keeports, Rohm and Haas
Pete Lodal, Eastman Chemical
Bill Marshall, Eli Lilly
Mike Marshall, OSHA
Neil Maxson, Bayer Material Science
Lisa Morrison, PPG
Jim Muoio, Lyondell
Karen Tancredi, DuPont
Tony Thompson, Monsanto
Bill Wheeler, Hughes
Roy Winkler, Sunoco, Inc.
Gary York, Rhodia
Mark Preston, BP Grangemouth

References

1. Center for Chemical Process Safety (CCPS). A Challenge to Commitment. New York, American Institute of Chemical Engineers. 1989.

2. Center for Chemical Process Safety (CCPS). *Guidelines for Technical Management of Chemical Process Safety.* New York, American Institute of Chemical Engineers. 1989.

3. Center for Chemical Process Safety (CCPS). *Plant Guidelines for Technical Management of Chemical Process Safety.* New York, American Institute of Chemical Engineers. 1992.

4. Center for Chemical Process Safety (CCPS). *Guidelines for Implementing Process Safety Management Systems.* New York, American Institute of Chemical Engineers. 1994.

Bhopal Accident and its Effects on Process Safety Worldwide

J. P. Gupta
Professor
Department of Chemical Engineering
Indian Institute of Technology
Kanpur-208016 (U.P.), India
Phone:+91+512+2597175, Fax: 2590104, e mail: jpg@iitk.ac.in

The sudden release of 41 T of deadly MIC and its reaction products on the cold and quiet night of December 2-3, 1984 in Bhopal was the worst-ever industrial accident anywhere. It resulted in the prompt death of approximately 4,000 and the subsequent death of another over 20,000, thus far and counting. Over 200,000 are undergoing lifelong sufferings. The genetic mutations due to the effects of MIC are affecting children born since 1984 and will affect several generations yet to be born. The pollution of underground water bodies due to chemicals discharged by the company during operations and the seepage underground of the chemicals left behind at site is affecting several hundred thousand more. A quote from Charles Perrow of Yale University describes it appropriately:

"...In catastrophes, the common cliché used is 'we were lucky it wasn't worse'. However, in case of Bhopal, it does not hold. Bhopal was that rare accident that could hardly have been worse; they are hard to arrange" [1]

The accident at Bhopal was a wake-up call worldwide about the dangers of process industry. A worldwide movement of making the chemical process industry safer started immediately thereafter. It has been 20 years since the tragedy and the process industry has moved a lot forward in safety. It is time to take a stock of the situation. In this paper, we will briefly describe the current scene in Bhopal ascertained by personal visits to the city and the plant. Then we will detail the steps taken by major international organisations, individual countries, trade and professional bodies as well as universities to make the process industry safer. We will further discuss the recommendation of the International conference 'Bhopal and its Effects on Process Safety', to be held on the 20th anniversary of the Bhopal Tragedy, December 1-3, 2004, at I.I.T., Kanpur, India. It is recognized that the time has come to look seriously into and resolve the issues involved rather than spending disproportionate effort in apportioning and denying blame for the tragedy.

Key words: Bhopal Gas Tragedy, Process Safety

[1] Perrow, Charles, 'Normal Accidents', Princeton University Press, Princeton, USA (1999)

My documents/CCPS - 2005 Atlanta conference Abstract

Two Hundred Years of PSM at DuPont

James A. Klein, C. Curtis Clements, and David E. Cummings

E. I. du Pont de Nemours and Company
Wilmington, DE, USA
james.a.klein@usa.dupont.com

Process Safety Management (PSM) is a comprehensive, integrated set of management systems designed to help manage chemical risk and other types of process hazards with the primary goal of preventing process-related incidents and injuries. The purpose of this paper is to discuss the history, current practices, performance, and future improvement of PSM at DuPont.

Since its founding as an explosives company in 1803, DuPont has had a core value for safety and process safety, with E. I. DuPont noting that "We must seek to understand the hazards we live with." The design of the first powder mills, for example, included consideration of such details as minimum safe distances between mills, proper siting and orientation of the production facilities, minimization of ignition sources, blast-resistant construction and over-pressure venting, all intended to help minimize the effects of any potential incidents. E.I. DuPont also established clear management accountability in terms of process safety in the operation of the powder mills, including a policy that line management was accountable and that "no employee may enter a new or rebuilt mill until a member of top management has personally operated it".

Safe operating procedures, management of change practices, and predecessors of other PSM systems date to early in the company's history, and management directives for a goal of zero injuries dates from at least 1915. In the 1950's, an early application of the company's first computers was to maintain the records for preventive maintenance of process equipment, foreshadowing today's mechanical integrity efforts. A process hazards review program was introduced company-wide in 1966, following a major site explosion in 1965, in effect, beginning the modern era of PSM at DuPont. Following a series of serious accidents and near-miss incidents in the 1970s, the PSM program was further developed, refined, and in 1979, formalized to include a total of 14 elements, organized under the categories of technology, personnel, and facilities.

With additional major industrial incidents, the resultant regulatory and industry guidance from the 1980's to the early 1990's, and application of the learnings from company experience, the PSM program at DuPont has continued to evolve. In the last ten years, the scope of PSM was expanded to include Lower Hazard Operations (LHO) and laboratory/pilot plant facilities, in addition to traditional Higher Hazard Processes (HHP), thus extending the principles of PSM to help improve process safety at all company facilities. More recently, a new focus on operational discipline has been introduced, helping to ensure that PSM systems are properly maintained and followed. DuPont's business portfolio continues to change by developing new technologies and through bolt-on acquisitions with a multi-faceted focus of traditional chemical manufacturing, biotechnology, and integrated science. PSM is an important program to help ensure effective risk management across all existing and new operations.

1. Introduction

DuPont was founded over 200 years ago in 1802. At the time, Thomas Jefferson was the third U.S. president, the Louisiana Purchase from France wouldn't occur for another year, and Lewis and Clark wouldn't leave St. Louis for their famous expedition exploring the West for two more years. In fact, President Jefferson encouraged the company's founder, E. I. du Pont, to manufacture gunpowder, since the United States at that time did not have reliable producers of high quality powder [1,2]. E. I. du Pont was perhaps uniquely qualified for this role, since he had apprenticed in France under Antoine Lavoisier, a famous French chemist who also was chief of the royal powders works. Of course, manufacture of gunpowder was a dangerous business, which E. I. du Pont readily recognized. Reflecting on these dangers, he noted that:

> We must seek to understand the hazards we live with,

establishing the strong commitment and core value for safety at DuPont which has remained constant for over 200 years.

While the hazards of 1802 are probably different than the hazards of today, the foundation of the strong current Process Safety Management (PSM) program and performance at DuPont rests in our corporate history. The continued dedication over 200 years to understanding the hazards present in our processes, and in successfully managing them to prevent serious injuries and incidents, has contributed significantly to current PSM priorities and practices. Learning from experience – what has worked and what hasn't – both in our company, and in the broader chemical industry, is essential for continuous improvement. Therefore, reviewing the past can contribute to a stronger and safer future.

2. Foundations of PSM

2.1 Powder Mills on the Brandywine

E. I. du Pont selected a location along the Brandywine River in Delaware for the first gunpowder mills. The river provided water power for the mills, local granite could be used for construction, and the local port in Wilmington could be used for shipping [3]. In his original plans, du Pont noted that "At the side of the stream, there must be sufficient land to distribute the different buildings of the plant at prudent distances from each other [4]." Spacing considerations were key to limiting the effects of possible explosions related to manufacture of gunpowder. Specific instructions included:

> The stamp mill should be at a considerable distance from… any other building. This machine may cause violent explosions and should be

> surrounded by high and thick walls... It [the graining mill] always contains a somewhat large quantity of finished powder and the necessary presence of the workers may cause most serious accidents, it is therefore important to isolate it as much as possible. [4]

Construction at the Brandywine location was started in 1802, and the first product was first sold in 1804 with the name Brandywine Powder (later changed to DuPont Powder) [1,5]. DuPont gunpowder quickly developed a reputation for high quality, and production on the Brandywine continued until 1921.

In addition to careful siting of buildings at minimum safe distances, much consideration was also given to the actual design of the mills. As shown in Figure 1, high, thick granite walls were constructed on three sides. The fourth side, which faced the river, and the light gauge roof were designed to direct any explosions towards the river and away from workers and other buildings. The success of the basic design is shown in Figure 2, where the aftermath of an explosion in the 1880's show that the granite walls remain relatively intact. Additional design features were intended to prevent powder contamination and potential ignition sources, which could lead to explosions, and to allow workers to exit quickly if needed [6]. Redundant safety valves were also used:

> Two safety valves are used – which are absolutely necessary, as when one only is employed it may by accident get choked and the works explode. [7]

Various barricade designs were also used at other locations as the business expanded, as shown in Figure 3.

Members of the du Pont family and supervisors were accountable for the safe operation of the mills, with corporate policy requiring that they be present to startup new equipment or manufacturing processes. The following was posted on the doorway to the mills:

> No employee may enter a new or rebuilt mill until a member of top management has personally operated it. E. I. du Pont. [2]

Management also gave personal instruction on how to safely roll powder kegs and how to avoid dropping them. By 1811, E. I. du Pont issued official safety rules to help ensure safe manufacture:

> As the greatest order is indispensable in the manufacturing as well as for the regularity and the security of works, than the safety of the workmen themselves, the following Rules shall be strictly observed by every one of the men employed in the factory. [8]

Family commitment and accountability was so strong that E. I. du Pont lived at the site. The original plan for the site, in fact, included the requirement that the director's house

Figure 1 – Powder Mills on the Brandywine River, ca. 1830

Figure 2 – Destroyed Powder Mills, ca. 1889

be constructed "in such a position that the whole plant may be seen from its windows [4]." The du Ponts raised seven children in the house, and on at least one occasion, his wife was injured and the house was severely damaged in an explosion. Ultimately, several family members died of work-related causes [1,2].

Despite precautions, explosions in this dangerous business did occasionally occur. An explosion in 1815 resulted in 9 deaths, which were the first company casualties, and another serious explosion that occurred in 1818 resulted in 40 deaths. All accidents were carefully investigated: "the accident cannot be attributed to any stone or nail or other foreign body mixed by chance with the materials [9]" and "This is the only cause which our knowledge of the business enables us to find for the accident [10]." Two years following the major explosion in 1818, changes had been made and it was noted that:

> We have last Thursday another accident in one of our mills which has been attended to with some satisfaction, as nobody was injured and it has proved a fair experiment that upon the plan on which our mills are now built we have not to fear any general explosion like the one which happened here two years ago. From the particular construction of the mill the effects of the explosion have been directed in such a way so as not to communicate to any other part of the works. The whole of our mills are now constructed on that safe plan. [11]

In another instance, it was noted that "3000 lbs. of powder have never before exploded together so innocently and with so little damage [10]."

Yet, manufacture of gunpowder was inherently dangerous and accidents occasionally occurred. As a result, "Men in powderhouses usually have an arranged plan of escape in their minds, and at the least unexpected noise, have not hesitated to plunge into the canal [6]." A related euphemism for possibly dying in an explosion was known as "going across the creek [1]." The conscientious efforts by the company to operate safely and to learn from accidents, though, was recognized in 1855 by the Wilmington Morning News:

> The number of persons killed in explosions has been reduced by the persistent efforts of the firm to discover and introduce safeguards to prevent catastrophes. [2]

Lessons from these early years in the company's history are today reflected in many elements of PSM, including safe work practices, training, incident investigation, pre-startup safety reviews, emergency response, and operational discipline.

Figure 3 – Explosion Barricades [12]

2.2 The Safety Division

The company safety effort was re-organized in 1911 [13]. Safety Commissions were formed to study and promote accident prevention measures. Safety Commission meetings of three operating departments in 1913, for example, required over two weeks and resulted in 300 recommendations. Rules were also issued governing:

- Operating procedures and practices
- Plant and equipment inspections
- Injury report forms and reporting requirements
- Standard drawings for construction and mechanical safeguards,

which were precursors to modern PSM elements such as Operating Procedures, Auditing, and Process Technology.

A corporate Safety Division was established in the Engineering Department in 1915 [12,13] to serve as central clearinghouse for evaluation and purchase of all safety items. The Safety Division acted in an advisory capacity for all plants, which included technical training, safety inspections, and review of all construction projects during the design stage. Safety groups were also established at most large plants. About this time, the goal of zero injuries was first mentioned in corporate directives [2]. Over the next 10 to 15 years, the Safety Division was very active making plant inspections, compiling and analyzing accident statistics, conducting special investigations of new processes, and participating in design reviews. In 1926, safety and fire protection were merged to form the Safety and Fire Protection Division. A set of Safety and Fire Protection standards was subsequently issued around 1930, including for example, Safe Practices in Entering Tanks. Detailed, periodic inspections based on these standards were also conducted at plants.

Safety groups were eventually established at all plants. Reflecting early DuPont history, though, responsibility for safety shifted from safety departments at sites to operations management, based on:

> The acceptance throughout the company of the fundamental principle that accident prevention is an integral part of the operating routine and therefore responsibility. [13]

Training programs were offered to all workers, since employee commitment to working safely – operational discipline – was recognized as essential for the success of safety programs:

> Recognition of the fact that the safety morale or attitude of the actual workers will mostly govern the success of the safety effort and consequently any program to be successful must be designed to reach every worker through regular organization lines. [13]

The success of these programs is evident from safety statistics collected during this period. Fatality rates were reduced from 5.66 per million hours of exposure in 1910 to

0.09 by 1931. Injury rates and days lost were similarly reduced by 90-95% from 1913 to 1931 [13].

2.3 Expansion into Chemicals

In the early 1900's, and especially in the 1920's, DuPont aggressively expanded into new chemical businesses, such as dyes, paints, plastics, and cellulose products [1]. A good source of nitrogen for explosives, fertilizers, and refrigerants was particularly desired, so DuPont began investigating production of ammonia following World War I. In 1924, a contract was completed to acquire the rights to the Claude process developed in France for high pressure catalytic synthesis of ammonia [1,14]. The first ammonia was made at the Belle, WV site in 1926, and Figure 4 shows some of the compressors required for the high pressure process. Some problems were encountered during startup, and a failure of a liquefaction column resulted in a 4-month shutdown and re-location of the columns near the river behind concrete barricades. Other safety improvements were made to the design, including the early application of rupture discs:

> Spring loaded safety valves were installed at the start on all stages of the compressors, but the performance at pressure about 3000 psi was so erratic, even with the greatest care in assembling and servicing, that we learned in mid-1926 of a type of safety or rupture disc that failed by shearing a thin disc [and] we lost no time in arranging with the originator for their use. It is unquestionably the most dependable relieving device above 5000 psi when correctly designed and mounted. [14]

It's interesting to note that one of the first applications of computers at the Belle site thirty years later (an IBM 650, available in the mid-1950's for approximately $500,000) was for preventive maintenance, since "former efforts took too many people too long to keep necessary records [14]." This is not surprising when considering the need for reliable, well-maintained equipment operating at several thousand psi with hazardous materials present. Mechanical integrity programs continue to be of great importance today for ensuring safe operations.

In support of ammonia manufacture, a Safety and Fire Protection standard on "Safe Practices in the Handling and Storage of Anhydrous Ammonia for Industrial Plants Other Than Refrigerating Plants" was issued at least by 1929 [15]. The standard included sections on Physical Properties, Hazards, Storage and Handling, and Protection of Employees. It was noted, for example, that inhalation of 0.2% ammonia for long periods will "produce a severe and consequently dangerous irritation of the respiratory organs and the mucous membrane of the eye… one per cent is liable to be fatal." Subsequently, "men should be excluded from work in this connection who have active tuberculosis or a history of tuberculosis, acute and chronic bronchitis, and asthma, as these illnesses are greatly aggravated by ammonia fumes." Workers were "to be instructed fully as to its hazards and the precaution they should take." Some additional protections included:

- Storage tanks and containers should be strong enough to withstand elevated pressures (200 psi working pressure), should be equipped with pressure relief valves (standard ammonia type set to relieve at 200 psi), and should not be filled to full capacity to allow for expansion from temperature variation (90.6% of capacity at 70°F).

- Tanks should be showered with water to keep them cool on hot days or if a fire occurs nearby.
- Employees should wear gloves and goggles, with canister type gas masks readily available for emergencies. Safety showers should be nearby.

Standards describing safe handling and storage of hazardous materials such as ammonia were early precursors to current Highly Toxic Material guidelines, which are discussed in the following section.

Lessons from these early years of the 20th century proved invaluable as DuPont continued to expand into Neoprene, Nylon, Teflon, and many other new products. DuPont expertise in engineering and safety eventually also led to involvement in the Manhattan Project during World War II. Today, these lessons are reflected in many elements of PSM, including process technology, management of change, mechanical integrity, pre-startup safety reviews, operating procedures, and emergency response.

Figure 4 – Ammonia Process, 1930 [14]

3. Modern PSM

3.1 A Comprehensive PSM Program

In many respects, the modern era of PSM at DuPont began with a serious process incident in Louisville, Kentucky in 1965. A mechanical failure in a vinyl acetylene compressor led to high temperature, which initiated a chemical decomposition resulting in a major explosion. 12 fatalities, 61 injuries, and a property loss of over $50 million occurred. Following the incident, in what is considered the beginning of DuPont's modern process safety management policy, corporate management asked each site to review their production facilities and procedures to assess the potential for catastrophic events and to take appropriate preventive measures. In particular, the location of storage tanks was reviewed to ensure that they were sufficiently isolated from process operations that could be sources of fire or ignition.

An annual review was also instituted to ensure that process additions or changes did not create new hazards. In following years, many plants conducted annual, in-depth process hazard reviews to evaluate the process safety of site processes. These studies led to many process improvements, but ultimately, were increasingly less effective after several years of reviews. By 1973, new guidance was issued that responded to this problem by providing suggested hazard review methodologies and frequencies.

In 1974, the cyclohexane explosion at Flixborough in the United Kingdom resulted in 28 fatalities and over $250 million in losses. In 1976, the dioxin release at Seveso in Italy potentially exposed many people to a toxic chemical. These and other serious incidents led to an increased industry and regulatory awareness globally of the potential for catastrophic process events.

In 1978, serious process incidents involving explosions at four different DuPont sites occurred, resulting in 1 fatality and over $40 million in property loss, despite the contributions of earlier process safety programs. The causes of these incidents were evaluated in detail and common factors – technology, human, and facility – were identified. The result of this effort was the recognition that a comprehensive, integrated process safety management program was required. A corporate guideline for Process Hazards Reviews was issued later in 1978, and a corporate Process Safety Management guideline was issued in early 1979.

The PSM guideline was intended to help prevent "serious, process-related incidents, which might affect plant personnel, off-site communities, the environment, or result in significant property loss or loss of business." Its purpose was to:

- Establish a framework to help focus management efforts on this important, serious, and complex subject.
- Comprehensively describe the principles and essential features of Process Safety Management for use by sites in managing Process Safety.
- Describe Corporate Business responsibilities and activities.

Business and site management responsibilities were clearly defined, and application of the guideline required that:

> Each site and/or manufacturing facility manufacturing, handling, using, or storing hazardous substances *will*, within the framework of this Guideline, develop a detailed Process Safety Management Program suited to its specific organizational structure and needs.

The PSM Program included specific requirements and recommended practices for 14 elements, including:

- Process Technology
- Process Hazards Analysis
- Operating Procedures and Safe Work Practices
- Management of Change – Technology
- Personnel Training and Performance
- Management of Change – Personnel
- Incident Investigation and Communication
- Emergency Planning and Response
- Quality Assurance
- Mechanical Integrity
- Pre-Startup Safety Reviews
- Management of "Subtle" Changes

Today, most of these elements are recognized as consistent with the OSHA PSM Regulation 1910:119, which was finalized later in the early 1990's. More details on the current PSM program are provided in Section 4.

3.2 Business Specific Programs

In addition to corporate PSM programs, individual businesses developed specialized process safety programs, often related to reactive chemicals, to help ensure the safety of site operations. In many cases, these programs resulted from serious process incidents, leading to formation of global teams from multiple manufacturing sites with similar processes to help prevent further incidents. These teams shared technical and process safety information, developed specific guidance and standards for promoting safe operations, developed hazard evaluation methodologies, provided training, and developed audit protocols. In several cases, these programs pre-dated the issuance of corporate PSM guidance and are still active today, continuing to contribute to the prevention of serious process incidents. Currently, representatives from three teams – the TFE Committee, the Reactor Safety Program, and the Intersite Neoprene Monomer Safety Committee – are working together on the DuPont Chemical Reactivity Leadership Team to identify common practices to provide a template for other chemical reactivity teams that may be formed in the future.

The DuPont Tetrafluoroethylene (TFE) Committee has been active since 1968. TFE is a colorless gas used to make polytetrafluoroethylene (PTFE), various co-polymers, and agriculture chemicals. TFE is highly flammable, and when heated, can decompose violently leading to fire or explosion. The mission of the TFE Committee is to function as an advisory board and a communication network on all matters relating to handling and using TFE in a way that:

- facilitates the exchange of technical information about TFE between users
- develops standard procedures for handling and using TFE

- provides guidance for the design and operation of TFE facilities
- reviews new processes and facilities using TFE

so that TFE facilities are operated in a safe manner. The agenda from a recent annual meeting included standards development, review of industry incidents, evaluation of new equipment designs, and discussion of explosion and reactivity testing,

Following two serious incidents in the coatings business in 1975, the Reactor Safety Program [16] was started with a goal of completely safe resin manufacturing. Used for automotive and refinish coatings applications, these resins are polymerized via various methods including free radical (e.g., acrylic resins) and condensation polymerization (e.g., alkyd and polyester resins). Runaway reactions are possible, and use of large amounts of flammable solvents can also lead to fire or explosion. A survey of resin-producing facilities worldwide was made in 1976 to re-evaluate if existing resin reactor systems were adequately designed. The survey resulted in recommendations to:

- Better define resin-related process safety hazards and re-assign production based on process capability
- Develop equipment design and operating standards to ensure safe operations
- Upgrade process equipment and instrumentation, including layered process safeguards such as shortstop systems, improved emergency relief system designs, and catch tank systems
- Establish periodic process safety audits to ensure proper implementation of the new standards.

The Reactor Safety Program has continuously evolved over the last 30 years, with team representatives from all global resin-manufacturing plants meeting regularly to share technical and incident information. Key program activities currently include:

- Hazard evaluation, including computer simulation and experimental testing, to ensure safe resin formulation, scale-up, and manufacture [17,18]
- Design change authorizations, requiring safety review of all equipment changes that could affect process safety by Reactor Safety specialists
- Continuous improvement of equipment design and operating standards
- Vapor cloud and runaway reaction prevention programs
- Development of training programs, including computer-based training
- First and second party audits using specialized audit checklists.

The Intersite Neoprene Monomer Safety Committee (INMSC) was formed in 1984, as a result of continued serious process safety incidents following a 1983 explosion, which resulted in 3 fatalities, numerous injuries and major equipment damage. Neoprene is a chlorinated elastomer. Chlorobutadiene and dichlorobutadiene monomers are produced, then reacted via free radical emulsion polymerization to form the elastomer.

These monomers are highly reactive, flammable, and exothermally react at room temperature. Unchecked, these reactions can lead to runaway conditions, loss of containment, fires, or explosions. The purpose of the INMSC is to lead the effort to continuously improve the safety of manufacturing, handling, and using chlorobutadiene and dichlorobutadiene by:

- Recommending procedural and equipment design standards
- Providing secondary audits against standards, management systems, and metrics
- Identifying performance gaps that lead to the development of new and improved training materials, systems, auditing, standards, and practices.
- Identifying technology gaps that necessarily lead to new emerging safety technology development
- Consolidating all available information on monomer chemistry relevant to managing process safety
- Sharing responsibility for implementation of safety guideline improvements
- Providing an effective communication link for the exchange of information, ideas, problems, and recommendations among plants.

3.3 Response to Bhopal

The incident in Bhopal, India in 1984, involving the release of acutely toxic methyl isocyanate (MIC), resulted in about 3000 deaths and thousands of injuries and significantly raised the awareness of both industry and regulators about the potential for off-site catastrophic incidents. The response in DuPont was immediate. At the time, a product in the agrichemicals business used MIC as a key ingredient in the manufacture of an insecticide. DuPont was purchasing MIC and shipping it in railcars to a remote manufacturing location, and then storing fairly large amounts until needed [21]. Following Bhopal, the R&D organization developed a new process using inherently safer process principles to allow use of a less hazardous material as the starting raw material, eliminating the risk of shipping and storing MIC. Process upgrades allowed this new raw material to be converted to MIC in the reactor, which was then immediately reacted as part of the next step, limiting the amount of MIC in the process to low levels.

A Highly Toxic Materials Subcommittee was also formed to review DuPont operations worldwide. This review found that DuPont's traditional emphasis on management responsibility for managing safety and process safety management was effective in minimizing risks both on- and off-site. Additional safeguards were also recommended, though, to provide an extra margin of safety, providing additional guidelines for facilities, storage, operations, transportation, detection, and community protection. The corporate Safety, Health, and Environmental Quality Policy was also modified to include:

> DuPont will inform employees and the public about the safety and health effects of its products and workplace chemicals; and will provide

leadership in establishing community programs to respond to emergencies involving hazardous materials.

In 1985, three new corporate guidelines were issued on Off-Site Risk Assessment, Community Preparedness, and Management of Highly Toxic Materials. The Off-Site Risk Assessment guideline provided "management with information to aid in identifying and assessing potential off-site public exposure," and included risk reduction measures such as the following, incorporating principles of inherently safer processes:

- Material substitutions
- Increase distance to exposed populations
- Reduction of total quantity of material
- Smaller process equipment and storage containers
- Upgrading equipment and material specifications
- Secondary containment
- Emergency de-inventory
- Leak detection
- Neighborhood alerting/response
- Transportation route selection considerations.

The Community Preparedness guideline endorsed strong support of the CMA CAER program and similar efforts in other countries, and it provided guidance on how sites could establish and renew appropriate community organizations to plan and respond to chemical emergencies.

The Management of Highly Toxic Materials guideline identified 15 hazardous materials, such as ammonia and chlorine, where special consideration was required to ensure safe storage, processing, and transportation. Highly Toxic Material (HTM) Safety Guardian Committees were established for managing each hazardous material globally, and specific principles and guidance were provided on facilities, storage, operations, transportation, detection, and community protection. For example, one principle for detection was:

> Sites handling highly toxic materials will include use of automatic or continuous specific chemical detection systems, where appropriate, as an integral part of minimizing the potential for harmful exposure on- and off-site.

Each HTM Safety Guardian Committee used this guidance to develop specific standards and practices to help ensure safe use of each material. A HTM Safety Guardian Leadership Team was formed in 1988 to "provide continued global focus, leadership, coordination, and leveraged support for the safe management of Highly Toxic Materials." Upgrades to HTM guidance over the last 15 years have included audit requirements, product stewardship reviews, and additional guidance on minimizing storage and preventing, detecting, controlling, and containing releases. HTM Safety Guardian Committees now exist for 24 substances.

3.4 Renewed Focus on the Goal is Zero

In late 1996, the corporate ZIP (Zero Incidents… Period!) Team was formed to focus on eliminating process-related injuries and incidents, while at the same time, building business value for PSM. The ZIP Team included members from all businesses and regions, ranging from senior corporate leadership, to plant managers, to operators and mechanics, and eventually included 70 members. The ZIP Team reaffirmed the vision of "The Goal is Zero" for process incidents and communicated these key messages:

- There is no silver bullet, no single answer, to improving process safety.
- The goal has to be ZERO process-related injuries and incidents. The question is, not IF, but HOW, this goal can be achieved, just as it is for personal safety.
- PSM is a business issue, not only a manufacturing issue, requiring the contributions of a broad cross section of the organization to ensure success.
- Businesses therefore need to establish a sustainable continuous improvement process toward achieving the goal of zero PSM injuries and incidents that is embedded in business planning.

All businesses and regions were requested to have implementation plans by the end of 1997, which would be included in executive level discussions on performance and planning. ZIP implementation plans for each business or region needed to:

- Establish the VP as the PSM Champion to promote business/region accountability.
- Ensure that systems were in place to understand current process safety performance and for building value and awareness for achieving zero process incidents through specific goals and metrics.
- Ensure that a process for allocating resources to process safety was part of the business planning process.
- Apply PSM systems to all process operations, including lower hazard operations, with improved audit emphasis and metrics.
- Develop the necessary systems and training to further apply Inherently Safer principles to process and product development activities, including R&D, operations, and capital projects.
- Develop a program to drive operational discipline – the deeply rooted dedication and commitment by every member of an organization to carry out each task the right way each time – at all sites.

The result was a renewed corporate effort to drive continuous improvement activities that led to upgraded standards and practices, training programs, evaluation tools, and business metrics that has greatly impacted PSM performance in DuPont. This is reflected in the current PSM program, discussed in the next section.

4. The Current PSM Program

PSM is the application of management systems and controls (programs, procedures, audits, evaluations) to a manufacturing or chemical process in a way that process hazards are identified, understood, and controlled, so that process-related injuries and incidents are prevented. While a comprehensive program is essential for achieving excellent results in preventing process incidents, it is only one of four key steps required to establish and maintain an effective PSM effort [19]:

- Step 1 – Establishing the Safety Culture: A PSM program will only be as effective as the underlying safety culture permits. If an organization has a weak safety culture, PSM efforts are doomed to conflicting priorities, such as production needs or lowering costs. A strong safety culture, based on a commitment to core values on safety, health, and environmental issues, and as evidenced by organizational policies, goals, metrics, and day-to-day decision-making, supports establishing safety systems where safety priorities are recognized, not as conflicting with other priorities, but rather as inherently necessary for completing any task the right way. DuPont's fundamental policy on safety was established by the company founder and has remained a core value for over 200 years, creating a strong overall safety culture.
- Step 2 – Providing Management Leadership and Commitment: Management leadership and commitment which is visible throughout the organization is required for continued excellence in PSM. Beginning with the formation and fostering of a strong safety culture, continuing with policies that set high expectations for excellent safety performance, and implemented in day-to-day decisions that provide resources and establish high priority for safety-related activities, the actions of all levels of management must support and reinforce strong PSM programs and accountability.
- Step 3 – Implementing a Comprehensive PSRM Program: A comprehensive PSM program includes integrated managing systems for various elements of process safety, such as Process Hazards Analysis, Incident Investigation, and Mechanical Integrity. A good starting point for developing a strong PSM program is existing regulations, such as the OSHA PSM Rule and the EPA Risk Management Program in the United States. Regulations should generally be considered the starting point only, though, representing minimum essential levels of practice. Additional requirements should be considered that go beyond regulations and better meet corporate needs for process safety. Additional details of DuPont's PSM Program is provided below.
- Step 4 – Achieving Operating Excellence through Operational Discipline: Operational discipline is the deeply rooted dedication and commitment by every member of an organization to carry out each task the right way each time [20]. No matter how comprehensive and well-designed PSM programs may be, it is the day-to-day ability of everyone to practice operational discipline that successfully changes safety systems from concept to reality. Operational discipline not only helps prevent serious injuries and incidents, but also contributes to excellent

business performance through increased operating excellence, including higher productivity, higher quality, reduced waste, and lower costs.

Over the last 25 years, the PSM program at DuPont has been continuously upgraded and extended due to the leadership of corporate and business management, PSM professionals, and site personnel. Today, PSM requirements are defined by a series of corporate process safety standards that both reflect and extend federal and state regulations. The principles and essential features of the PSM program are described by 14 elements grouped by Technology, Personnel, and Facilities, as shown in Figure 5. This figure is typically called the PSM Wheel, with each of the 14 elements, such as Process Hazards Analysis, Operating Procedures and Safe Work Practices, and Personnel Training and Performance, arranged around the spokes of the wheel. Management leadership and commitment, necessary for implementing and maintaining strong PSM programs, is shown at the center of the PSM Wheel. Operational discipline is shown as the rim of the PSM Wheel, connecting all of the elements and translating the required managing systems into real results for preventing injuries and incidents.

Figure 5 – DuPont PSM Model

4.1 Beyond Minimum Essential Practice

Some examples of where DuPont PSM goes beyond minimal essential practice, as defined by federal and state regulations, include the following items.

Management of Change – Personnel

The Management of Change – Personnel element in DuPont PSM recognizes that people are an essential factor "interwoven through all the elements of PSM" and that "it is important to maintain a minimum level of (1) specific direct process experience and (2) knowledge and skill in managing process safety." Since the loss of experienced personnel through job transfer or organizational changes can significantly affect PSM performance, the following requirement has been established:

> Sites shall develop and implement criteria and guidelines regarding management of change of personnel to help ensure that minimum levels of specific, direct, process experience and minimum levels of knowledge and skill in managing process safety are maintained.

This requirement leads to sites evaluating and managing personnel changes (e.g., transfers, promotions, retirements, etc.) and job assignments (e.g., across different operating shifts) to help ensure that appropriate levels of experience in operating the process safely are available.

Operational Discipline

As previously discussed, operational discipline is the deeply rooted dedication and commitment by each member of the organization to carry out each task the right way, each time. Clearly, operational discipline is related to Management of Change – Personnel, but extends further to ensure that not only are appropriate levels of experience available, everyone is also motivated to work safely. The requirement for operational discipline is:

> Sites shall implement and maintain programs and systems to achieve and maintain a high degree of operational discipline for all PSM elements in a manner that supports business and operating objectives.

Organizations that have a high level of operational discipline usually exhibit an identifiable set of positive behaviors or characteristics. While it's possible that these characteristics may vary from one organization to another, DuPont experience has shown that 11 key characteristics seem to be most important:

- Leadership by example
- Sufficient and capable resources
- Employee involvement
- Active lines of communication
- Strong teamwork
- Common shared values
- Up-to-date documentation
- Practice consistent with procedures
- Absence of shortcuts

- Excellent housekeeping
- Pride in the organization

A more detailed discussion of the development of operational discipline practices at DuPont and tools used to help sites evaluate and improve operational discipline performance is available [20].

Application

PSM has traditionally applied to high hazard processes involving hazardous materials, such as defined in the OSHA PSM Rule, which presents requirements for "preventing or minimizing the consequences of catastrophic releases of toxic, reactive, flammable, or explosive chemicals [22]." While this has certainly been true at DuPont as well, PSM efforts have also been extended to include higher-hazard processes (HHPs), lower-hazard operations (LHOs), and laboratory and pilot plant facilities. The impact is that there are relatively few manufacturing or research facilities at DuPont where some level of PSM or related programs is not applied to help ensure that process hazards are appropriately evaluated and managed.

HHPs include processes covered by the OSHA PSM Rule. HHPs also include other processes where use of hazardous substances can potentially lead to death or irreversible health effects, significant property or environmental damage, or impact on the surrounding community. Examples include use of flammable, toxic, or reactive materials, which do not exceed OSHA Threshold Quantities for covered substances, such as many polymerization reactors, film drying lines, and pigment dispersion mills. The requirement for HHPs is:

> Each site and manufacturing facility operating an HHP shall develop, within the framework of this standard, a PSM program suited to its specific parameters. Such guidance shall embrace all the principles and essential features of PSM.

LHOs are processes where there is a low potential for death or irreversible health effects, significant property or environmental damage, or impact on the surrounding community. Examples of LHOs include many extrusion, palletizing, conveying, spinning, drying, packaging, and treatment operations. The requirement for LHOs is:

> Each site and/or manufacturing facility containing an LHO shall develop a PSM program suited to its specific parameters. Such guidance shall embrace the principles of PSM and incorporate those essential features deemed appropriate by site management.

The purpose of the LHO requirement is to gain the benefits of a PSM program for non-traditional PSM processes, with the program applied selectively to help manage the hazards of the process without unnecessarily applying requirements that provide few safety benefits. Typically, all 14 elements remain relevant for most LHOs, although the number of requirements is greatly reduced. Where a process falls into a "gray zone" between a HHP and a LHO, a consequence analysis must be completed to help identify and document process hazards and potentially hazardous events, allowing the process to be properly classified.

A new guideline for Safety and Risk Management for Laboratories and Pilot Plants was issued in 2003, based on the principles of PSM. Previously, many pilot plants had been (and continue to be) classified as HHPs or LHOs, depending on the process hazards present. The new guideline extends PSM further to remaining pilot plant facilities and to laboratories located at manufacturing and dedicated research sites, if significant hazards are present. The purpose of the guideline is to help prevent serious injuries and incidents at these locations through the flexible application of PSM systems. Additional flexibility is permitted to avoid inhibiting research, while also ensuring that any hazards present are identified and managed, as appropriate.

4.2 PSM Performance

Figure 6 shows the number of process-related Lost Workday Cases (LWCs) at DuPont over the last 30 years, using the 5-year period of 1975-79 as the base. While the corporate total recordable rate is low (1.2 in 2004), any number above the goal of zero is not acceptable. The corporate PSM guideline was issued at the end of the base period in 1979 and significant additional efforts were made following the Bhopal incident in 1984. The impact of these efforts was seen in 1985-89, where the number of process-related LWCs dropped by over 50%. The number of process-related LWCs then stayed relatively constant, although a drop was seen in the most recent 5-year period of 2000-04, possibly due to the impact of ZIP Team recommendations. As discussed previously, the ZIP Team reaffirmed the goal of zero process-related injuries, increased business accountability for PSM, extended PSM to LHOs, endorsed inherently safer process methodologies, and recommended new programs to improve operational discipline. Interestingly, most process-related LWCs over the last two years have occurred in LHOs, which is now the focus of additional investigations.

A similar story is seen in the number of process incidents, as shown in Figure 7. All process incidents at DuPont are scored based on actual and potential consequences and are evaluated by sites for key factors and trends. Process incidents scoring greater than 75 points are communicated broadly through the company, are used as a corporate metric, and are evaluated further for higher level learnings and trends. Figure 7 shows the number of these incidents reported over the last 10 years. A clear downtrend is observed in the number of incidents, partly due to a change to the scoring system in 2001, but also likely due to the same reasons that process-related LWCs have decreased. The number of process incidents in 2004, for example, is down 75% vs. the base year of 2001 using a consistent scoring system. In fact, the threshold for reporting incidents was reduced to 50 points in 2003 to provide additional data for improving evaluation of key incident learnings and trends to help identify continuous improvement opportunities.

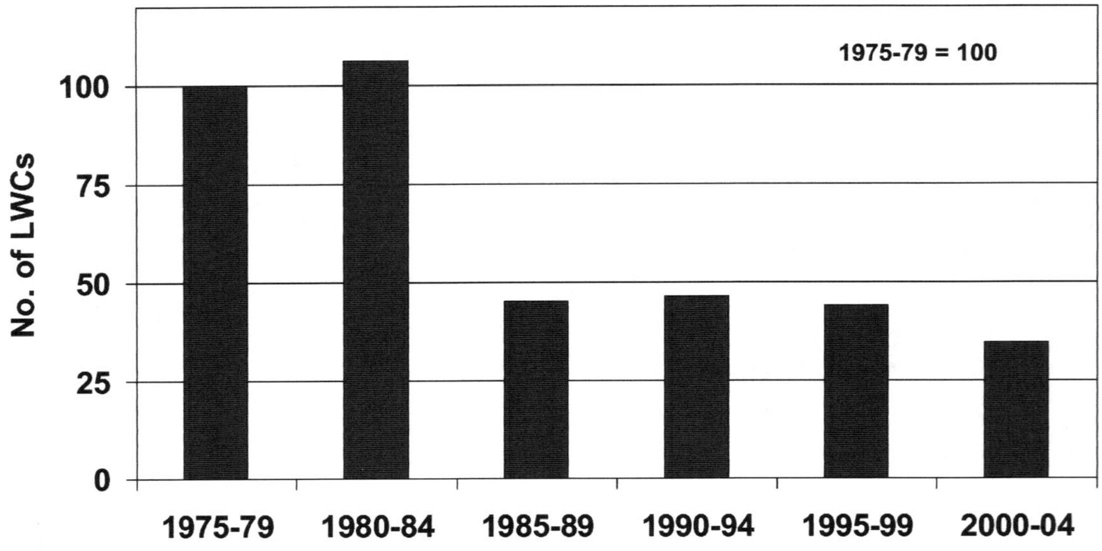

Figure 6 – Process-related LWCs (1975-79 = 100)

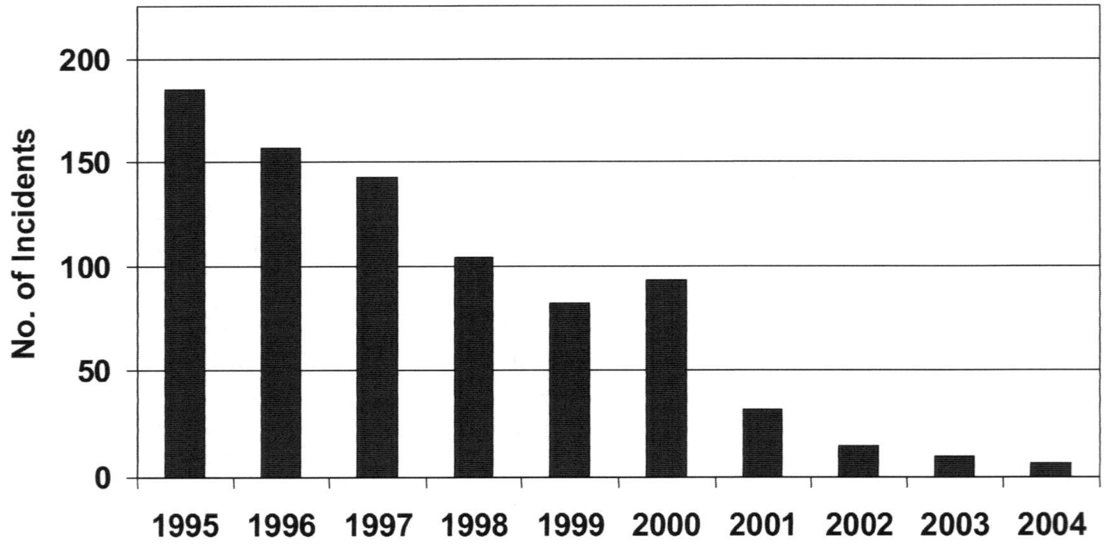

Figure 7 – Process Incidents (75 points and above)

5. Looking Forward

In thinking about the future, the one certainty is that change will be continual. Businesses will grow, acquisitions will be made, and new technologies will be invented. PSM will need to change as well. New acquisitions may present challenges due to different corporate or regional safety cultures. New technologies, perhaps in biotech, nanotechnology, or some other new innovation, may benefit from the application of

PSM, but at the same time, may require that new approaches be developed. Standing still and celebrating our successes, though, will not work. Continuous improvement will always be necessary:

> Systems and organizations continually experience change as adaptations are made in response to local pressures and short-term productivity and cost goals... A corollary of this propensity for systems and people to adapt over time is that safety defenses are likely to degenerate systematically through time, particularly when pressure toward cost-effectiveness and increased productivity is the dominant element in decision making. Thus, the redundancy and other precautions added to protect against human error often degenerate over time as other work practices adapt to increase efficiency within the local environment. The critical factor here is that such adaptation is not a random process – it is an optimization process... [23]

In effect, if we are not moving forward, we are falling behind. The challenge is to recognize this fundamental force and to ensure that management commitment and organizational safety culture function to maintain robust PSM systems.

A greater dependence on technology may not provide the complete solution. Where incidents in the past have generally related to equipment or human failures, greater reliance on digital technology for process design and control introduces new complexity:

> we are increasingly experiencing a new type of accident that arises in the interactions *among* components (electromechanical, digital, and human) rather than the failure of individual components. [23]

The result is a need for increased reliance on systems engineering approaches that evaluate individual component failures (e.g., equipment breakdown, software faults, human error), and also failures resulting from component interactions. As operators rely on software for process control, for example, they move further away from the process and an understanding of what can go wrong and what should be done to fix problems in the process or in the software. How information is communicated and how instructions are entered becomes more critical, as does the need to maintain experienced, knowledgeable personnel. Preventive maintenance of critical equipment, good design and testing of software, and high levels of operational discipline will always be important for safe operations. In the future, systems engineering capabilities to manage the growing complexity of our processes will also increasingly be important.

The goal now, and in the future, is to prevent serious injuries and catastrophic incidents – the goal is zero. Learning from experience and going beyond the minimum essential practice defined by regulations are essential if zero is to be achieved. At DuPont, our core belief in the value of safety – for our employees and for our other stakeholders – was established over 200 years ago by the founder, E. I. du Pont. Seeking to understand and manage the hazards of our processes to help ensure a safe future, despite increasing complexity, will always be our challenge and our commitment.

References

1 Kinnane, Adrian, *DuPont: From the Banks of the Brandywine to Miracles of Science,* E. I. Du Pont de Nemours and Company, 2002

2 Mottel, William J., Joseph F. Long, and David E. Morrison, *Industrial Safety is Good Business: The DuPont Story*, Van Nostrand Reinhold, 1995

3 Carisio, Justin, *History of DuPont*, DuPont video, 2003

4 du Pont, E. I, *The Location and Construction Necessary for Manufacturing Gunpowder*, Hagley Museum, Wilmington, DE

5 Johnson, Allen and Dennis Molene, *Dictionary of American Biography, Vol. 5*, Charles Scribner's Sons, 1930 (Hagley Museum, Wilmington, DE)

6 Cassell's Magazine, "In a Powder Mill," Hagley Museum, Wilmington, DE

7 Du Pont, B. G, *Life of E. I. du Pont, Vol. 6, 1802-04*, University of Delaware Press, 1925 (Hagley Museum, Wilmington, DE)

8 du Pont, E. I., opening paragraph to "Rules" for his manufactory, Hagley Museum, Wilmington, DE, 1811

9 DuPont, *Letter Books, 1820-21*, Hagley Museum, Wilmington, DE

10 DuPont, *Letter Books, 1825-27*, Hagley Museum, Wilmington, DE

11 du Pont, E. I, letter to a customer in Virginia, Hagley Museum, Wilmington, DE, 1820

12 Burton, Jehu and Janet Dill, *DuPont: Celebrating 100 Years of Engineering, 1903 - 2003*, E. I. Du Pont de Nemours and Company, 2004

13 DuPont, *Development of Accident Prevention Work in E. I Du Pont De Nemours and Co.*, Hagley Museum, Wilmington, DE, 1932

14 Denham, Charles J. (editor), *Sentimental Journey: The DuPont Belle Works, A 75 Year History, 1926 – 2001*, E. I. Du Pont de Nemours and Company, 2001

15 DuPont, *Safe Practices in the Handling and Storage of Anhydrous Ammonia Recommended for Industrial Plants Other Than Refrigerating Plants*, Hagley Museum, Wilmington, DE, 1929

16 Balchan, A. S., J. A. Klein, and F. G. Klein, "Process Safety of Polymer Resin Manufacturing: A 20-Year Perspective," in Loss *Prevention and Safety Promotion in the Process Industries*, Vol. 1, Elsevier Science, 1995

17 Klein, J.A. and A. S. Balchan, "Safe Formulation and Manufacture of Acrylic Resins," in *International Conference and Workshop on Process Safety Management and Inherently Safer Processes*, Center for Chemical Process Safety, 1996

18 Balchan, A. S., D. A. Paquet Jr., and J. A. Klein, "Emergency Relief Adequacy for Acrylic Polymerization Processes," Process Safety Progress, Summer, 1999

19 Burk, Arthur F., "Achieving Excellence in Process Safety and Risk Management," Managing Safety and Health in the Workplace, Chicago, June 13-14, 1994

20 Klein, James A., "Operational Discipline at DuPont," Paper TH005-2, Plant Process Safety Symposium, Atlanta, April 11-13, 2005

21 Clements, C. Curtis, "Application of Inherently Safer Process Concepts in DuPont," Paper TH002a, Plant Process Safety Symposium, Atlanta, April 11-13, 2005

22 OSHA PSM Rule 1910:119

23 Leveson, Nancy, "A New Accident Model for Engineering Safer Systems," Safety Science, Vol. 42, P. 237-270, 2004

Nurturing a Strong Process Safety Culture

Shakeel H. Kadri
Air Products and Chemicals, Inc.
Allentown, PA 18195 USA
kadrish@airproducts.com

David W. Jones
ChevronTexaco
Richmond, CA 94801 USA
DWJones@chevrontexaco.com

ABSTRACT

Process safety management introduced a formal, structured, management systems approach to accident prevention that represented a "step change" improvement in identifying and reducing the potential for major process plant accidents. While the process safety management elements have proven effective, the overall sustainability remains a challenge. This paper will highlight the importance of a strong process safety culture that is needed to sustain a strong process safety performance. The authors are using the Columbia Shuttle disaster incident as a backdrop to focus the attention on the significance of not having an optimal "Safety Culture." Examples have been included from our industry to further exemplify its broad applicability. The paper proposes an approach that will help organizations to self-assess the state of the process safety culture in their organizations, which can be an aid in their ongoing cultural improvement effort.

1. INTRODUCTION

Process safety management introduced a formal, structured, management systems approach to accident prevention that represented a "step change" improvement in identifying and reducing the potential for major process plant accidents. The previous prevention systems in place, based largely upon standards, procedures, and "engineering judgment," had begun to prove ineffective in the face of more complex operations and mounting public scrutiny/reaction to increasingly "smaller" events.

While the process safety management elements have proven effective, the overall sustainability remains a challenge. As company personnel change and corporate memory fades (along with the reasons why certain activities must always be performed), accident rates again increase. This pattern suggests a new "path forward" may be in order to help ensure the ongoing sustainability of process safety management.

In evaluating safety culture, NASA's experience with the Challenger and then the Columbia tragedies offers well-analyzed and relevant information for the process industries. Although NASA is a unique organization with a focused mission, the cultural failures that led to the Columbia disaster have counterparts in any operation with a potential for significant incidents.

2. DEFINING PROCESS SAFETY CULTURE

There are many definitions of culture, including those found in Webster's Dictionary, which characterizes culture as "the concepts, habits, skills, arts, instruments, institutions, etc., of a given people." Common terms for company culture include "patterns of beliefs," "shared values," or "written and unwritten norms." The Columbia Accident Investigation Board offered the following definition of organizational culture:

> "Organizational culture refers to the basic values, norms, beliefs, and practices that characterize the functioning of a particular institution. At the most basic level, organizational culture defines the assumptions that employees make as they carry out their work; it defines "the way we do things here." An organization's culture is a powerful force that persists through reorganizations and the departure of key personnel."
>
> *Columbia Accident Board Report, Volume 1, Page 101*

For process safety management purposes, we propose the following definition for process safety culture:

> **"The combination of group values and behaviors that determine the manner in which process safety is managed."[1]**

An important characteristic of all cultures is that they are supported and promoted by those who have survived in it. New members to the culture observe the actions, successes, and failures of those before them and act accordingly. Written and unwritten organizational rules evolve that become part of day-to-day business. In this way, support or violations of process safety management fundamentals can be promoted and passed down through generations of plant personnel.

3. COLUMBIA SHUTTLE INCIDENT

> "In our view, the NASA organizational culture had as much to do with this accident as the foam."
>
> *Columbia Accident Board Report, Volume 1, Page 97*

Much has been written about the Columbia Shuttle disaster incident. We are not planning to duplicate the incident review and its findings. However, we have included a brief overview of this incident and its learnings taken from the *Columbia Accident Report*[2] to highlight the importance of the underlying causes of this incident that were significant contributors. The authors wanted to raise the awareness that a well-respected organization like NASA, who is a leader in the advancement of science, engineering, and technology and has a very strong safety performance but had a blind spot around their safety culture that led to this disaster.

Many organizations in our industry have a strong safety performance. However, past safety successes can create complacency and potentially generate "blind spots." As part of this paper, we have included screening indicators that will help raise awareness for identifying "culture" blind spots contributing to your significant incidents. This will also aid in developing leadership actions and commitment that organizations need to continue expanding process safety culture and minimizing significant incidents.

3.1 The Physical Cause of the Incident

On February 1, 2003, the Space Shuttle Columbia disintegrated upon re-entry into the earth's atmosphere, killing all seven crewmembers aboard. The physical cause of the loss of Columbia and its crew was a breach in the thermal protection system on the leading edge of the left wing, caused by a piece of insulating foam that separated from the left bipod ramp section of the external tank at 81.7 seconds after launch and struck the wing. During re-entry, this breach in the Thermal Protection System allowed superheated air to penetrate through the leading edge insulation and progressively melt the aluminum structure of the left wing, resulting in a weakening of the structure and break-up of the Shuttle.

3.2 The Underlying Causes of the Incident

Soon after the physical cause was confirmed, the Board was specifically trying to understand one key issue … Why was it that serious concerns about the integrity of the Columbia, raised within one day after the launch, were not acted upon in the two weeks available between launch and return? With little corroborating evidence, management had become convinced that a foam strike was not, and could not be, a concern. What the Board was most interested in finding out in this part of the investigation was the state of the "Organizational culture" at the time – what NASA employees were assuming as far as the basic values, norms, beliefs, and practices as they carried out their work.

3.3 The Underlying Causes Expanded

- The Board was already familiar with the problems rooted in the Space Shuttle Program's history and culture, including the original compromises that were required to gain approval for the Shuttle, subsequent years of resource constraints, fluctuating priorities, schedule pressures, and mischaracterization of the Shuttle as operational rather than developmental. This had allowed several cultural traits and organizational practices detrimental to safety to develop.

- Some of these problems had continued at NASA and had become key contributors in degrading its organizational values. The aggressive launch schedule to attain the deadline of 19 February 2004 to complete the preliminary phase of the

International Space Station was heavily promoted by senior-most NASA officials. This had created an environment at NASA in which some workers felt reluctant to raise safety issues for fear of delaying the launch timetable.

- To elaborate on the decision-making process at NASA, the Board goes in detail to highlight several missed opportunities when the management could have done something about the debris strike. After every shuttle liftoff, NASA engineers reviewed the liftoff imagery to see if any debris had struck and developed any damage to the shuttle. On the second day after the liftoff, when the engineers reviewed the liftoff images, they noticed a large piece of debris had struck the shuttle's left wing. However, because of the resolution and the camera angle, the engineers could not assess the severity of the damage caused by the debris from the launch film. Their requests for outside images were the first step toward discovering the extent of the damage and mounting an effort to rescue the shuttle crew, if required. Despite a fear of losing their jobs when they spoke up, NASA engineers made numerous attempts to communicate their concerns to their management. However, their appeals were consistently ignored. When the engineers presented their findings to the mission managers, the managers disputed their findings and identified gaps in their analysis.

3.4 Underlying "Culture" Themes

At least five underlying themes emerge that helped create NASA's "culture" blind spot leading to the tragedy.

- **Maintaining a Sense of Vulnerability**

  Since catastrophic accidents involving hazardous materials or activities are not very common, most organizations never have the unfortunate opportunity to experience one. This can create a false sense of security and complacence. At NASA, even after it was established that significant debris had struck the shuttle during its initial ascent, the perception that it could lead to a disaster had been diminished due to the prior 17 years of successful operation.

- **Combating Normalization of Deviance**

  When pre-established engineering or operational constraints are consciously violated without any resulting negative consequences, an organizational mindset is encouraged that more easily sanctions future violations. At NASA, early in the Space Shuttle Program, foam strike was considered a dangerous problem for the fragile Thermal Protection System. However, over the 113 Shuttle missions flown, foam shedding and debris impacts had come to be accepted as routine and maintenance concerns only because it had not created any safety to flight issue. This normalized the fundamental deviation from the intended design.

- **Performing Appropriate and Timely Hazard/Risk Assessment**

Without a complete understanding of risks and the options available to mitigate them, management is hampered in making effective decisions. At NASA, there was a lack of consistent, structured approaches for identifying hazards and assessing risks; and most identified concerns were simply labeled as "acceptable." Also, many of the action items emerging from these studies were not addressed.

- **Establishing an Independent and Unassailable Role for Safety**

  By not having an independent critical view, an organization can miss valid safety concerns. Senior management strongly stressed that budget cuts will not impact the safety of the Shuttle. Over the years at NASA, the safety organization had degraded and had ultimately been relegated to "rubber stamping" critical safety-related decisions. Also, the safety staff was put in the untenable position of having to prove that an operation was unsafe, rather than requiring the engineers and managers to prove that the operation was safe.

- **Ensuring Open and Frank Communications Across all Levels**

  NASA's organizational culture did not encourage "bad news," rather it encouraged consensus and tactfully suppressed the dissenting views. Their "chain of command" emphasis in communications actually stifled communication or diluted the content of the message, and participants were afraid to speak up.

4. PROCESS INDUSTRY PARALLELS

While the Columbia and Challenger shuttle accidents are specific to NASA, the lessons are broadly applicable. Some examples from our industry are included here to further illustrate applicability of these lessons:

4.1 Piper Alpha – 1988 (Lord Cullen Report)[3]

On the night of July 6, 1988, the Piper Alpha oil production platform burned to the sea, killing 167 people. The initiating cause was a permit-to-work breakdown, resulting in a release of light hydrocarbon through an open pipe where a relief valve had been removed for maintenance. The Lord Cullen Report offered the following detailed findings. We have added possible cultural themes:

Permits *(Normalization of Deviance)*

- Work permits on the same, or related, systems were normally not cross-referenced. The contractor supervisor and most platform personnel had received no training on the permit-to-work system.
  - The work permit required blinds to be installed when the relief valve was removed. Blinds were not installed, per accepted platform practice.

- Work permits were often signed without "field" checking the job. It was the practice to lay the work permits on the head operator's desk without informing him of the status of work.

Engineering and Operations *(Lost Sense of Vulnerability and Lack of Hazard/Risk Assessment)*

- The managers of the adjacent platform assumed the Piper Alpha would bring their fire under control and, therefore, that they did not need to shut down.
- There had been no training or emergency drills conducted for inter-platform emergencies.
- The consequences of a high pressure gas pipeline failure had been predicted 12 months earlier in an engineering study. No action had been taken.
- Fire walls had been installed on Piper Alpha, but blast walls were not considered.
- The Piper Alpha platform manager had never been trained in the emergency response procedures for the platform.

Auditing *(Lack of Critical Independent Review for Safety)*

- The Piper Alpha platform permit-to-work system had undergone a significant number of audits. Every day, one operator had the only task of monitoring the work permits. There was never a report to management indicating the work permit system was not being followed properly.
- Six months before the accident, an internal company audit looked at the permit-to-work systems and found no deficiencies.

4.2 Longford – 1998 (Longford Royal Commission Report)[4]

Two workers were killed when a heat exchanger failed by brittle fracture in a gas processing facility. The combination of process upset, failed alarms, and a lack of understanding by management and supervisors about cold temperature effects on carbon steel led to the accident. The Longford Royal Commission Report published the following findings. We have added possible cultural themes:

Leadership Accountability (*Lost Sense of Vulnerability*)

- Management was not ensuring operating discipline and adherence to procedures.
- A culture was established that emphasized production ("not walking the talk").
- Process safety systems, while structurally present, were weak in content.
- Management was ignorant of the hazard, and did not know they were ignorant.

Behaviors - Failure, Across all Shifts, to Adhere to Basic Operating Practices. (***Normalization of Deviance***) For example:

- Appropriate checks not made before operations.
- Chart recorders were not maintained.
- Common operation in alarm mode
- Violations of procedures were accepted.

Programs without Follow-Through ***(Lack of Critical Independent Review for Safety)***

- Procedures did not address fundamental hazards.
- Gaps in training.
- Poor change control.
- Inadequate hazard identification.
- Failure to perform incident investigations.
- Lack of management commitment to established programs and procedures.

The above reports were referenced because they are available for public review. However, most major incidents can often be traced back to multiple organizational culture failures, as illustrated above.

5. INDICATORS THAT YOUR PROCESS SAFETY CULTURE IS NOT STRONG

NASA's experience may or may not be applicable to all organizations; however we believe that it provides significant learning opportunities to improve process safety management culture especially to those organizations that are involved in highly hazardous activities. We have included indicators that will help these organizations as a starting point to self-assess the state of the process safety culture in their organizations and can be an aid in their ongoing cultural improvement effort.

Indicators that you are NOT Maintaining a Sense of Vulnerability

- You assume your safety systems are good enough.
- You treat critical alarms as operating indicators.
- You allow backlogs in preventative maintenance of critical equipment.
- Lessons from related industry disasters are not routinely discussed at all levels in the organization.
- Actions are not taken where similar deficiencies have been identified.

Indicators that you ARE Allowing Normalization of Deviance

- You allow systems to operate outside design safe operating limits without detailed risk assessment.
- You allow deviations from established procedures without MOC review and approval.
- Willful, conscious violation of an established procedure is tolerated without consequences for the persons involved.
- Staff cannot be counted on to strictly follow procedures when supervision is not around to monitor compliance.

Indicators that Appropriate and Timely Hazard/Risk Assessment are NOT Being Done

- Availability of experienced resources for risk assessments is limited.
- The recommendations from risk assessments are not meaningful.

- The recommendations from risk assessments are not implemented in a timely manner.
- The actions taken differ from the intent of the original recommendations.
- Bases for rejecting risk assessment recommendations are mostly subjective judgment or based upon previous experience and observation.

Indicators that the Safety Role is NOT Independent and Unassailable

- People monitoring safety-related decisions are not sufficiently technically qualified and independent.
- Key process safety management positions have been downgraded over time.
- Recommendations for safety improvements are resisted on the grounds of cost or schedule impact.
- No system in place to ensure an independent review of major safety-related decisions.
- Audits conducted by people not technically competent and are regarded as negative or punitive.

Indicators that Open and Frank Communication at All Levels is NOT Happening

- The bearer of "bad news" is viewed as "not a team player."
- Safety-related questioning "rewarded" by requiring the suggestor prove that he/she is correct.
- Critical, safety-related news that circumvents official channels not welcomed.
- Communications get altered, with the message softened, as they move up the management chain.
- Employee cannot speak freely to anyone else about their honest safety concerns, without fear of career reprisals.

6. IMPROVE YOUR PROCESS SAFETY CULTURE BY TAKING ACTIONS

The five organizational culture themes identified above can serve as a starting point for helping companies identify potential cultural weaknesses leading to serious incidents. Specific steps might include:

1. **Create Awareness and Buy-in About Process Safety Culture Themes**
   Utilizing NASA's findings, along with the industry parallel, can be a starting point to raise the awareness and buy-in of your management and employees to the importance of having a strong process safety culture in your organization. It is critical to generate a dialogue around the question, "Could this happen to us?"

2. **Adapt Process Safety Culture Themes to your Company Experience**
   Review your own serious process safety incident reports for key culture themes. Look for related themes, based upon the NASA and other examples provided.

Identify lessons learned and follow-up actions for both management and employees to take to strengthen their process safety culture.

3. **Utilize Suggested Indicators to Identify Specific Areas of Improvements**
To determine specific contributors to potential process safety culture breakdown, utilize the suggested indicators in Section 5 above. This will point to opportunities to address both tactical and strategic cultural issues. Ensure that you have a buy-in from your senior management on your proposed high-level improvement plan.

4. **Develop a Strategic Improvement Plan to Strengthen and Sustain your Process Safety Culture**
Based on the identified improvement areas, create a campaign that keeps process safety culture issues on the forefront. Start your meetings with process safety culture topics. Popularize process safety culture themes through posters and other company communication tools. Periodically assess your process safety culture against the proposed indicators to ensure that the process safety culture improvement progress is sustained.

7. REFERENCES

1. Dave Jones, "Turning the Titanic, Three Case Histories in Cultural Change," presented at CCPS International Conference at Toronto, October 2001.
2. Columbia Accident Investigation Board Report, August 2003.
3. Lord Cullen Report on Piper Alpha Disaster, November 1990.
4. Report of the Longford Royal Commission, Government Printer for the State of Victoria, June 1999.

The Right People – Key to a Successful Hazard Review

William Bradshaw, P.E.
ABSG Consulting Inc.
Knoxville, Tennessee
wmbradshaw@absconsulting.com

Jonathan Babcock
Eli Lilly and Company
Indianapolis, Indiana
jbbabcock@lilly.com

ABSTRACT

At Eli Lilly and Company (Lilly), the means used to identify and manage process hazards have evolved over time. Although the rigor of process hazard review (PHR) methods, the use of technology, and the availability of information have noticeably changed, one critical input remains constant: identifying and effectively managing hazards depends on people. In this paper, we address the importance of having (1) highly qualified PHR team leaders, (2) depth of experience and diversity of viewpoints among PHR team members, and (3) highly visible management involvement, from the initial planning stages through resolution of PHR team recommendations to completion of action items.

In addition to describing what has worked well at Lilly's Biosynthetic Human Insulin plant in Indianapolis, Indiana, this paper includes a comparison of data from PHR revalidations that were conducted between 1998 and early 2002 to PHRs for the same processes that were redone when they came due for their next revalidation. These data provide insight into the increased effectiveness and efficiency of the current approach, which includes (1) using leaders who are experts in PHR techniques rather than experts in the process being reviewed, (2) a more structured method of assigning team members, and (3) a higher degree of management oversight and involvement. In addition, the paper briefly examines how these issues differ for fill/finish operations that typically involve lower inventories and lower hazard chemicals than bulk pharmaceutical manufacturing processes.

1. INTRODUCTION

There are many factors that contribute to the success of a process hazard review (PHR) and to the overall effectiveness and efficiency of a site's PHR program. These include:

- Clearly defining the scope of the PHR, the types of consequences that are of interest to the PHR team, and the threshold severity for the various consequences that have been identified
- Having complete, current, and accurate technical information on the process
- Using an appropriate PHR technique
- Assigning the right people to each PHR team
- Producing well-organized and user-friendly reports
- Having management support for the PHR program
- Generating enthusiasm among operators, maintenance personnel, and engineers about participating on PHR teams and addressing PHR recommendations

At Eli Lilly and Company (Lilly), we have found that involving the right people in roles that maximize their effectiveness contributes to the success of the PHR program. In addition, by involving the right combination of Lilly personnel and expert PHR team leaders, we have transformed what was once considered by some to be a burden on the operations team to what is now generally considered to be an opportunity for improvement.

2. BACKGROUND

Traditionally at Lilly, the primary responsibility for conducting the PHR, deciding on actions to address PHR team recommendations, and tracking the completion of those actions fell on the operations group, with significant support from process engineering. Standardization was achieved by (1) having a detailed written manual that addressed all aspects of the PHR program, starting with PHR planning and extending through documentation of completion of PHR recommendations, (2) sending all PHR leaders to a 3-day course, and (3) conducting a series of internal technical and quality reviews of PHR reports (based on detailed checklists that were developed to help standardize the review process).

This approach led to inefficient use of people in the following ways:

- The PHR process consumed a large amount of staff resources, primarily because:
  - "Process experts" rather than "PHR experts" were used to lead PHR teams
  - PHR teams tended to be relatively large
  - To mitigate the effects of inefficiencies caused by using relatively inexperienced PHR team leaders, PHR meetings were conducted in 1- to 2-hour sessions; processes were finely divided in order to complete

a PHR in a timely manner using this approach (some PHRs included a single unit operation or process step)

- Resolving PHR team recommendations into actions and completing the actions also consumed significant resources because:
  - —Some PHR team recommendations were not related to process safety
  - —The relatively large number of PHRs contributed to sets of recommendations that were not always well aligned to be efficiently implemented (e.g., numerous redundant/overlapping recommendations)

Additional background on each of these factors is provided in this section.

Process engineers were typically assigned to lead PHR teams and document the results. There was a general belief that depth of process knowledge was a more important attribute for a PHR team leader than knowledge and experience in leading PHRs, and no cadre of highly experienced PHR leaders was readily available. Therefore, a process engineer would typically be assigned to lead one or two PHRs. In fact, from 1998 to early 2002, 18 different people were assigned to lead a total of 26 PHRs for the Biosynthetic Human Insulin (BHI) plant located at the Lilly Technology Center (LTC) in Indianapolis, Indiana.

Lilly used relatively large PHR teams. On each of the 26 PHRs conducted from 1998 to early 2002 (described above), an average of seven PHR team members were assigned on a full-time basis. One or more PHR team members were typically appointed to represent each of the following groups/disciplines:

- Production (typically one or more operators; sometimes an operator and a first-line supervisor)
- Maintenance
- Process engineering (the process engineer for the area was assigned to lead 19 of the 26 PHRs, 3 were led by people with significant operational experience in the process area, and 4 were led by engineers who were not specifically assigned to the process area)
- Technical service (the BHI group that typically has the most knowledge of process chemistry)
- Automation (the group that looks after programmable control systems)
- Project engineering (if the PHR involved new equipment)
- Safety and health (on a limited basis)

In some cases, four or more operators were assigned to a PHR team. There is no clear evidence that this increased the information available to the team. Also, the process chemistry is fairly simple for some processes and the control system is well documented throughout the BHI facility, resulting in relatively light participation on the part of the technical service and automation representatives.

PHR meetings extended over weeks or months. Since all of these team members had full schedules, PHR meetings were held on a regular basis in small increments over several weeks or months (typically 1- to 2-hour meetings held 1 to 2 weeks apart).

Production areas were finely divided. To help limit the duration of the string of PHR team meetings, the BHI plant was divided into many different processes over the years. At one point, there were 68 different defined process areas within the BHI plant. During the 1998 to early 2002 period, there were 29 BHI process areas that were classified as "high risk" (i.e., processes that are regulated by OSHA's process safety management [PSM] regulation and/or EPA's risk management program rule, or processes that Lilly chooses to include in its PSM program based on the results of a hazard screening tool). During the 1998 to early 2002 period, the PHRs were revalidated for 26 of the 29 BHI high-risk processes, requiring over 8,000 staff-hours for PHR preparation, PHR meetings, and documentation of PHR results. (The 8,000 staff-hours do not include the time required to address actions that resulted from PHR recommendations.)

Consequences of interest were not always well defined. For a variety of reasons, some actions resulting from the recommendations contained in these 26 PHR reports tended to be difficult to resolve. For example:

- Some of the recommendations were redundant (and not always well aligned) with similar recommendations generated by other PHR teams charged with reviewing other parts of the same production area
- Some recommendations probably should have been managed via another system (e.g., submit a maintenance work order, generate a request to correct a simple error or omission in a procedure that was unrelated to process safety)
- Some of the recommendations appeared to be good opportunities to improve operations but were unrelated to process safety

Presumably, the PHR teams that identified these opportunities for improvement had one means to communicate that information – make a PHR recommendation. Unfortunately, this practice resulted in significant effort to either (1) document why a recommendation should not be implemented or (2) manage and track completion of an action that could have been managed by a simpler management system (e.g., the maintenance work order system) or was not directly related to process safety.

PHR recommendations tended to lead to redundant or inefficient actions. Implementing actions resulting from PHR recommendations can be very resource-intensive. In fact, it is likely the biggest (albeit hidden) cost of a PHR program. A PHR program that involves a large number of teams, processes, and leaders can lead to numerous redundant or overlapping actions, as was the case for the BHI plant.

3. CHANGES THAT HAVE IMPROVED THE PHR PROGRAM

Lilly has implemented several changes to its PHR program over the past several years to improve the program's effectiveness and to better utilize the time of people associated with the entire PHR program.

Centralize coordination of the PHR program. The responsibility for the PHR program for all bulk pharmaceutical manufacturing facilities at LTC (which includes the BHI plant) was assigned to one PHR coordinator. This role included "owning" the PHR procedure, scheduling all PHRs, identifying process boundaries, selecting PHR team members, coordinating final report review/approval, developing the initial plan for implementing actions resulting from PHR team recommendations, coordinating final resolution of PHR recommendations, and tracking PHR metrics.

A PHR Oversight Team was established and chartered to improve the PHR program and monitor its effectiveness. This team also established metrics that were tracked and communicated broadly on a monthly basis. These metrics included the number of open recommendations, the number of open recommendations that are past due, the number of recommendations coming due in the next 90 days, and a detailed listing of the action item owners and their supervision. In addition, the recommendations were given a risk ranking, and special emphasis was put on resolving the highest priority items first.

As the BHI management team became more aware of the overall status of the PHR program, additional emphasis was put on addressing the backlog of PHR recommendations and ensuring that PHR revalidations were completed in a timely, effective, and efficient manner. The use of divisionwide metrics and consistent/broad communication tools made these metrics more visible and understandable, resulting in an increase in ownership and accountability throughout the organization. This and other improvements described in this paper have led to a 90% reduction in open action items over the past 2 years.

Expand process boundaries. Process boundaries have been expanded so that a process generally begins with a well-controlled set of raw materials (e.g., materials purchased to a specification and certified by the supplier, recovered/recycled materials that have been tested against an internal specification, in-process inventories that have been similarly tested or certified) and ends with finished products, byproducts, recycled materials, and waste materials. Dividing the processes so that they begin and end with predictable/specified inputs and outputs allows the team to focus on the process under review – the process with which team members are normally most knowledgeable – rather than searching upstream or downstream processes for causes and consequences of upsets. This is not to say that teams do not consider the full range of process upsets that could affect the process that is being reviewed. In fact, the current approach provides structure that allows teams to quickly identify and examine deviations such as high flow, low flow, reverse flow, and high concentration of contaminants in upstream or downstream processes that might impact the process being reviewed. This also reduced the total number of BHI high-risk processes from 29 to 9.

In our experience, the PHR team's efficiency in the first day of PHR meetings is often 50% or less of the efficiency of the same team by the second or third day. Expanding process boundaries helps take advantage of the increase in efficiency that naturally occurs during the course of the meetings.

We have also noted that the meeting time required for the PHR team to address the questions in most checklists (e.g., the human factors and facility/stationary source siting checklists found in the book titled *Revalidating Process Hazard Analyses* [CCPS, ©2001]) is largely independent of the number of nodes or process sections in the scope of the PHR. Thus, the time required for checklist-based reviews increases linearly with the number of "processes" that are identified for a particular process area. Also, using the same checklist multiple times for a single, finely divided process area is likely to result in the identification of similar hazards or issues, but the different teams may suggest alternative strategies for managing the hazards or addressing the issues. This can lead to overlapping or redundant action items, which tends to be less efficient than a single, coordinated approach to managing the hazard(s) identified by the team.

Schedule the PHR meetings in a block format. PHR meetings no longer extend over weeks or months – they are now typically scheduled for 6 hours per day in 2- to 4-day blocks. If a PHR cannot be completed in a single week, a second week of meetings is scheduled, normally within 2 weeks of the initial set of meetings. Using this schedule, most PHRs at BHI are completed in 1 week, with a few extending up to 3 calendar weeks. This minimizes the loss in efficiency that was observed when meetings were strung out over several months, while minimizing the impact on team members.

BHI has specifically avoided (1) scheduling meetings for 8 hours each day, 5 days per week, and (2) scheduling meetings for consecutive weeks if the PHR cannot be completed in 1 week. Although a compressed schedule is often necessary for a new plant (where the PHR meetings are often a critical path for completing plant design and construction), this is generally considered suboptimal at the BHI plant. Personnel who have a solid understanding of the process, valuable experience, and the respect of their peers are normally very busy. Scheduling them for 8-hour meetings ignores the important role they play in supporting day-to-day operations. Expecting that team members will devote 40 hours to PHR meetings in a single week can result in (1) team member burnout due to very long days (i.e., 8 hours of PHR meetings followed by 4 hours of their "day job"), (2) compromising on team composition/team membership, and/or (3) team members being constantly called out of meetings at inopportune times, which can significantly impact the efficiency of the meetings. These problems are minimized by providing time each day for each team member to adequately plan for other activities (e.g., most PHR team members attend the morning production meeting prior to the start of the PHR meetings and hand off any important tasks to other members of the production team).

Appoint a limited number of expert PHR team leaders. BHI has made the decision to use PHR team leaders with significant expertise and experience in leading PHRs but with limited process-specific knowledge, rather than vice-versa. Prior to 2002 there was a perception that it was vital for the PHR team leader to be an expert on the process being reviewed. Clearly, understanding the process helps the leader; however, we have found that team members can provide all of the knowledge and expertise required while the team leader helps facilitate and direct the discussion. Using a small number of expert PHR leaders has led to improvements in:

- PHR team efficiency
- PHR recommendations that are easier to implement and provide more "bang for the buck"
- Timeliness in completing PHRs
- Opportunities to use more effective PHR methods
- More consistent PHR reports

As noted previously, 18 different people were assigned to lead 26 PHRs performed between 1998 and early 2002. Since mid-2002 all of the BHI PHRs have been led by one engineer. Lilly uses another engineer in a similar manner for other units at LTC, and these two PHR leaders have substantial knowledge of Lilly's PHR process and have been identified as each other's alternate. Using experienced PHR leaders helps ensure good productivity during PHR meetings. For example, when meeting times are compared at BHI, experienced leaders have been able to redo PHRs in approximately 45% of the meeting time required by the previous PHR team.

The "ideal" PHR team member must be diverted from one of many opportunities to improve operations in order to help with the PHR. Maximizing the efficiency of each PHR team member's time helps further improve the PHR program because:

- Key personnel, who typically have significant informal influence within the organization, are much more likely to agree to serve on the PHR team and actively participate in the PHR meetings because they see it as a good use of their time
- Attracting team members who have the respect of the entire production team not only improves the quality of the PHR, it helps establish credibility for the team's recommendations
- Often, the same key personnel play a key role in completing actions resulting from their recommendations, and this ownership helps ensure successful implementation of the actions

As we mentioned previously, good recommendations are the most important output from a PHR. And, resolving and completing the recommendations is normally the most costly step in the PHR process. Therefore, it makes sense to find ways to maximize the efficiency of this step. We have found that expert leaders have helped achieve this efficiency by ensuring that:

- The recommendations are well documented. Recommendation statements that used to be short statements with little context have been replaced with stand-alone statements that include the hazard, the team's rationale for making the recommendation, and, in many cases, suggestions for addressing the hazard.
- Opportunities for finding synergy between recommendations developed by various PHR teams are noted. This requires that the PHR team be keenly aware of related efforts, which is difficult in a plant with eight different control rooms managed by five production team leaders, and process engineers normally dedicated to a single production area. The simplest and most efficient way we have found to provide this knowledge of potential overlapping issues between various PHR teams is by using a common PHR team leader.

Prior to 2002 PHRs at Lilly were typically led and documented by process engineers, and the teams met periodically over several months. However, the same process engineers are intimately involved in process troubleshooting and providing other day-to-day support to operations tasks that are time-critical. Although a process engineer may plan to use all of "next Wednesday morning" to prepare for a PHR meeting (or to work on documenting the PHR report), an equipment breakdown at 7 a.m. on Wednesday tends to refocus priorities and the PHR may be set aside. Setting aside a longer-term task (like conducting a PHR meeting or documenting a PHR report) when an immediate and important operational need arises provided a constant challenge to process engineers and sometimes delayed completion of PHR meetings and/or delivery of PHR reports.

Another benefit of using expert PHR team leaders is that an expert generally has more tools available. In other words, an expert leader (1) is more likely to recognize that a particular process section might be better analyzed using a particular hazard review technique and (2) can quickly orient the team and move forward using that technique. As the saying goes, if your only tool is a hammer, every problem looks a lot like a nail.

A final benefit that comes from limiting the number of PHR leaders is that it improves the consistency of PHR reports. At BHI, detailed PHR procedures (nearly 50 pages) are in place to help ensure consistency in conducting and documenting PHRs. It's similar to the tax code; written instructions and checklists help ensure consistency, but long, detailed written documents are not always user friendly, particularly to someone who only refers to the document once a year (or maybe once in their career). It is generally difficult, and certainly inefficient, to relearn these procedures for each PHR.

For some time, Lilly has used PHR reports to identify safeguards for inclusion in its mechanical integrity program and to determine operating limits, consequences of deviation, and steps to avoid or correct deviations from operating limits. Our vision is that (1) the reports will play an expanding role in the identification and evaluation of hazards associated with changes to the process, (2) incident investigation teams will look to the reports more often for possible causes of process upsets, and (3) information contained in the reports will be more directly integrated into operator training. However, the PHR reports tend to be quite long (often over 100 pages), and it is vital that a standard format be logical and strictly adhered to if users are going to be able to quickly find information in the report. This goal has been accomplished, in part by not having a multitude of process engineers writing PHR reports with various layouts, formats, and so forth.

Allow for part-time or on-call PHR team members. Assigning certain PHR team members to serve on a part-time basis helps (1) improve the efficiency of the PHR meetings, (2) make the best use of the team member's time, and (3) focus PHR team recommendations.

As stated previously, prior to 2002 an average of seven people were assigned to PHR teams at the BHI plant. We now assign an average of five "core" team members who are well supported by a larger group of specialists who attend meetings as appropriate and are readily available to support the PHR team in other ways.

In many cases, PHR team members have specific questions that can be quickly verified by a subject matter expert. For example, they may want to know what preventive maintenance work orders are in place for a specific pressure relief device. While this information is normally not immediately available in the PHR meeting, it can be quickly provided by the area maintenance technician (MT), particularly if the technician is sitting in his office and talking to the team on the speakerphone. We have found that assigning the MT to the team on a part-time basis (1) improves the quality and timeliness of this sort of information and (2) allows the part-time team members to be much more productive in supporting the PHR team while minimizing the impact that the PHR meetings might otherwise have on ongoing production and maintenance operations.

Likewise, part-time participation by automation engineers and technical service representatives has proven effective because these people have the time to research a specific technical question and report back to the PHR team, often later the same day. Based on the information provided, the team can then document a safeguard, explain why an apparent cause is not credible, or make a recommendation based on facts rather than speculation. This also helps teams avoid making recommendations that begin with "Investigate the hazard associated with . . . and take appropriate action," and causing BHI to resolve the recommendation and track/document completion simply because the PHR team members were not aware if the hazard actually exists (or what safeguards are actually in place).

Finally, in reviewing recommendations from previous PHR reports, one finds that many times a PHR recommendation addressed an industrial safety issue, an equipment reliability issue (where the team clearly documented that the equipment would fail in a safe state), or a product quality issue. This may have been due to the lack of discipline in defining consequences of interest (which will be addressed later), but it was also partly a result of the makeup of the PHR teams.

Use the most suitable PHR methodology. Prior to 2002 the "what-if/checklist" was the most common PHR method used at the BHI plant. This is partly because the 3-day PHR leader training tended to focus on this technique, and partly because there was a perception that the hazard and operability (HAZOP) method took much more time than the what-if method. Both of these limitations are closely related to the point made earlier that expert PHR leaders have a wider set of tools.

A review of previous PHR reports clearly indicates that wide use of the what-if method led to inefficiencies. Specifically, PHR teams would brainstorm what-if questions, and then answer them over the course of the 1- to 2-hour meeting. This process was repeated in subsequent meetings, many times several weeks later, and the teams ended up asking and answering similar questions multiple times. Also, since there is limited structure to the what-if method, teams failed to identify consequences of one mishap as a cause of another mishap and "link" them together to help improve efficiency. Because of this, when addressing a what-if question that might be a cause of high flow to a fixed roof storage tank, teams would identify consequences and safeguards associated with knock-on effects (high level, high pressure, and loss of containment). When addressing another question that might be a cause of high level, the team would end up going down the same path, resulting in analyzing high pressure and loss of containment several times. As mentioned previously, the time required for PHR meetings has been reduced by an average of 55%, partly due to the efficiency of using the HAZOP method, which helps provide more focus and structure to the PHR.

Lilly engineers have also noted other benefits from using the HAZOP method, including (1) much more structure to the tables, which allows an occasional user to quickly find a particular scenario, (2) a means to quickly identify causes or consequences of scenarios that involve process equipment that is outside the scope of the process being reviewed, and (3) a general belief that the PHR teams are less likely to miss a hazard using a library-based method of generating process deviations (such as HAZOP versus the more free-form, what-if brainstorming method).

Focus on consequences of interest. Once a PHR team makes a recommendation, the natural assumption is that addressing the recommendation should reduce the likelihood or mitigate the consequences of a loss event related to process safety. In other words, it's difficult to justify doing nothing with a PHR team's recommendation. At a minimum, management must document the basis for rejecting a PHR team recommendation (e.g., the recommendation may not be necessary to protect the health and safety of employees and contractors, the public, and the environment because . . .). Since 2002 PHR teams have been encouraged to maintain a list of non-PSM-related suggestions, and these lists are passed along to the appropriate groups for further consideration (to improve operations, economics, product quality, etc.). This helps capture good ideas while not creating significant extra work to track the completion of actions associated with these good ideas.

4. RESULTS

Although it is sometimes difficult to directly identify benefits from PSM programs, we have compiled a few quantitative and qualitative results that have been observed at the BHI plant. Although we believe that these data are accurate and are important to making sound management decisions on staffing of PHR teams, *effectiveness*, not efficiency, is the primary goal of Lilly's PHR program. These data are primarily provided as benchmarks – to help readers evaluate the potential impact of implementing some of the changes described in this paper at another facility.

Quantitative results. Data related to efficiency are provided in Table 1. Efficiency has improved in the following ways:

- Meeting hours are fewer for each process
- Lilly staff time for participation in PHRs has been reduced considerably
- Reduction in the number of recommendations generated by PHR teams reflects efforts to:
  - — Eliminate redundant recommendations
  - — Sharpen the focus on predefined consequences of interest
  - — Use the best means to communicate the need for changes (i.e., submit a maintenance work order rather than make a PHR recommendation when appropriate)

Table 1. Comparison of the Impacts of Changes on the BHI PHR Program[1]

Current Process (number of processes in previous PHR)	Meeting Hours		Lilly Staff-hours		Number of PHR Team Recommendations	
	Previous	Current	Previous[2]	Current[3]	Previous	Current
Processes 1A and 1B (7)	220	90	2,000	435	307	63
Process 2 (2)	27	24	560	124	26	10
Process 3 (4)	82	30	1,762	160	81	24
Process 4 (4)	88	30	1,573	160	72	7
Process 5 (1)	30	24	760	136	89	10

[1] This analysis is limited to six of the nine currently defined processes; the data on meetings and staff-hours are not available for some of the PHR revalidations conducted between 1998 and 2002 that correspond to two recently completed PHRs (the ninth PHR is scheduled for revalidation in 2005)

[2] Includes the time to document the results of the PHR

[3] Does not include the time for the non-Lilly personnel who led the PHR meetings and documented the PHR

Qualitative results. The qualitative improvements are much more important than the quantitative results described above. An "efficient" PHR program is not the goal; if a team fails to recognize a hazard and make an appropriate recommendation where safeguards are less than adequate, the PHR team has failed. There is a general belief among BHI engineers and managers that PHRs that have been redone when they came up for revalidation over the past 3 years are more comprehensive than the previous PHRs and that the PHR program is delivering better results.

For example, a recent incident resulted in high temperature in a tank due to a high flow rate of acid to the tank (and a control system that did its best to keep the pH within range by feeding caustic at a high rate). The process operators were able to identify the cause of the incident, and the situation was brought under control without injury or equipment damage. The incident investigation team noted that the scenario had not been identified by previous PHR teams (i.e., there was no question that forced the team to consider this scenario). The incident investigation team was also concerned about the potential for other, unrecognized hazards involving the tanks in this process area, and recommended that the PHR be revalidated 2 years ahead of schedule. Management supported the recommendation and the PHR was recently redone, illustrating a vote of confidence in the PHR program.

PHRs are now seen as value-adding activities, and there is a marked increase in the willingness of very busy people to serve on PHR teams. This willingness to participate and the belief that the overall quality of the PHR program has markedly improved are probably the most significant factors that have led to improvements in the completion of action items.

5. CONDITIONS THAT FACILITATED IMPROVEMENTS TO THE PHR PROGRAM

As with most significant improvements, no single factor or point in time can be clearly identified as the event that caused the organization to "turn the corner." An incident involving a release of toxic gas at the BHI facility in 1997 precipitated a step-change increase in focus on process safety (there were no injuries, but the incident resulted in a 2-month business interruption). This incident could arguably have initiated the improvements to the PHR program. However, other factors were equally, if not more, important.

Management oversight and leadership. Since 2001 BHI has established a standing team of senior managers who are charged with oversight of the PSM program. This team meets monthly to review metrics, discuss progress, help clear any obstacles that are reported, and generally assess the PSM program.

Ownership at all levels. By any measure, ownership of BHI's PSM program, including activities associated with PHRs, has moved from the health and safety group to the engineering and operations group. The discussion has largely moved from "What do we have to do?" to "How can we best do this?" and "How can we leverage this effort to . . .?" The safety and health group has transitioned from the role of doer/enforcer to helper/coordinator.

Synergy with quality improvement initiatives. During the same time frame, considerable effort has been focused on quality initiatives. The entire BHI plant is much less tolerant of discrepancies between stated plans/procedures and results; a past-due PHR action item is fundamentally a discrepancy between a plan and a result. This intolerance has helped focus the PHR program at all levels, from planning to completion of action items.

6. PATH FORWARD

Based on the results presented in this paper, Lilly has embarked on the following initiatives to further parlay its recent BHI PHR experiences into improvements at other facilities.

Expand the "limited number of expert PHR team leaders" concept to other plants. While the success of utilizing expert PHR team leaders was recognized by many on a qualitative basis, the quantitative improvements represented in this paper will be useful in promoting similar changes at other Lilly sites. This is not to say that there is only one approach to effectively perform PHRs; however, consideration will be given to the advantages experienced in the BHI plant.

Consider duplicating the "limited number of expert team leaders" concept within incident root cause analysis. PHRs are not the only initiatives that require experienced leadership. Another effort that is very analogous to the PHR is root cause analysis (RCA) following an incident. Applying the same approach of using a limited number of expert RCA leaders should have similar effects: more efficient use of RCA team member time (coupled with increased willingness to participate), better use of tools, more focused (and less redundant) conclusions, and improved documentation of results and recommendations.

Build PSM expertise into process engineering. Since the changes were made to increase the expertise in the PHR leader position, the BHI plant has also chosen to further integrate PSM into process engineering. Four senior process engineers at BHI have been assigned to be technical reviewers of process safety information and PHR reports. They also serve as reviewers/approvers of hazard reviews performed in conjunction with management of change. In the future, the senior process engineers will be given additional PSM responsibilities, such as reviewing/approving process safety information and implementing actions to resolve PHR recommendations. Finally, one of the senior process engineers has been designated by management to expand his knowledge of PSM requirements and practices to help ensure that the process engineering discipline continues to play an integral role in the application of PSM principles.

Replicate this PHR process in fill/finish operations. The greatest process safety considerations in pharmaceutical manufacturing exist in bulk chemical operations. But there are still safety concerns, even of a process safety nature (though on a lesser scale), in the final formulation and packaging phase, referred to as "fill/finish operations." While the consequences of interest may change, the general tools and approaches described previously still apply. The ability to transfer the PHR process with the least cost and greatest benefit is viewed as being highly dependent on the use of an expert PHR leader. One such fill/finish area at Lilly has begun to perform PHRs in an effort to continuously improve its safety performance. The BHI PHR coordinator provided guidance on the expert PHR leader approach and the accompanying PHR management systems used in bulk pharmaceutical manufacturing to managers of fill/finish operations. Additional replication of the PHR process used in the BHI plant may be pursued in other fill/finish areas.

The Dow Chemical Company's Reactive Chemicals Resource Center

Marabeth Hofelich, Ph.D.
The Dow Chemical Company
Midland, Michigan

Harold Johnstone
The Dow Chemical Company
Freeport, Texas

Brenda Prine
Dow Chemical Canada Inc.
Sarnia, Ontario, Canada

Joerg Steinmetzer
Dow Deutschland GmbH & Co OHG
Stade, Germany

ABSTRACT

The Reactive Chemicals Resource Center (RCRC) was launched from the Core R&D Function of The Dow Chemical Company in May of 2002 to improve the delivery of Reactive Chemicals information within Dow. While resolving and documenting client requests for data and information for thermal and flammability hazard evaluations, the RCRC simultaneously developed and implemented a knowledge management system for organizing and making shared information easily accessible. Although Dow has operated Customer Technical Support Centers for several years to service external business customers, the RCRC represents the first technical contact center to serve internal Dow clients from a leveraged capability. This initiative was driven by a desire to improve speed of response in solving Reactive Chemicals requests and has achieved a productivity gain of over 10% since startup.

1. INTRODUCTION

When the Analytical Sciences Laboratory within The Dow Chemical Company's R&D function globalized in the mid-1990's, the Reactive Chemicals (RC) Technology group began an initiative to improve technical information management in hazard evaluation. [1] The business climate of decreasing expert resources coupled with a strong drive to high productivity levels led to the development of the "Reactive Chemicals Technology Package", a living document used extensively by plants and project teams for Reactive Chemicals/Process Hazard Analysis (RCPHA) reviews.

In 1999, Knowledge Management (KM) became a strategic element for the Reactive Chemicals group and was driven and facilitated by the technical leader team. This led to the creation in late 2001 / early 2002 of the first "virtual contact center" for Dow-internal technical support named The Reactive Chemicals Resource Center (RCRC). While the overall KM strategy is beyond the scope of this paper, the main elements of the RCRC system (people resources, technical infrastructure, and work processes) are described below.

1.1 Challenges for the Reactive Chemicals Group – a Case for Change

In 1997, discussion about Knowledge Management started within the global R&D Analytical Sciences Reactive Chemicals team. We realized we faced current and future challenges, such as:

- need to serve a growing client base with the same or fewer RC resources
- clients need more complete and more complex problem solutions
- 80% of our "product" is made of Knowledge
  (paradigm shift: 15 years ago, 80% of our product was Lab testing data)
- Our workload is growing (with growth of Dow) but our workforce size is not keeping pace
- We must free-up time for value-creating opportunities, e.g., growth projects, and maintain balance in our personal lives
- We have to change to meet the RC requirements now and in the future
- It is critical that we improve the Knowledge Management work process

Predicted effects of resource over-commitment and of not changing:

- Productivity per project declines
- Lack of delivery on critical safety issues
- "Fire-fighting" efforts consume resources
- Key Subject Matter Experts (SME's) lose clarity of purpose
- Failure at retaining experienced resources
- RC resources and functions become a bottleneck for projects
- Cycle times for projects lengthen

1.2 The Vision

The vision we created was:

1) a KM system to enable the RC experts to capture and mine problems and solutions.
2) a self-help system for internal clients .
The realization of this vision would empower RC Experts and clients to leverage RC information faster, for lower costs and with more impact than ever before.

2. PEOPLE RESOURCES

2.1 Experienced Subject Matter Experts

Three experienced SME's from the USA, Canada, and Germany formed the initial nucleus of the RCRC. The inclusion of personnel from multiple countries improves global support over multiple time zones and various national holidays.

The three founding SME's have over 80 years of combined experience in Reactive Chemicals technology. Their familiarity with many Dow processes and their depth of RC expertise have been key to the success of the RCRC. They are able to provide quick answers to many client questions through the knowledge they have gained over the years. After about 18 months of operation, two additional SME's were also added to the RCRC to guarantee sustainable service levels when an SME is out due to extended illness or vacation.

Figure 1

Dow's Reactive Chemicals Resource Center Locations

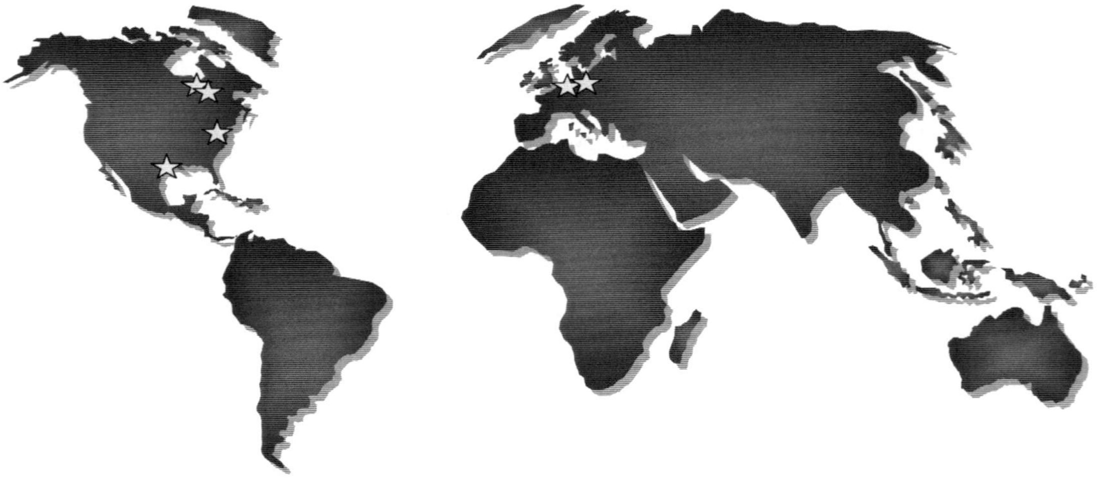

2.2 Supporting Staff

The role of the RCRC Assistant was created and filled on a part-time basis with two experienced RC technologists (whose primary job responsibility was still to do thermal stability testing in a Reactive Chemicals lab). Their role overlapped that of the RCRC Admin's (part-time office professionals providing support to RCRC) with respect to helping manage the flow of service requests. However, their role added the dimension of actually owning and resolving certain "routine" requests (e.g., helping clients find relevant data in Dow's Reactive Chemicals database).

3. TECHNICAL INFRASTRUCTURE

Many people think about technical infrastructure first when they hear the term knowledge management (KM). This is, in fact, the *last* element to think about. It must be designed to support the other elements of an overall KM strategy (e.g., People/Culture, Knowledge Content, Knowledge Flow). The Dow RCRC team chose the following primary hardware and software tools: one global telephone system and one global number to call (operated in conjunction with external support), Intranet (internal) web pages, a global functional email account, a certified Windows™ NT File Share with indexing and searching capability, and some special client relationship management software - Integrated Desktop (IDT from Siebel). These tools were leveraged from Dow's Customer Technical Service Center (CTSC) organization which has been operating Call Center/Help Desk services to external customers in most Dow business units for several years.

3.1 Telephone System

For most CTSC's, the personnel work together in one physical location. Since the RCRC is a virtual call center (each of the SME's at a separate location), the telephone needs are somewhat unique. We were able to establish a single phone number based in Midland which linked the original three SME's in the USA, Canada, and Germany. This number was programmed with some logic which routes calls by time of day and country of origin based on rules set up by the RCRC team.

In the event that clients are connected to the phone menu (no RCRC personnel immediately available), they will be given options which also vary by time of day and country of origin. If the client is in Europe, for example, choosing to speak with an assistant will connect them to CIG (a third-party service organization operated by Electronic Data Systems) in Belgium, (for multiple language support) during European business hours or to CIG in the USA. The phone menu (at all hours) also tells clients that another option is to send an email to the functional mailbox.

3.2 Intranet Web Pages

Since numerous web pages related to Dow's Reactive Chemicals technology already existed on the Dow Intranet, it was a natural fit to link to these with a special home page for the RCRC. This provided an avenue to advertise the services of the RCRC and a link to send a message or service request by email. The functionality of this page was later improved with a link to a pop-up form which provided a template for clients to use in submitting a service request. By building this form with required entries for key fields, the number of requests with incomplete information was greatly reduced. A companion form was also added for submitting information about samples whenever a service request involved a need for testing.

3.3 Network File Share

Discussions were held with several Dow businesses to select the best methodology for sharing and retrieving information. These discussions with users indicated that the use of a certified NT file share (CNTFS) would be the simplest approach (and one which was already being used successfully by several CTSC's). The purposes of the CNTFS were as follows:

- to simplify information access and security management.
- enable RC to work with only one instance of each document on the network, so that everybody is using the most current, up-to-date information.
- enable RC to include links to documents with Dow Confidential information with full (NT) security, thus keeping business confidential information out of IDT.
- capture and leverage documented explicit knowledge, related to Dow RC needs.
- enable searching of a shared global pool of RC documents to speed up necessary and mandatory RC work process in the businesses and functions.
- enable publishing of selected files on a web page for RC internal self-help.

4. WORK PROCESSES

4.1 Quick Resolution for Small Requests

Regardless of how a new client request enters the system (phone, regular email, web form), an SME in the RCRC will attempt to resolve it immediately if possible. Because these SME's are highly experienced, they can answer many "frequently asked questions" based on their accumulated knowledge. Other questions can often be answered relatively quickly, also, after consulting any of a variety of resources such as existing internal or external published data, reference books such as Bretherick's Handbook of Reactive Chemical Hazards, or internally documented resolutions to similar problems.

4.2 Escalation to Other SME's

If RCRC SME's have a heavy workload or if a new project is large or complex, then team members can escalate the request to other RC practitioners (Dow SME's outside the RCRC). A communication protocol was developed to handle these cases to insure a clear hand-off of responsibility. The receiving SME is responsible for re-negotiating the need date and deliverables with the client (if needed), resolving the request in a timely manner, and documenting the results with communication back to the RCRC.

5. RESULTS

5.1 First-Contact Fulfillment

Since one of the main purposes of the RCRC is to provide fast resolution to client Reactive Chemical problems, a team goal was established around First-Contact Fulfillment. This goal measures how many requests are resolved within 24 hours. IDT records the creation and completion date and time for each service request. Since RC testing is performed for 30-50% of all requests, and can have up to a two-week turnaround time or longer, this goal is not expected to reach 100%. Nonetheless, we are tracking these data to drive improvements in our performance. At the end of 2002, we provided <24 hour resolution for 38.3% of all requests. In 2003, we improved to 40.9%. In 2004, we continued to improve to 45%.

5.2 Meeting Client Need Dates

Requested need dates from clients are entered and tracked in IDT. When re-negotiated, these deadlines are updated with appropriate documentation that also describes the consequences of missing the need date. Historical data set the baseline for this team goal at 76% of clients' need dates met. By the end of 2002, after six months of operation, the RCRC measured 79.4% met. By 2003 year-end, 91.2% of clients' requested need dates were met on time. By 2004 year-end, the RCRC improved the rate again to 93.5% met on time.

5.3 Development of Self-Help Resources

In 2004, we started a Six Sigma Project to develop a work process to consolidate and verify the documented resolutions of client questions on the CNTFS. The objective is to internally publish the best available RC Knowledge as self-help resources, accessible globally via the Dow Intranet. To help prioritize these efforts, 1400 service requests from IDT were analyzed to determine what RC topics Dow clients ask about most frequently. The top issue from this analysis was chemical compatibility. A team was therefore assembled to carry out a pilot project on this subject and implement a Quality System. This pilot is still in progress as of the writing of this paper.

5.4 Productivity Improvement

While hard dollar savings are difficult to quantify, the RCRC has definitely created productivity improvements in the delivery of Reactive Chemicals technical support within Dow. For example, in the past, the same (or similar) questions might be posed to two or more RC SME's at different sites independently (the clients might be in the same or different business groups). This created a duplication of effort which was usually not discovered until after the fact. Now, the RCRC serves as a "clearing house" which can eliminate such occurrences. In addition, the population of the RC Knowledge Base is far enough along that it is starting to pay dividends. Recycling of information from the CNTFS has been a contributing factor in improving the rates of First-Contact Fulfillment and Meeting Client Need Dates. Other productivity improvements that are realized because of the existence of the RCRC are easy vacation relief coverage for all Reactive Chemicals SME's, reduction of interruptions for RC SME's working outside of the RCRC and facilitation of work load leveling to assist individuals in meeting critical deadlines. Continuing productivity improvements are expected as we further develop our RC Knowledge Management system and begin publishing self-help resources.

6. REFERENCES

1. L. Whiting, M. LaBarge, and J. Steinmetzer; "Global Technical Information Management in Thermal Hazard Evaluation", *Journal of Thermal Analysis*, vol 49 (1997) 1617-1624.
2. [Working knowledge : how organizations manage what they know], Thomas H. Davenport, Laurence Prusak. ISBN 3-478-36470-1].
3. Schreiber et al; Knowledge Engineering and Management: The Common KADS Methodology; MIT Press (1999). ISBN 0-262-19300-0
4. D. M. Amidon; Innovation Strategy for the Knowledge Economy, The Ken Awakening; Butterworth-Heinemann (1997). ISBN 0-7506-9841-1
5. P. Lyman and H. Varian; University of California at Berkeley Project: How Much Information? 2003; www.sims.berkeley.edu/research/projects/how-much-info-2003/
6. J. B. Quinn; Intelligent Enterprise: A Knowledge and Service Based Paradigm for Industry; The Free Press (1992). ISBN 0-02-925615-1
7. U.S. Chemical Safety and Hazard Investigation Board (CSB): Hazard Investigation: Improving Reactive Hazard Management. Report No. 2001-01-H (2002) www.csb.gov/completed_investigations/docs/ReactiveHazardInvestigationReport.pdf
8. The 2004 European research project CIKM (Creation of Innovation through Knowledge Management); funded by the European Commission's Information Society Technologies-Program (IST-2001-34442). www.knowledgeboard.com/cgi-bin/item.cgi?id=105428 (See also www.cikm.org/)

Using the Adiabatic Flame Temperature to Predict the Flammability of Lower Alkanes, Carboxylic Acid and Acetates

By

R. Ervin[1], M. Palucis[2], T. Glowienka[2], V. Van Brunt[2], W. Chastain[3], R. Kline[3] and P. Lodal[3]

[1]Vanderbilt Univ.(REU@ U. South Carolina), [2]U. South Carolina, [3] Eastman Chemical Company

Abstract

A Gibbs Free Energy minimization procedure is used to predict the flammability envelopes of alkanes, carboxylic acids, and acetates. In addition to providing the calculated adiabatic flame temperature (CAFT), the product profiles reveal regions of incomplete combustion products and the onset of cracking. Temperatures at the predicted onset of cracking appear to be closely related to the temperature at the upper flammability limit.

The predictions also reveal local maxima and minima in the concentrations of reaction products. Maps of incomplete combustion products for the flammability region show that incomplete combustion products are produced at fuel/oxygen ratios below the fuel/oxygen stoichiometric ratio. This region decreases as the limiting oxygen concentration is approached.

I. Introduction

Flammable substances, defined under EN 1127-1 [3] as gases, liquids, solids or a mixture of the three that undergo an exothermic reaction in the presence of air when exposed to an ignition source, are prevalent in today's chemical and petrochemical industries. Most solvents and hydrocarbons are extremely volatile under relatively normal operating conditions making for potentially dangerous situations in even the most routine of tasks, such as taking a vessel out of service or filling an empty tank. Thus, in order to prevent workplace explosions of such flammable vapors during these tasks, detailed knowledge of a particular compound's flammability region as a function of fuel, oxygen, and nitrogen concentrations is extremely useful. Most commonly this region is depicted on a triangular diagram, seen in Figure 1 [1], which identifies all three constituents of interest simultaneously. Each leg of the triangle is marked in a specific direction and indicates a constant concentration of a component. The flammability zone on the triangle is the main area of interest because all compositions within that region are potentially flammable given the presence of an ignition source. All regions outside of that zone are considered non-flammable and thus it is desirable to run under those conditions. The flammability zone may be divided into two subzones, a lower one between the stoichiometric line and the lower flammability limit and an upper one between the stoichiometric line and the upper flammability limit.

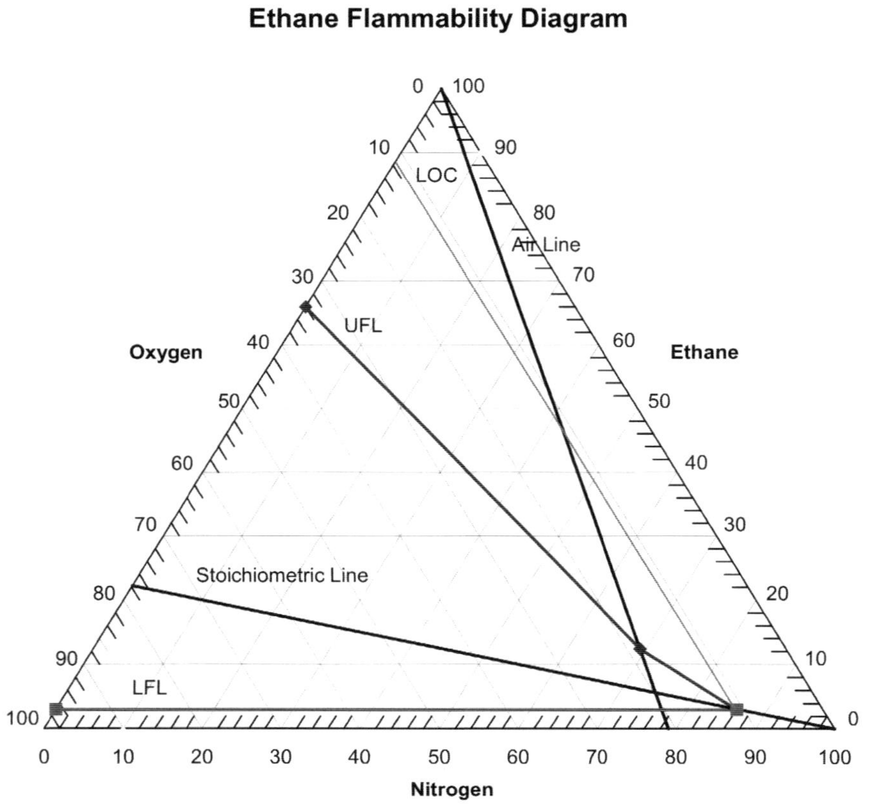

Figure 1. Ethane Flammability Triangle

Currently, there are several methods for determining this region for a given fuel. The most reliable is to conduct laboratory testing, which requires a highly automated apparatus of at least 20-L in volume to reduce quenching effects. In this vessel, hundreds of attempted reactions are performed at varying fuel, nitrogen, and oxygen concentrations [2,3,7]. Unfortunately this method is very costly and time-consuming. A second option is to use literature data to approximate the flammability zone, as long as the experimental data for the compound of interest is available. If the upper and lower flammability limits (UFL and LFL) in air, the minimum oxygen concentration (MOC), and the limits in pure oxygen are known then these points can be connected to approximate the flammability zone. While this method is a conservative one, the actual region will have some curvature at the "nose" (i.e., the low-fuel, low-oxygen portion of the diagram), especially in the upper flammability zone [3].

In order to avoid conducting experimental tests for the determination of a compounds flammability range, it has been suggested that there may be a correlation between the explosion limits and other parameters characteristic of the mixture. One such parameter is the adiabatic flame temperature (AFT). The theoretical determination of this temperature, referred to as the Calculated Adiabatic Flame Temperature (CAFT), is based on the thermodynamic principles of Gibbs Free Energy Minimization at constant temperature and pressure, which states that in the absence of work and appreciable kinetic and potential energy effects the energy liberated upon combustion is transferred to the combustion products [4]. The maximum temperature achieved by the products in the limit of such adiabatic operation is thus called the adiabatic flame temperature. Studies have proposed that there is a constant CAFT value adequate for predicting the limits of the entire explosion range of fuel-air-nitrogen systems with a generally accepted value of 1200K. It is believed that mixtures with flame temperatures higher than 1200K are flammable, and those below 1200K are non-flammable. This was first applied to the lower flammability limit and then extended to the upper flammability limit [3, 4, 5, 7, 8].

For this paper, the commercially available simulation package HYSYS 3.2 was used to perform a Gibbs Free Energy minimization to calculate chemical equilibrium and predict the flammability zone of three chemical families: the lower alkanes, carboxylic acids, and acetates. The Gibbs Free Energy minimization results from HYSIS 3.2 were compared with those generated using a similar methodology with Aspen Plus version 12.1. The reaction product composition and temperature results were comparable for selected alkanes and carboxylic acids. The favorable comparison of the results from two independent simulation packages validates the results of the investigation.

The product speciation and the CAFT at various oxygen-nitrogen-fuel lines were examined for each of the chemical families. Three representatives were chosen for detailed analysis: ethane, acetic acid, and methyl acetate. The validity of using a constant adiabatic flame temperature to predict flammability limits was examined. The findings, however, have indicated that the CAFT is not a conservative method for predicting the flammable region, especially within the range of 1000-1200K. For the three homologous series studied the CAFTs at the UFL, LFL and LOC varied considerably with composition and had temperatures lower than the accepted value of 1200K, thus under-predicting the flammability zone.

II. Computational Procedure

The simulation program utilized an algorithm based on the principles of Gibbs free energy minimization, which assumes that a reactor at steady-state in the absence of work or potential and kinetic energy effects will transfer all energy to the products or the surroundings. As the amount of energy transferred to the environment is reduced, the temperature of the products increases until the limit of adiabatic operation is achieved [4]. In order for this to be done in HYSYS, a flow sheet was constructed that contained a Gibbs Free Energy reactor with one input stream and two output streams, one liquid and one vapor. For each individual simulation, the following was entered into the input stream: the molar flow rate, a heat duty of zero, an initial temperature of 25° C, an initial pressure of 1 atm and the feed composition of fuel, oxygen and nitrogen. For the output streams, all possible partial combustion products were manually input into the system because complete combustion could not be assumed. This list included methane, hydrogen, water, carbon monoxide, carbon dioxide, nitrogen oxide, nitrogen dioxide, nitrous oxide, carbon, and formaldehyde. HYSYS then calculated the final product composition and CAFT of each outlet stream. As expected, the main difficulty in using this model was in deciding what species to include in the product stream. The final decision was based one assuming products from all possible combustion and decomposition reactions could be present in the final mixture.

HYSYS 3.2 is able to calculate the CAFT by setting the change in Gibbs free energy (G) to zero, meaning the reaction is in equilibrium. HYSYS then solves for this equilibrium using $\Sigma\ G_{products} - \Sigma\ G_{reactants} = 0$, where $G = H - T\Delta S$, by searching for a temperature that will make the change in Gibbs energy as close to zero as possible. Because the heat duty was specified as zero, this final temperature is the CAFT for the combustion reaction.

Two major assumptions were made concerning the simulation program. First, it was assumed that the fuel and the air were perfectly mixed in one stream before entering the reactor at 25°C. It could have been assumed that the fuel and air were in separate streams, each at 25°C, and then mixed and sent into the reactor. This assumption is valid as long as the molecule of interest has a boiling point less than 25°C, because at that temperature it is already a vapor before coming in contact with the air stream. For this study, only ethane, propane and butane had normal boiling points less than 25°C, while the remaining fuels studied had boiling points above the specified inlet temperature. For those, mixing the fuel and air streams prior to entering the reactor affects the CAFT because the fuel stream enters as a liquid, and it begins to vaporize only when it comes in contact with the air stream. This change in state causes the mixed stream to have a temperature between 5 and 50°C lower than the 25°C assumed in the first method, mainly due to the energy requirements necessary to vaporize the fuel. Therefore, the first method, which assumes one pre-mixed stream, was used for all fuels studied.

A second assumption made involved the Equation of State (EOS) that was used. A study of the different EOS's available in HYSYS was conducted to determine the affect that each had on the final CAFT at the LFL and UFL for each fuel. Those tested were the Ideal Gas, Peng-Robinson, Soave/Redlich/Kwong (SRK) and Lee-Kessler Equations of State. Of these, the Lee-Kessler Equation of State was the only one that gave drastically different results for the final value of the CAFT. All others gave temperatures within one to three degrees of each other. Therefore, the SRK EOS was used exclusively.

III. Results and Discussion

3.1 Alkanes

Product Speciation

The flammability range of five alkanes, namely ethane, propane, butane, pentane, and hexane, were analyzed using HYSYS 3.2 for various fuel-oxygen-nitrogen systems ranging from oxygen-depleted (i.e, at the Limiting Oxygen Concentration, or LOC) to oxygen-rich (Pure O_2) environments. The fuel composition ranged anywhere from 0.5% to 50% in order to map the entire flammable region along the fuel-oxygen-nitrogen line being studied. From HYSYS the CAFT was recorded as well as the composition of the main products, specifically, methane, hydrogen, carbon dioxide, carbon monoxide, and water. Other products appeared only in trace amounts. Of main importance were the onset of cracking indicated by methane production and the limit of complete combustion along each fuel-oxygen-nitrogen line, determined to be when hydrogen and carbon monoxide started to appear in appreciable amounts (10^{-4} mole%), respectively. Also noted were the maximum formation of water and carbon dioxide.

Since similar trends were observed among the five lower alkanes studied, the results for only one representative alkane, ethane, will be analyzed in detail, as will acetic acid and methyl acetate in later sections. The observed trends within and between the three chemical families will then be examined. For each alkane examined seven different fuel-oxygen-nitrogen lines were studied. Each line was a hypothetical mixture of an oxygen and nitrogen binary mixture with varying quantities of fuel. The O_2- N_2 mixtures being at the LOC, 15% O_2 85%N_2, 17% O_2 83%N_2, 19% O_2 81%N_2, 21% O_2 79%N_2 (Air), 60% O_2 40%N_2 and Pure O_2. In the case of ethane, the LOC is known to be 11.2 vol% O_2 [6] and thus the initial oxygen-nitrogen binary mixture was 11.2% O_2 88.8%N_2. The fuel compositions ranged from 1 to 15%, corresponding to phi (ϕ) values of 0.32 to 5.46, where phi is the ratio of the fuel to oxygen concentration to the stoichiometric fuel to oxygen concentration [9]. As shown in Figure 2, at the LOC the limit of complete combustion occurs when phi is equal to one, meaning the fuel to oxygen concentration equals the stoichiometric fuel to oxygen concentration. After this point incomplete combustion occurs, during which water and carbon dioxide formation begin to decline and there is the onset of hydrogen and carbon monoxide formation. The maximum predicted concentration of hydrogen occurred approximately at a phi of 3.1.

In moving left across the flammability triangle from the LOC to pure oxygen, thus increasing the amount of oxygen from 11.2 to 100 vol%, there were several trends observed, as can be seen in Figures 2 through 8. As the environment becomes more oxygen-rich, the value of phi at the limit of incomplete combustion decreases from 1 at the LOC to 0.18 in pure oxygen. Conversely, the phi value at which the onset of cracking occurs increased from 2.35 at the LOC to 3.5 in pure oxygen. The peak formation of hydrogen, water, carbon dioxide, and carbon monoxide also increases in magnitude as the environment becomes more oxygen-rich.

The product speciation resulting from ethane combustion along the seven observed fuel-oxygen-nitrogen lines (LOC to pure O_2) were also plotted on a flammability triangle, along with the upper and lower flammability limits as predicted from literature data [1]. This provided an approximate speciation map for the mixtures. Several trends can be seen. Examining Figure 9, the limit of complete combustion falls along the stoichiometric line between the LOC and air, known as the nose of the flammability envelope; however, it then deviates below the line as the oxygen percentage is increased. This is contrary to the common belief that the limit of complete combustion falls along the stoichiometric line for the entire range of the flammability envelope. Although the limit of complete combustion did not fall along the stoichiometric line as expected, it is interesting to note that the point of maximum water formation fell along the stoichiometric line for the entire range, from the LOC to pure O_2. Examining the point at which cracking begins, indicated by the onset of appreciable amounts of methane formation, within the realm of the predicted upper flammability region shows that from the LOC to air, the UFL lies inside the cracking line. Using this as a prediction tool would thus be a conservative method, except for the fact that as the environment becomes more oxygen-rich past air the cracking line falls below the UFL.

CAFT

Along with the composition of the combustion product stream, the theoretical CAFT was also found for all limit mixtures with the aid of HYSYS 3.2. This was done in order to determine its potential use in mapping the flammable region. For each fuel-oxygen-nitrogen line examined between the LOC and pure O_2 lines, a plot was constructed showing the CAFT versus phi, which is shown in Figure 10 for ethane. It can clearly be seen that as the combustion mixture becomes more oxygen-rich going from the LOC to pure O_2 the maximum value obtained for the CAFT increases. The curvature of the peak also changes going right to left across the flammable region. At the LOC there is a clearly defined sharp maximum at the stoichiometric composition. As the amount of oxygen is increased the peaks begin to spread more and the maximum is less clearly defined. This trend is to be expected and consistent with the speciation results because while at the LOC the region of complete combustion is the greatest, in pure oxygen it is the smallest. The rounding of the CAFT curves is consistent with incomplete combustion products being formed.

As mentioned earlier, current work has suggested that a constant CAFT value of 1200K may be a conservative approach for estimating the flammability envelope of a fuel. The validity of this argument was assessed by constructing temperature contour plots for each fuel. This was done by selecting six CAFT values, namely 1000K, 1250K, 1500K, 1750K, 2000K, and 2250K, and plotting them along with the LFL and UFL limits, as shown in Figure 11 for ethane. Mixtures whose CAFT's fell inside the LFL and UFL are flammable, and those outside are non-flammable. In the lower flammability region for ethane the LFL falls almost exactly along the 1250K contour line, indicating that this value would be an appropriate estimate. However, when looking at the upper flammability region 1200K is no longer a conservative estimate. In the nose of the envelope 1200K is a conservative estimate, but after this point the 1250K contour falls below the UFL, indicating that even at this CAFT a flammable mixture is still present. The 1200K contour underpredicts the flammability for oxygen concentrations higher than air.

The Lower Alkanes

In comparing product speciation for the entire family of lower alkanes it was seen that the same trends observed for ethane were true for propane, butane, pentane, and hexane. In validating the rule of a constant CAFT equal to 1200K as a conservative estimate for the prediction of flammability limits, propane was very similar to ethane, though in very oxygen-rich environments the 1250K contour line rose above the UFL. Butane showed that the UFL and 1250K contour line fell exactly on top of one another, indicating that 1200K would be an appropriate estimate for this species. Pentane followed the trend of ethane and propane with its behavior between the LOC and air and in oxygen-rich environments the UFL being below 1250K. but otherwise the UFL was below 1250K.

Other than in the nose region of the flammability envelope, the UFL for hexane was above the 1250K contour line. For all five of the alkanes investigated, the LFL could be conservatively estimated using a constant CAFT of 1200K; however, it appears that the upper flammability region would be under-predicted.

3.2 Lower Carboxylic Acids

Product Speciation

Product speciation of six carboxylic acids was predicted using the same method used for the alkanes. Similar to the alkanes here acetic acid is used as a representative of the family before trends of product speciation and CAFTs were compared.

To better understand the way the products changed in the acetic acid combustion reaction, the following oxygen-nitrogen ratio acid lines were used: the LOC, 14% O_2 86%N_2, 17% O_2 83%N_2, 20% O_2 80%N_2, 21% O_2 79%N_2 (Air), 60% O_2 40%N_2 and pure O_2. The LOC is known to be 10.8 vol% O_2 [6]. The speciation diagrams of the product compositions versus phi are given in Figs 12-18. In Figure 12, the limit of complete combustion is clearly defined at phi equal to one by the water and carbon dioxide compositions, the two complete combustion reaction products. In moving from the LOC to pure O_2, which equates to an increase in oxygen concentration from 10.8 to 100 vol%, several trends can be drawn from the product speciation figures. First, as was the case with the alkanes, the phi value that corresponds to the limit of complete combustion decreased substantially. This limit is illustrated by the rounding of the carbon dioxide concentration profiles shown in the speciation graphs. Also, as the environment becomes more oxygen-rich, the value of phi at the limit of complete combustion decreases from 1 at the LOC to 0.22 in pure oxygen. Conversely, the phi value at which the onset of cracking occurs increased from 2.15 at the LOC to 2.9 in pure oxygen. The peak formation of hydrogen, water, carbon dioxide, and carbon monoxide also increases in magnitude as the mixture composition becomes more oxygen-rich.

Product speciation for acetic acid was also examined on a flammability triangle in conjunction with the LFL and UFL. Figure 19 shows the flammability triangle for the combustion reaction of acetic acid at each of the seven fuel/nitrogen/oxygen mixtures. As the figure suggests, the limit of complete combustion appears far below the stoichiometric ratio, except in the nose region of the flammability envelope. Also, the UFL lies below the cracking line in the region from the LOC to air. Therefore, using the temperature associated with the point of cracking could potentially be an alternative conservative measure for predicting the UFL for acetic acid.

CAFT

The CAFT for all limit mixtures, ranging from the LOC to pure oxygen, was calculated in order to examine the potential to map the flammable region of a using CAFT. Figure 20 is a diagram of the CAFT versus phi for acetic acid. This graph clearly illustrates that the CAFT increased as the amount of oxygen in the mixture is increased. The curvature of each temperature profile changed in moving from the bottom to the top, similarly to that seen with the alkanes. In pure oxygen, the curve is broad, but as the concentration of oxygen in the mixture decreases, the temperature peaks become more defined, with the sharpest temperature peak corresponding to the LOC. Again, like the alkanes, the temperature maximums line up with one another around the stoichiometric concentration, $\phi=1$.

The validity of using a constant CAFT of 1200K was also assessed for acetic acid. Figure 21 shows temperature contours at 1250, 1500, 1750, 2500, and 2750 K, using data from the seven fuel/nitrogen/oxygen line calculated. As this figure illustrates, in this case, 1200K would be a conservative measure for predicting the flammable region of acetic acid because the CAFTs associated with the UFL fall below the 1250K contour line. The LFL also clearly lies within the 1250 K line.

The Lower Carboxylic Acids

The product speciation for the remaining five carboxylic acids, namely formic, propionic, butyric, valeric, and hexanoic acid, all displayed the same trends discussed above for acetic acid. In fact, the trends seen for the carboxylic acids were very similar to those seen for the alkanes. Using a constant value of 1200K as a conservative approach for predicting the flammability region for the remaining carboxylic acids, however, was shown to be not conservative. For formic acid assuming a constant CAFT of 1200 K is not conservative for nitrogen compositions greater than 0.68. Further, for propionic, butyric, valeric, and hexanoic acid respectively, 1200 K is not conservative in the upper flammability region; the UFL clearly lies outside the 1250 line. Table 1 shows that the required temperature is far below 1000K.

3.2 Esters

The trends for methyl acetate are shown in the accompanying Figures 22-28. The same oxygen-nitrogen binary mixtures as the carboxylic acids were used. As observed for the alkanes and acids, the speciation plots revealed a small region of complete combustion that diminished as pure oxygen is approached. The flammability triangle shown in Figure 29 shows lines indicating the complete combustion limit, maxima for carbon monoxide production and water formation and the onset of methane formation. The CAFT contours shown in Figure 30 show the same trends as previously observed. When translated to the triangular diagram in Figure 31, it can be seen that the 1250K contour under-predicts the UFL and the upper flammability zone, following the same trend as in Figure 11 for ethane.

IV. Conclusions

Both products and adiabatic flame temperatures were calculated using a Gibbs Free Energy minimization model assuming chemical equilibrium across the flammable region. It was shown that any particular CAFT value chosen may not be a conservative predictor of the flammable region even within the alkane, carboxylic acid and ester chemical groups.

Table 1 shows that although the CAFT at the LFL is fairly constant the CAFT for the UFL in both air and pure oxygen vary with carbon number and the presence of oxygen on the molecule. Although the onset of methane production (i.e., cracking) appears to be an alternative conservative measure of flammability for oxygen concentrations from the LOC to air, like the fixed CAFT temperature, it under-predicts the flammable region for several species.

Therefore the results of this investigation suggest that care should be taken to validate the extent of the upper flammable region with experiment particularly in the region from air to pure oxygen, as no single CAFT value can be chosen to provide a prediction that is both accurate and conservative.

V. References

[1] Crowl, Daniel A.; Louvar, Joseph A. *Chemical Process Safety: Fundamentals with Applications*. Prentice Hall, Inc, Upper Saddle River, NJ, 2002, 243.

[2] Mashuga, Chad V.; Crowl, Daniel. *Process Safety Progress*. **1998**, *17(3)*, 176-183.

[3] Razus, Domnina; Molnarne, Maria; Fuss, Oliver. *Chemical Engineering and Processing*. **2004**, *43*, 775- 784.

[4] Balzhiser, Richard E.; Samuals, Michael R.; Eliassen, John D.; *Chemical Engineering Thermodynamics*. Prentice-Hall, Inc, Englewood Cliffs, NJ, 1972, 516-517.

[5] Mashuga, Chad V.; Crowl, Daniel. *Process Safety Progress*. **1999**, *18*, 124-134.

[6] Zabetakis, Michael G.; "Flammability Characteristics of Combustible Gases and Vapors", United States Department of the Interior, Bulletin 627, Bureau of Mines (1965)

Rowley, R.L., Wilding, W. V., Oscarson, J. L., Yang, Y., Zundel, N. A., Daubert, T. E., and Danner, R. P., DIPPR Data Compilation of Pure Compound Properties, Design Institute for Physical Properties, AICHE, New York, NY (2004).

[7] Mashuga, Chad V. "Determination of the Combustion Behavior For Pure Components and Mixtures Using a 20-Liter Sphere." Ph.D. dissertation, Michigan Technological University, 1999.

[8] Melhem, G.A., "A Detailed Method for Estimating Mixture Flammability Limits Using Chemical Equilibrium" Process Safety Progress, 16, 203-218 (1997)

[9] Glassman, Irvin, *Combustion, 3rd ed.,*Academic Press, New York (1996)

Appendix

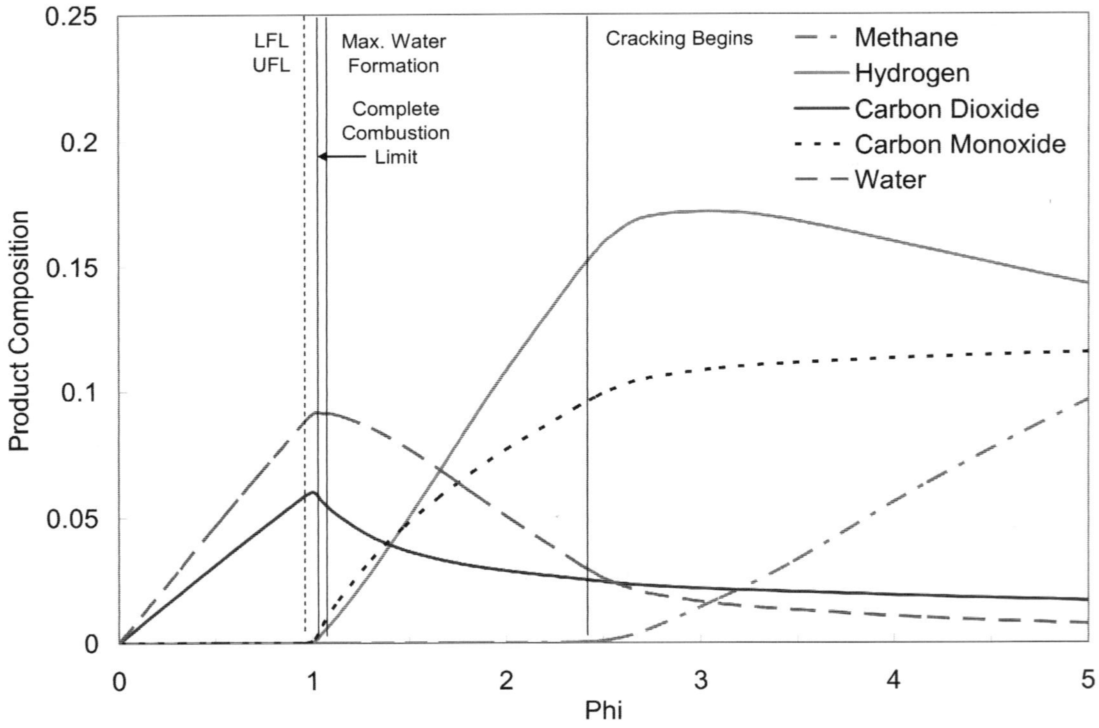

Figure 2. LOC – Ethane

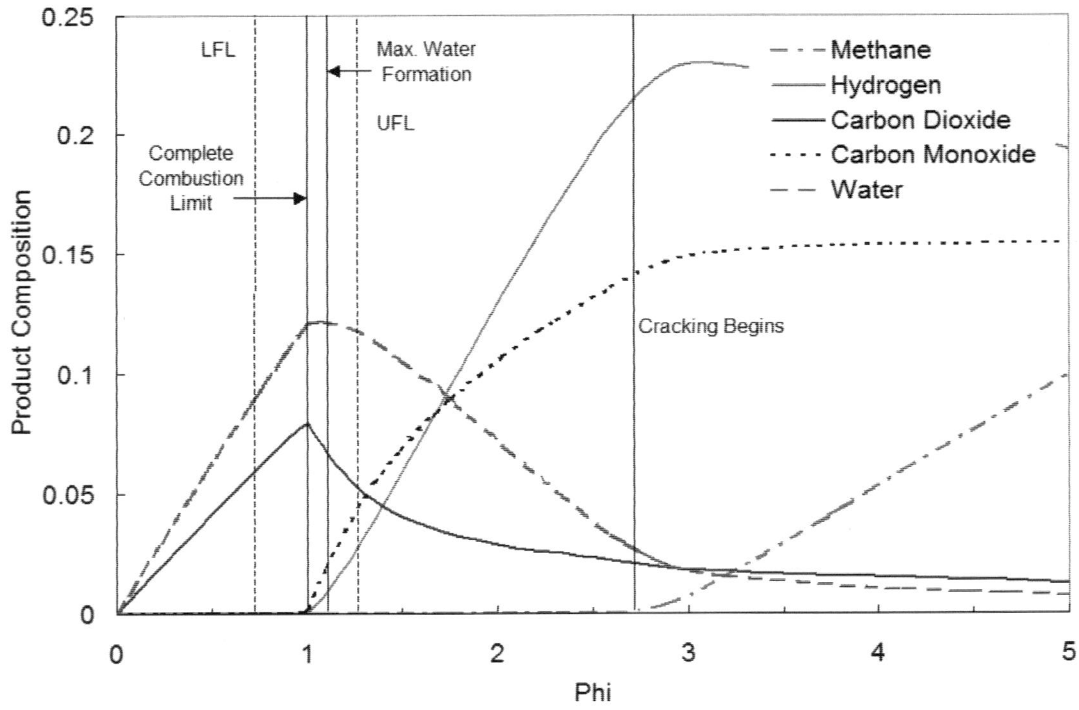

Figure 3. 15% Oxygen – Ethane

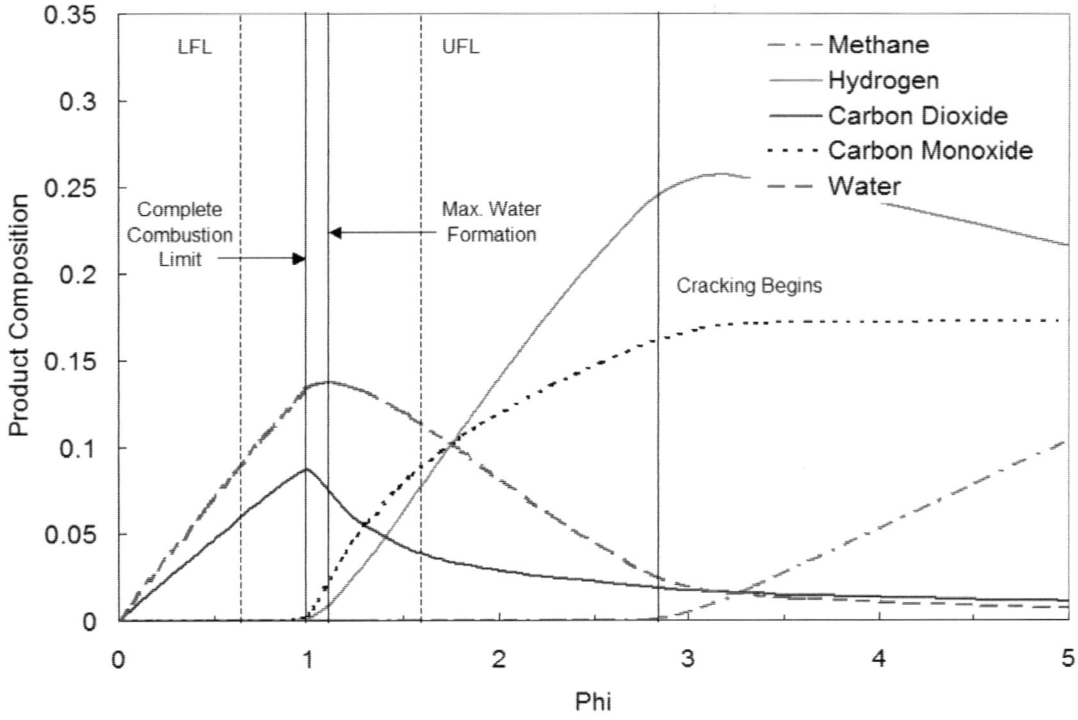

Figure 4. 17% Oxygen - Ethane

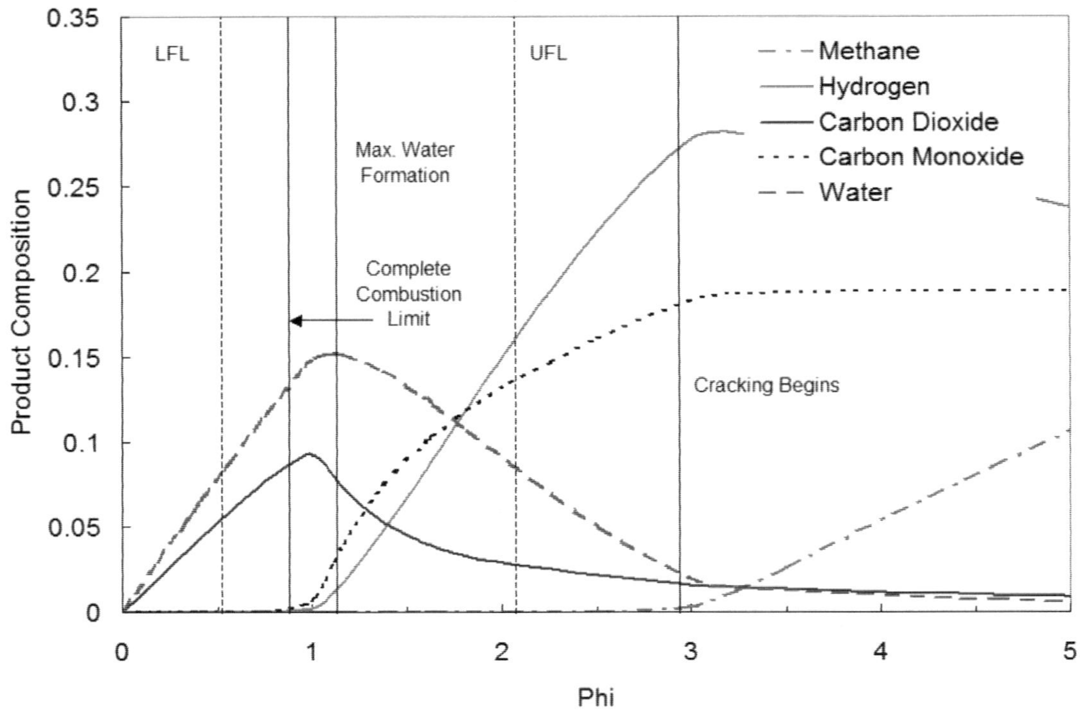

Figure 5. 19% Oxygen – Ethane

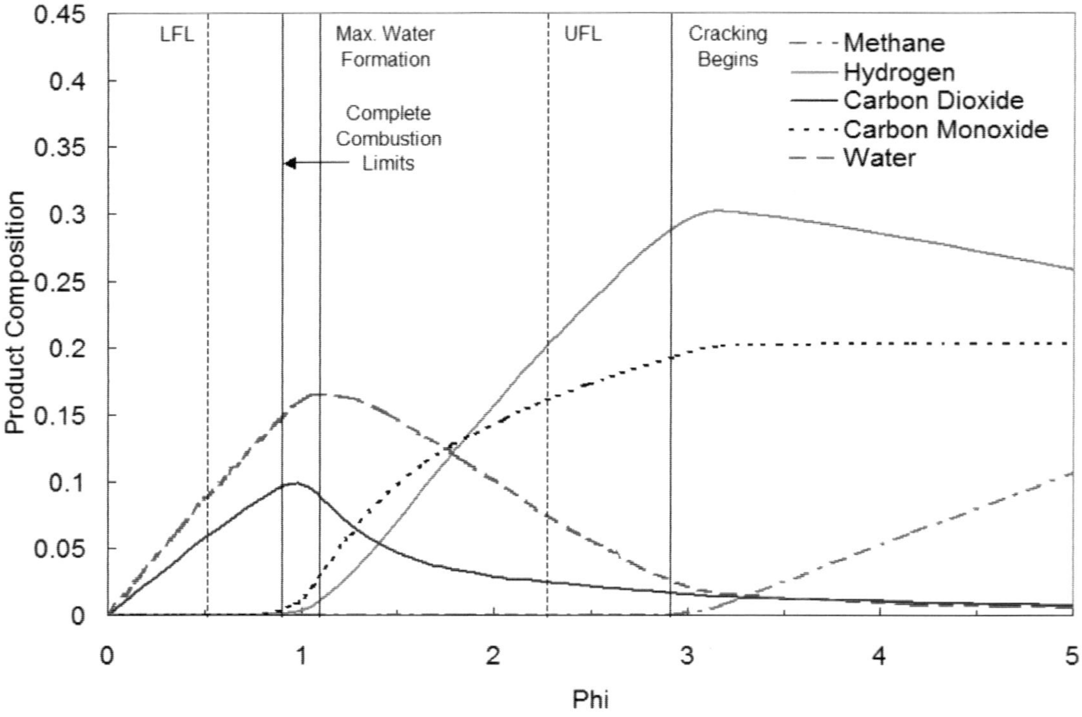

Figure 6. Air – Ethane

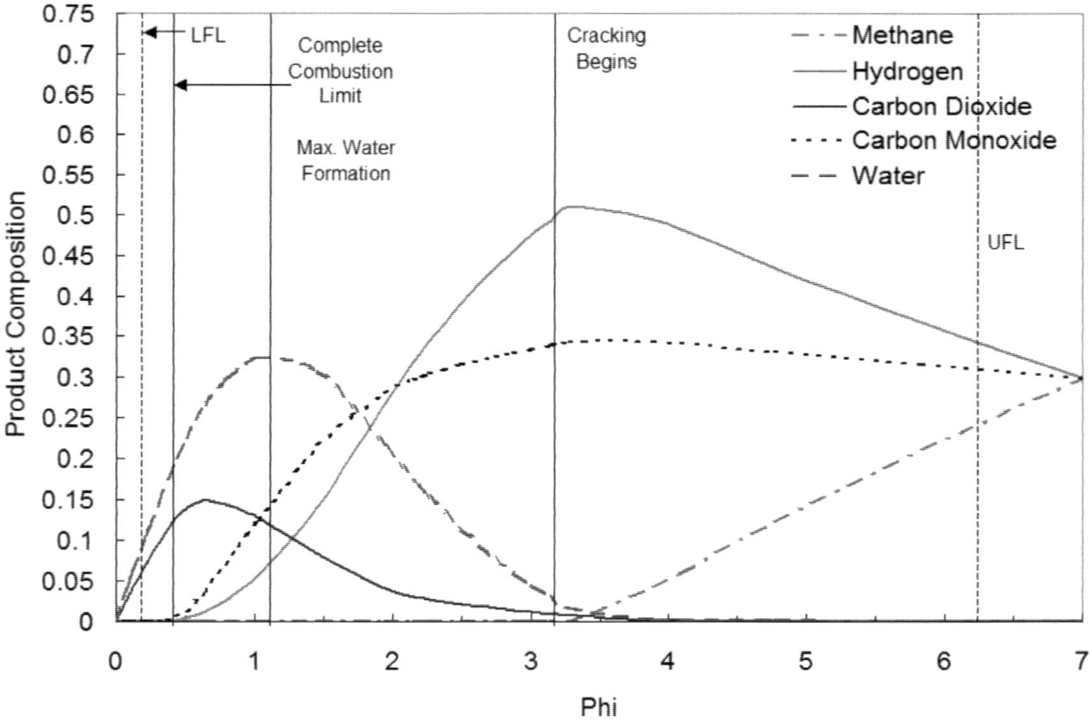

Figure 7. 60% Oxygen – Ethane

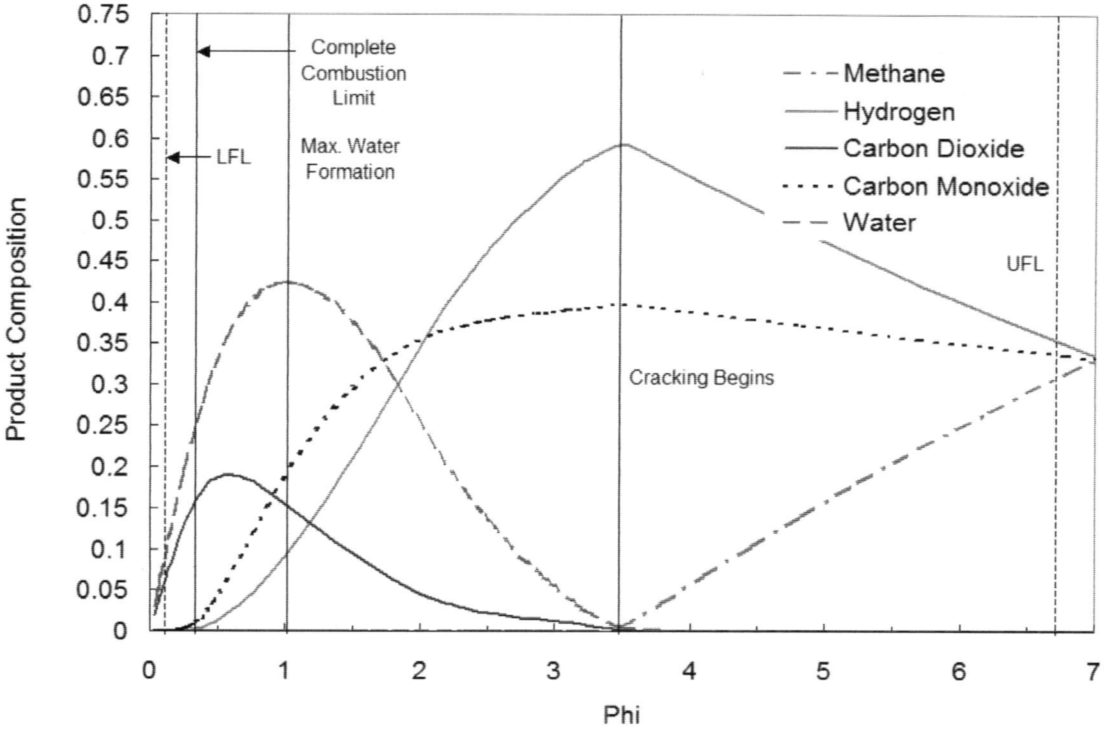

Figure 8. 100% Oxygen – Ethane

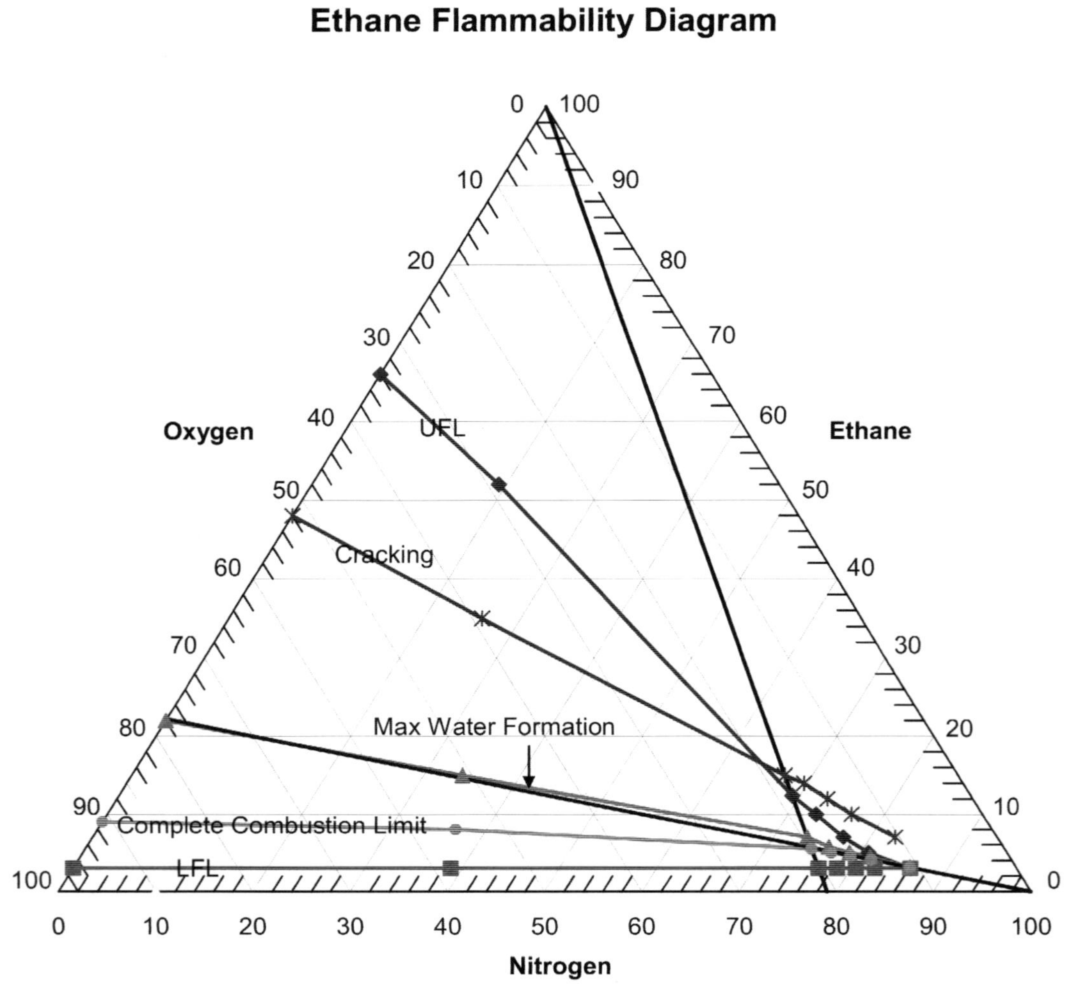

Figure 9. Product Speciation – Ethane

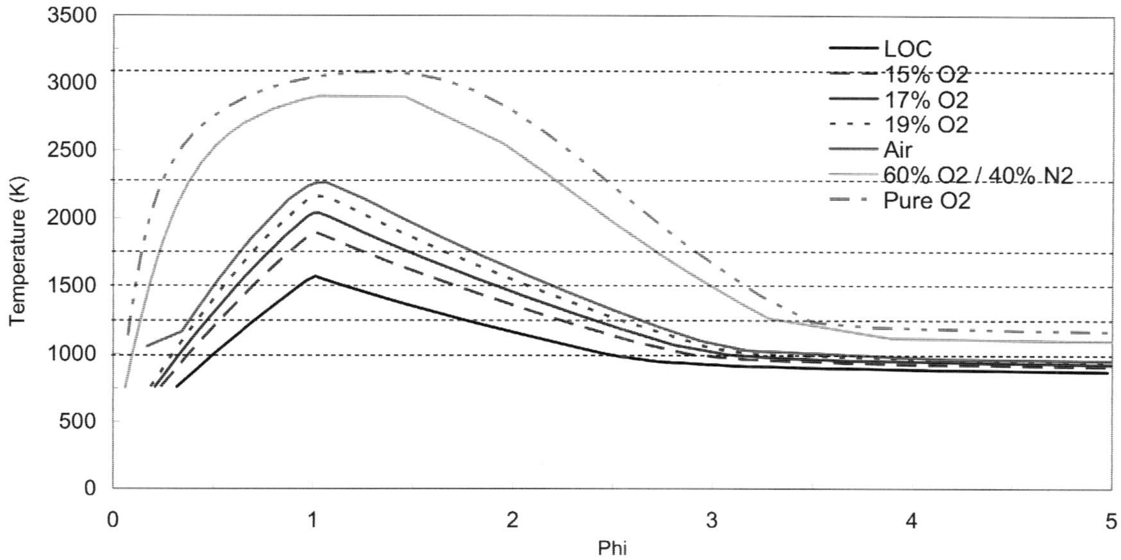

Figure 10. CAFT versus Phi - Ethane

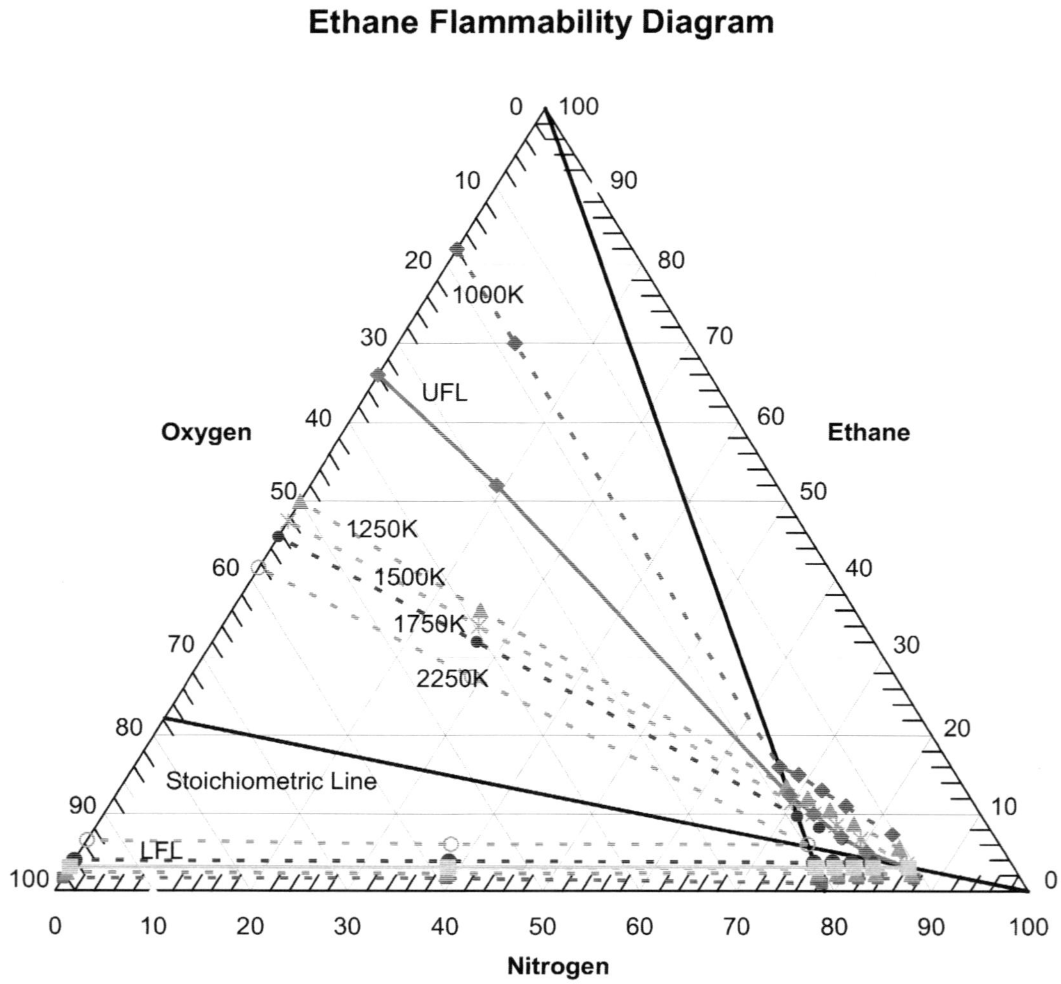

Figure 11. Temperature Contours with UFL and LFL – Ethane

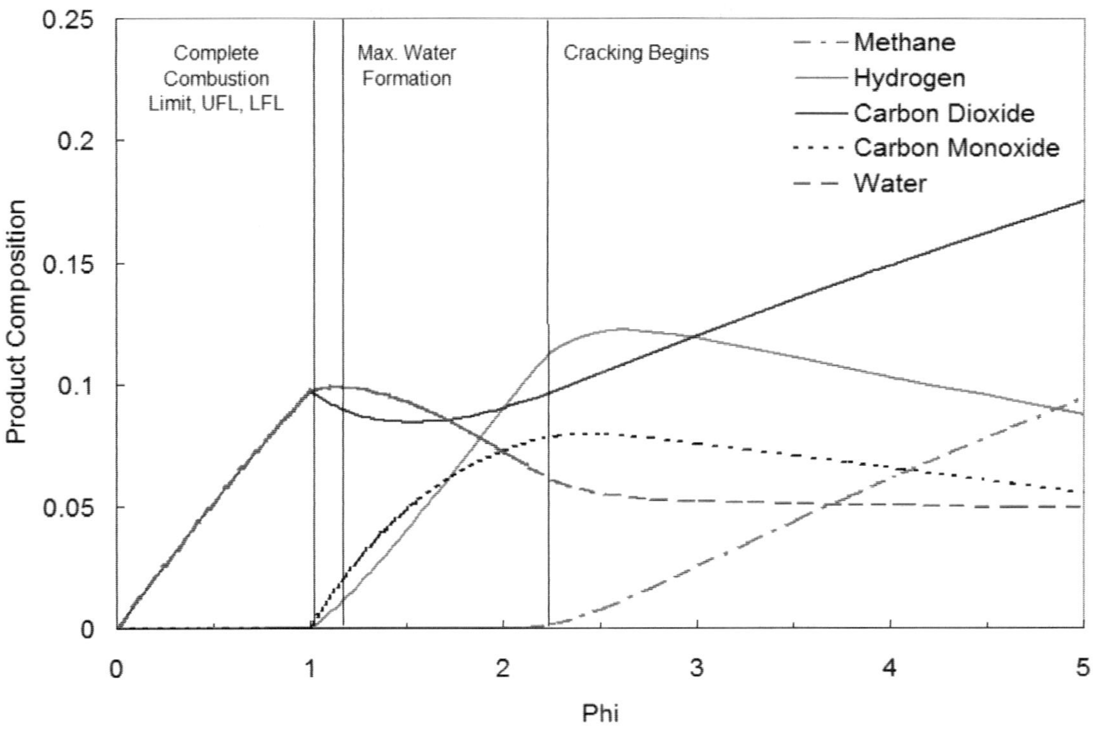

Figure 12. LOC – Acetic Acid

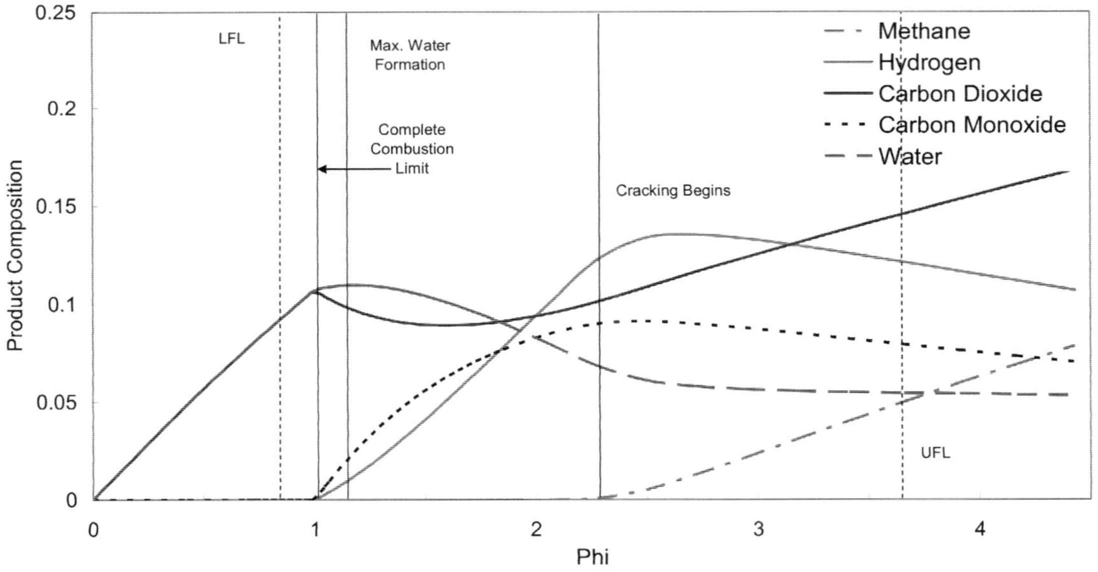

Figure 13. 12% Oxygen – Acetic Acid

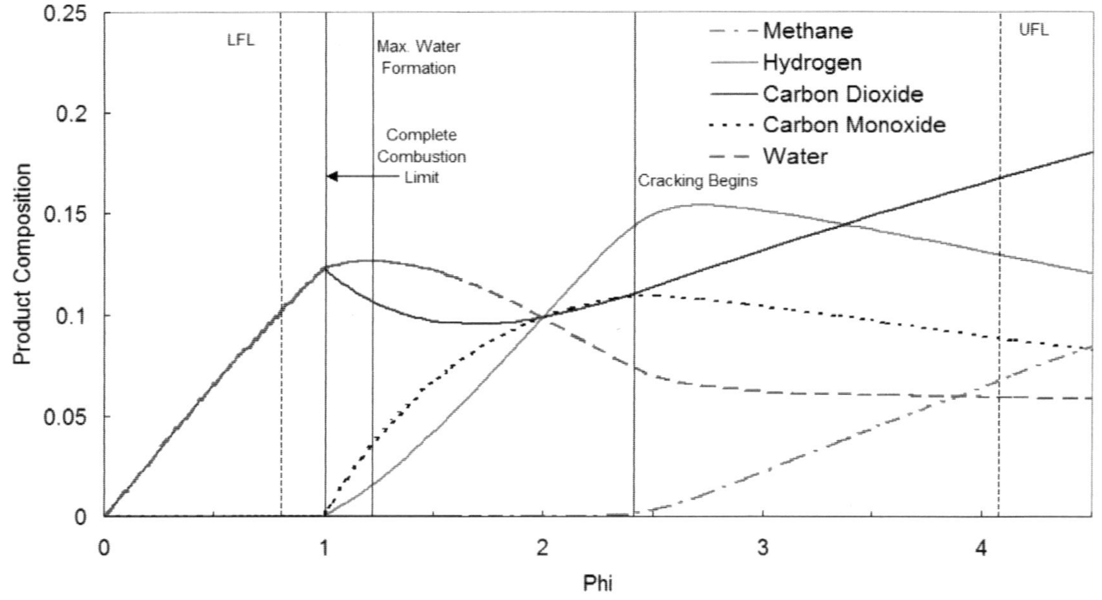

Figure 14. 14% Oxygen – Acetic Acid

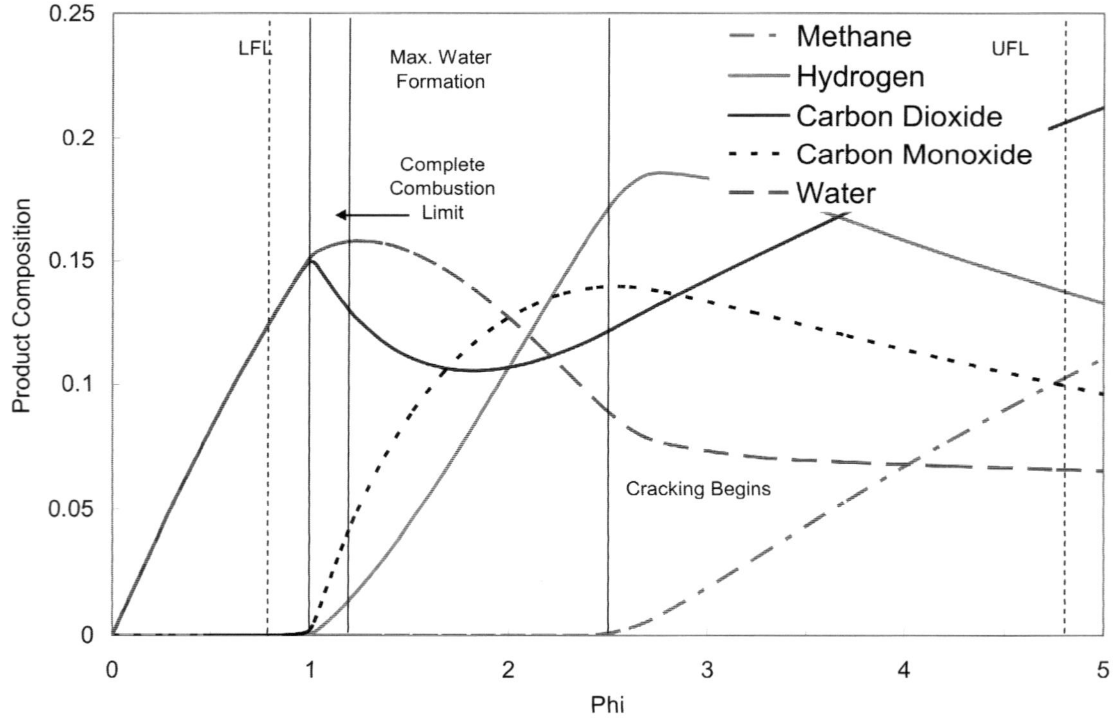

Figure 15. 18% Oxygen – Acetic Acid

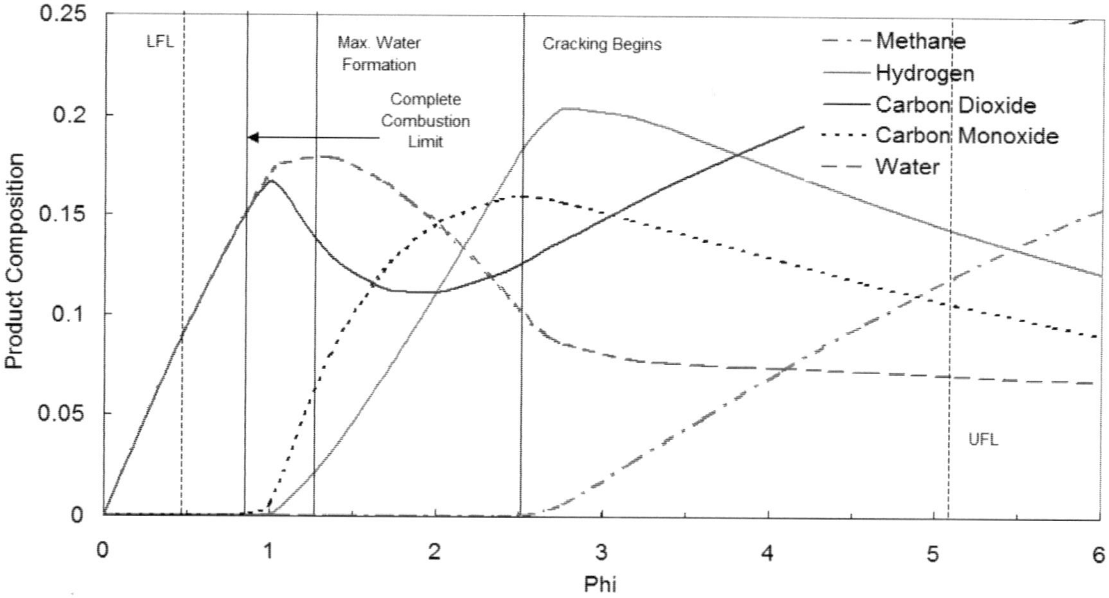

Figure 16. Air – Acetic Acid

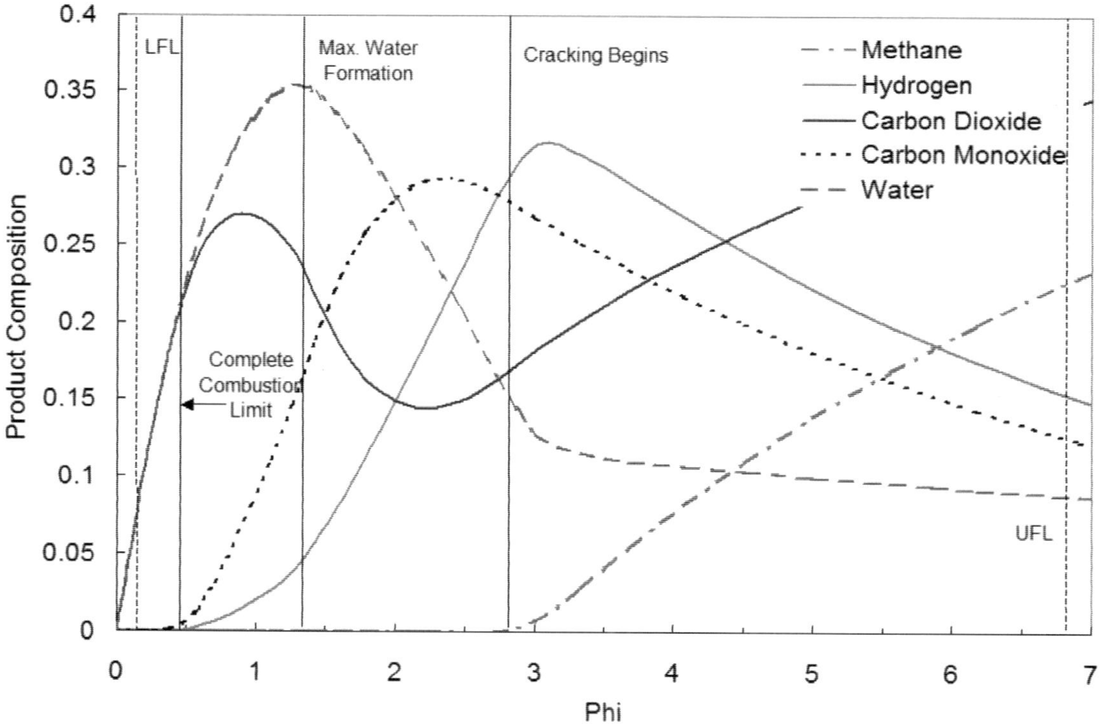

Figure 17. 60% Oxygen – Acetic Acid

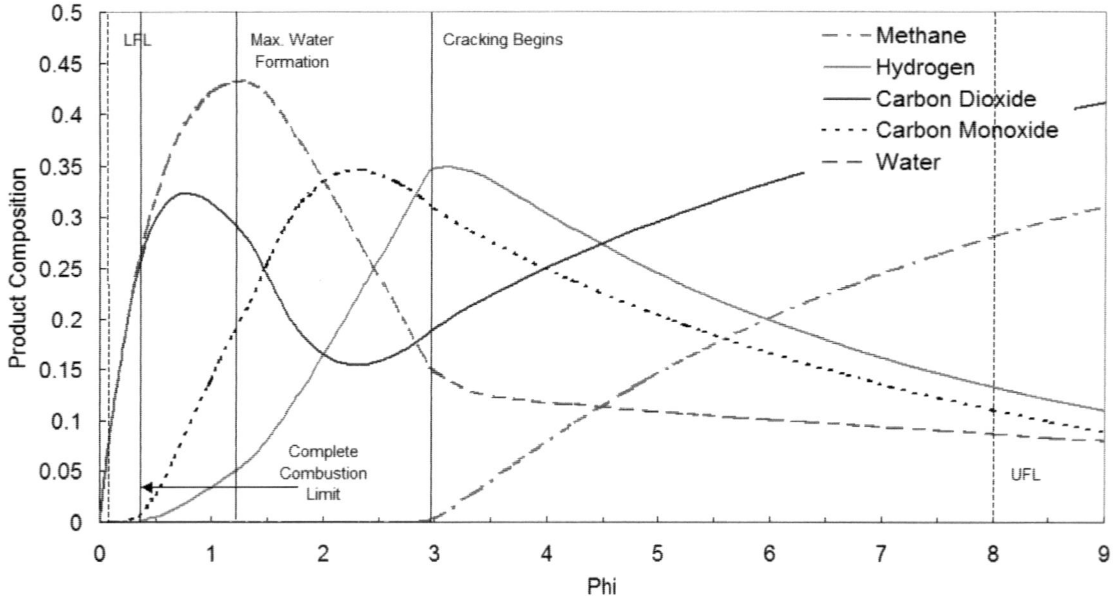

Figure 18. 100% Oxygen – Acetic Acid

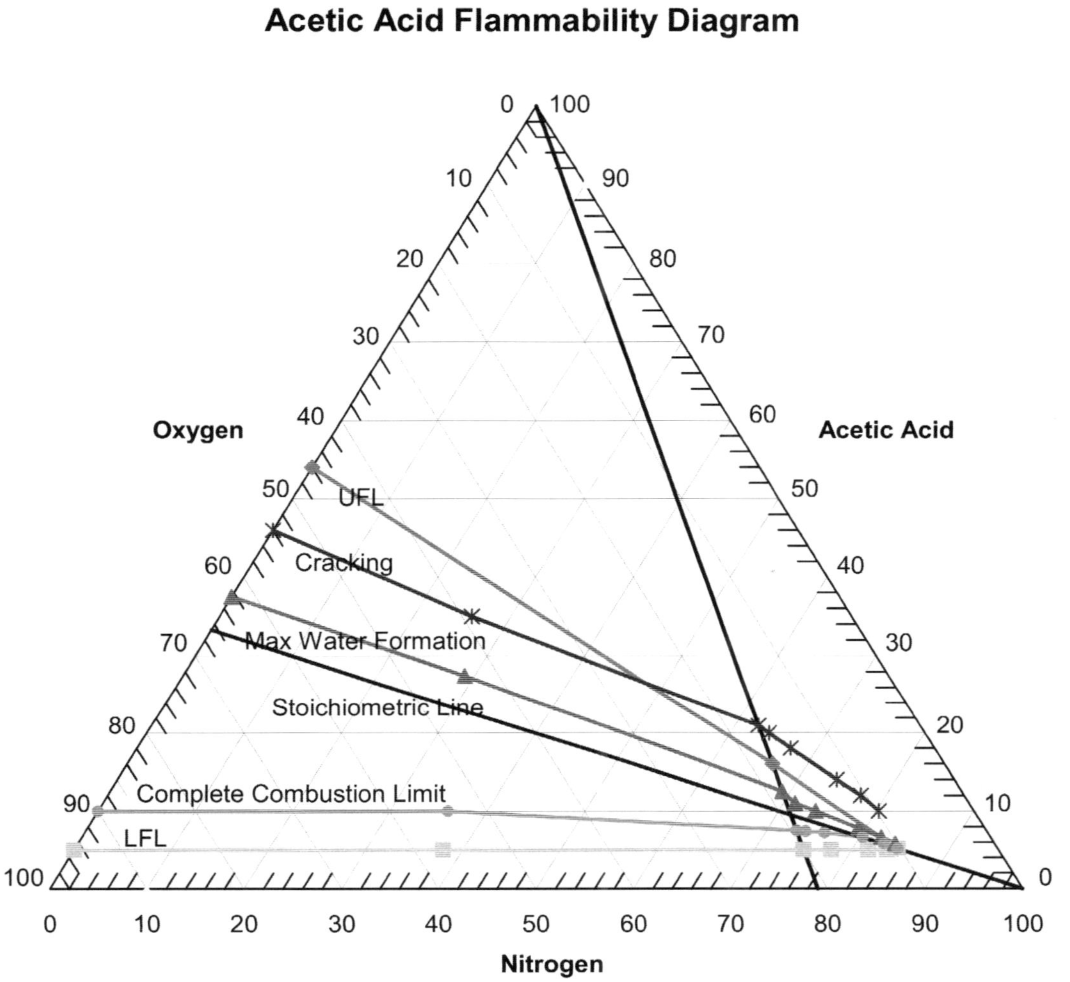

Figure 19. Product Speciation – AcOH

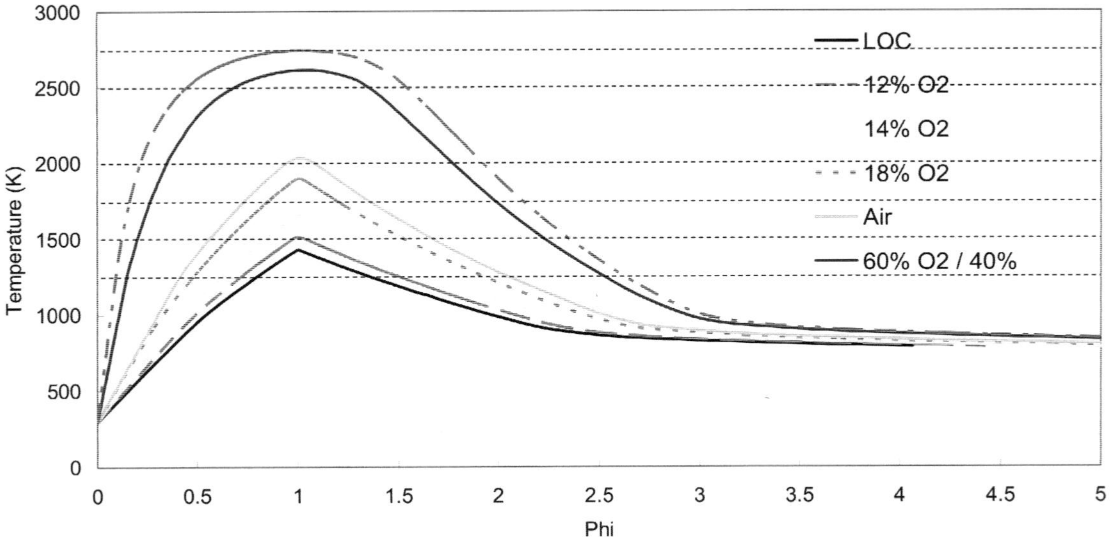

Figure 20. CAFT versus Phi – Acetic Acid

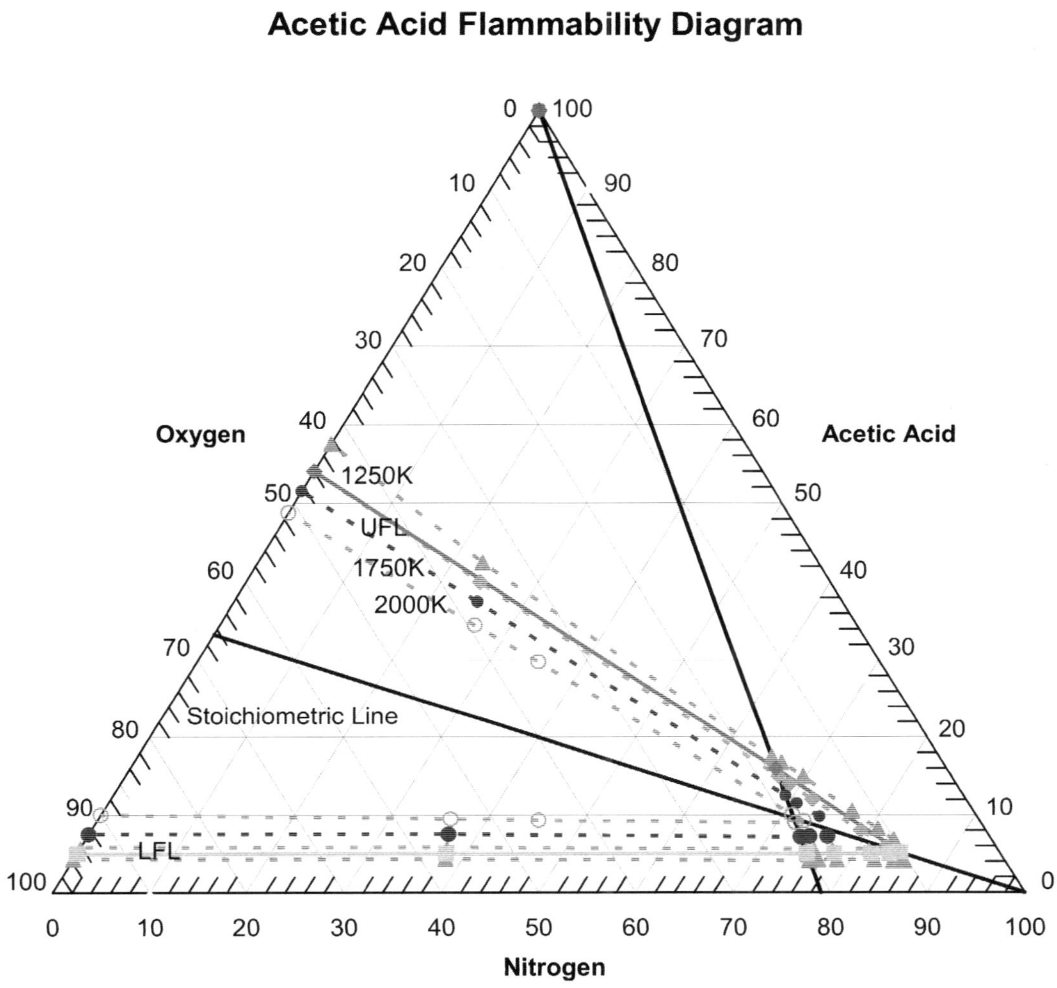

Figure 21. Temperature Contours with UFL and LFL – AcOH

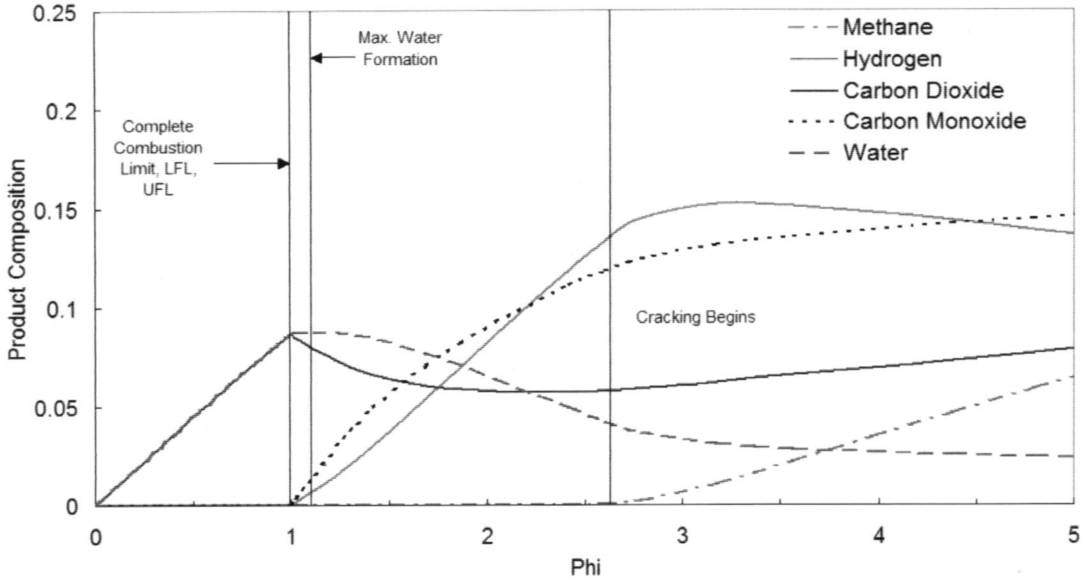

Figure 22. LOC – Methyl Acetate

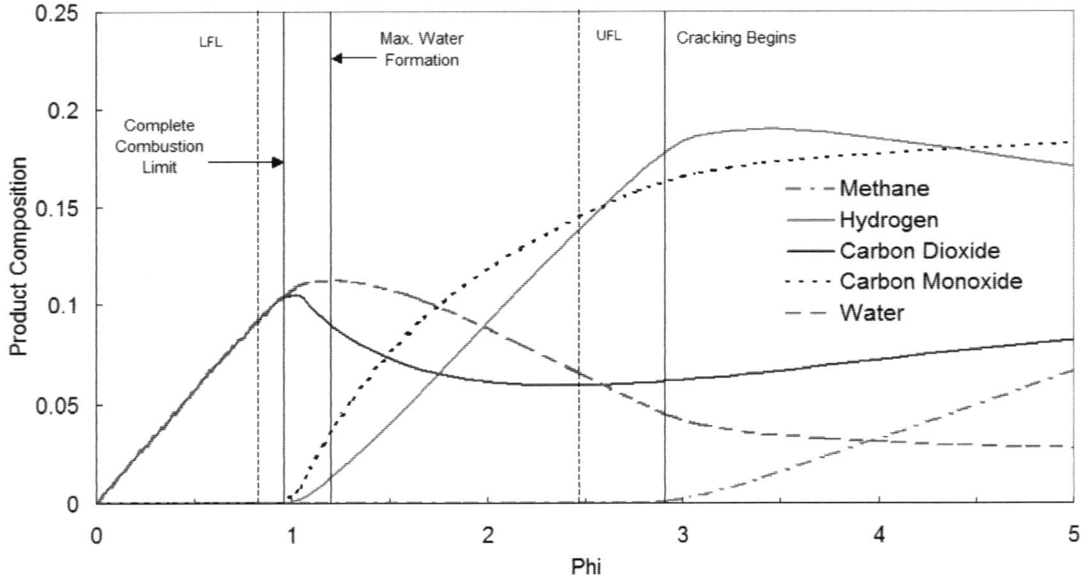

Figure 23. 14% Oxygen – Methyl Acetate

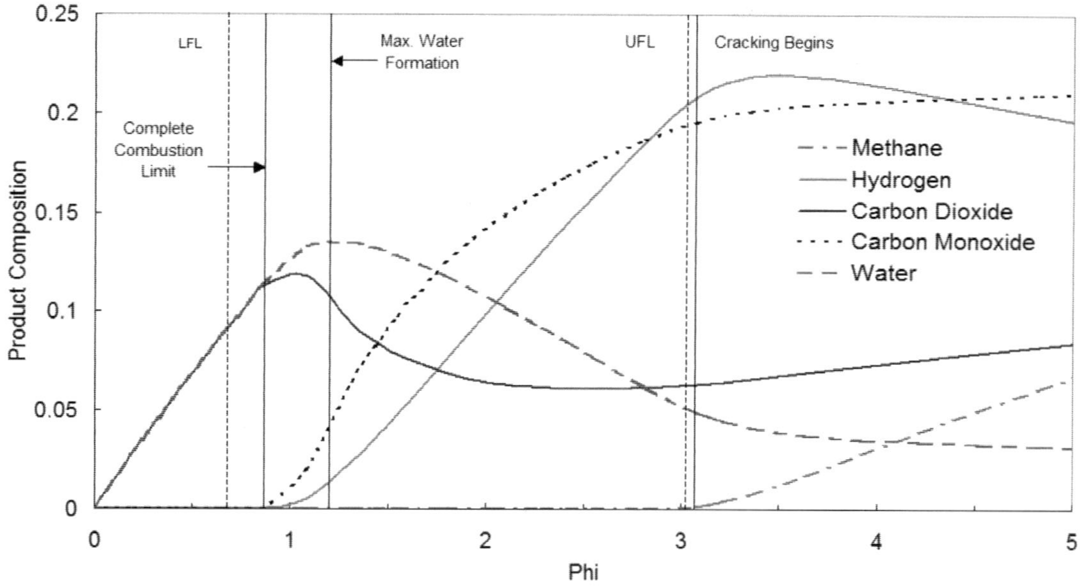

Figure 24. 17% Oxygen – Methyl Acetate

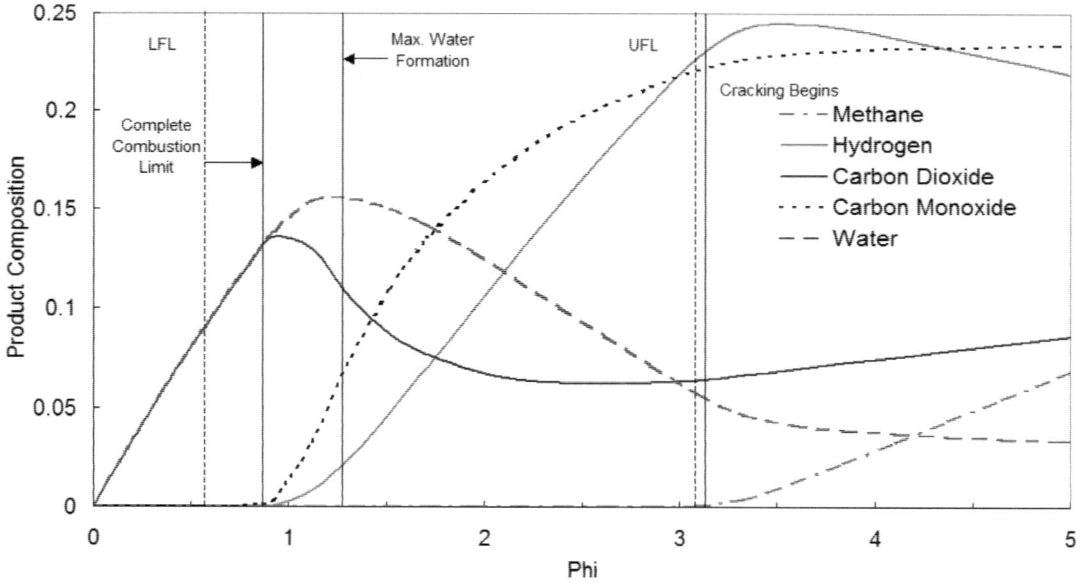

Figure 25. 20% Oxygen – Methyl Acetate

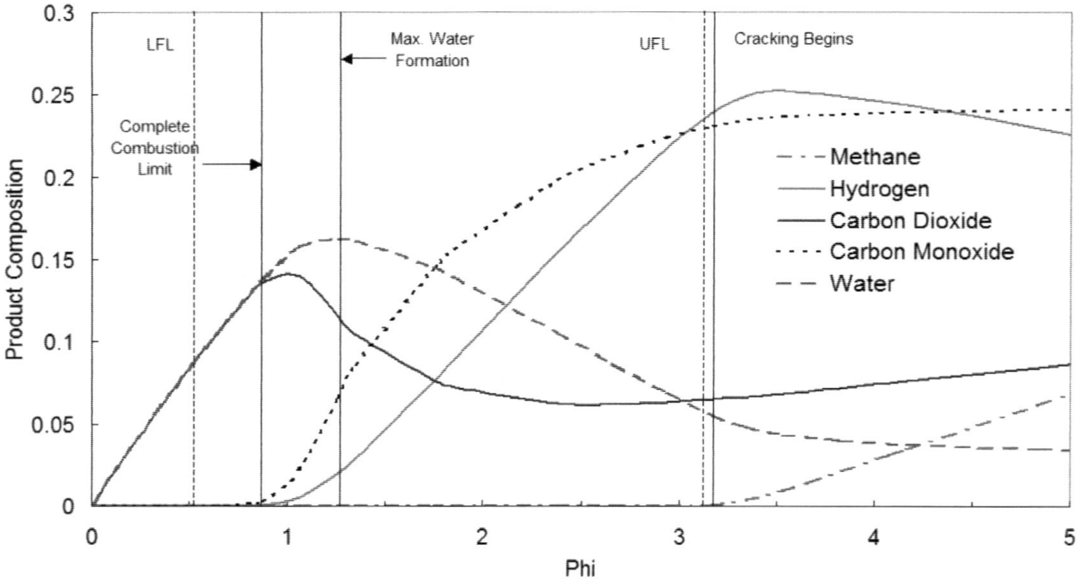

Figure 26. Air – Methyl Acetate

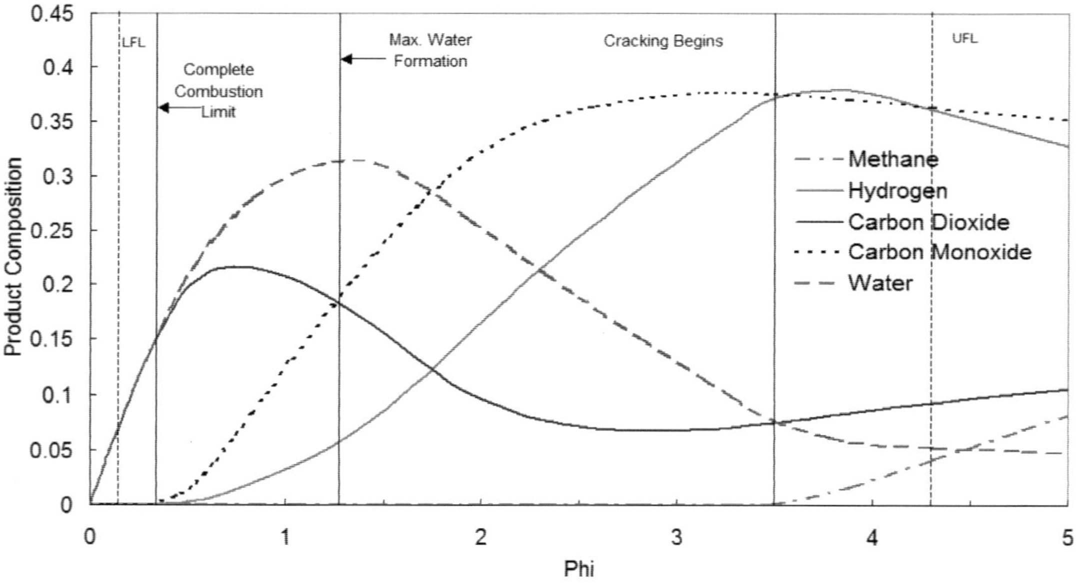

Figure 27. 60% Oxygen- Methyl Acetate

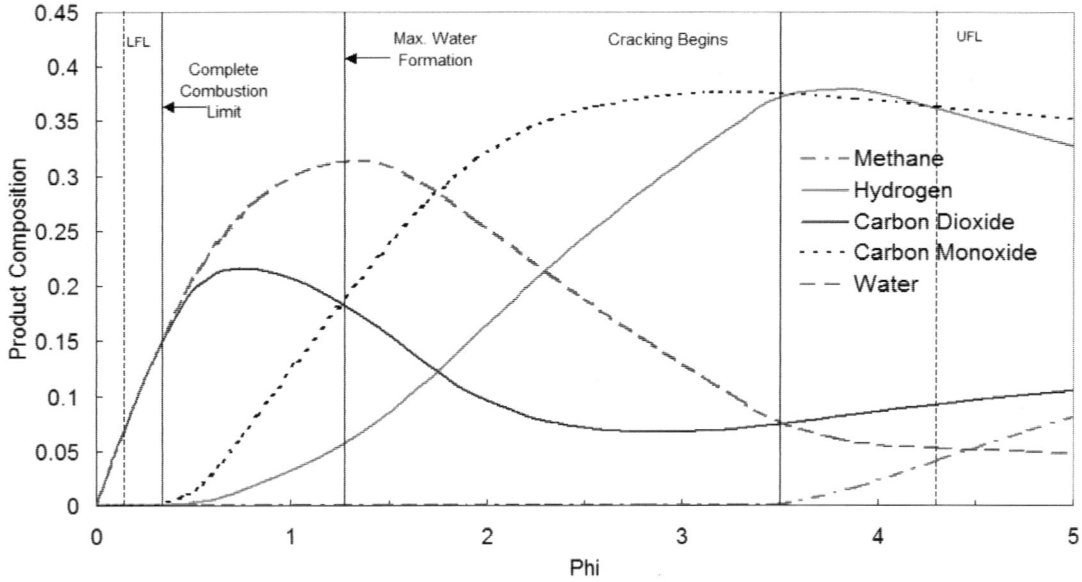

Figure 28. 100% Oxygen – Methyl Acetate

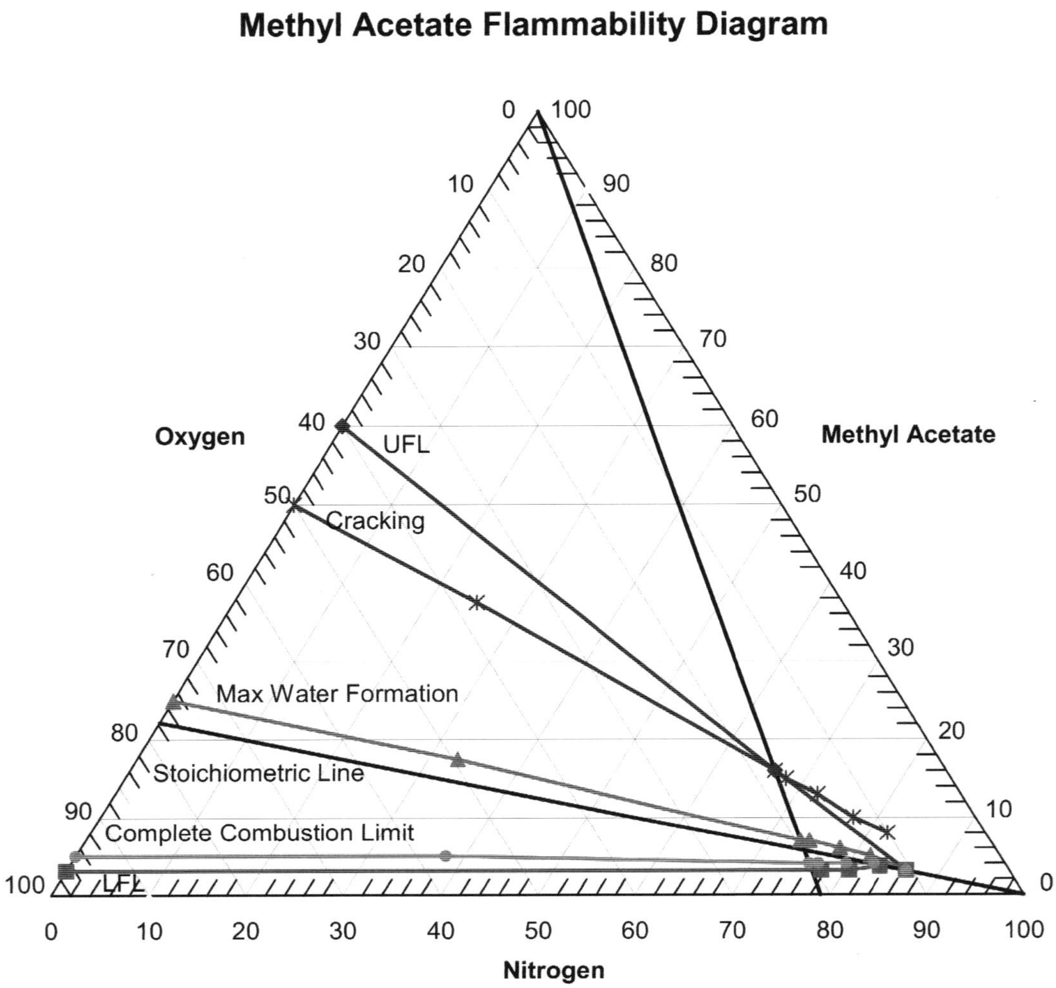

Figure 29. Product Speciation – Methyl Acetate

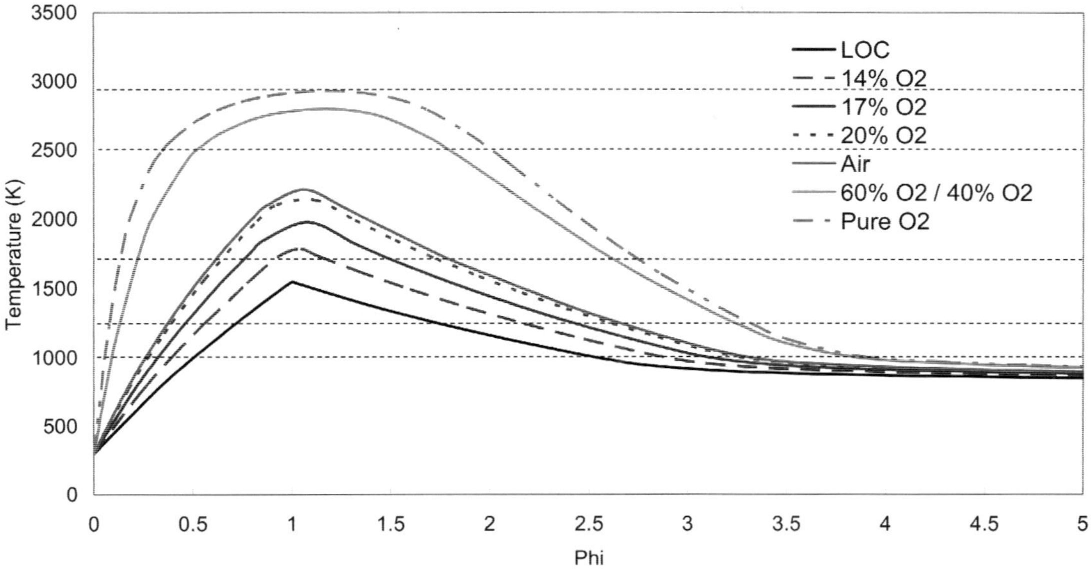

Figure 30. CAFT versus Phi – Methyl Acetate

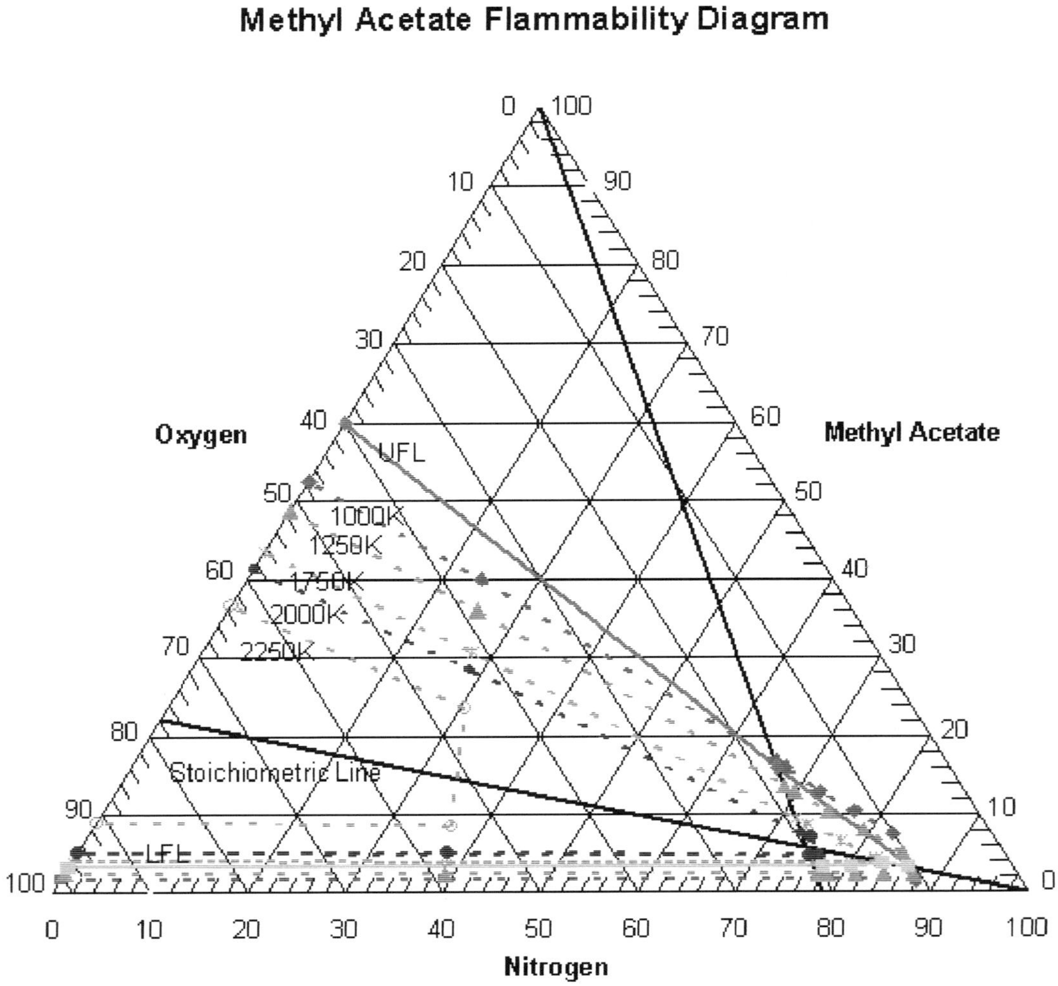

Figure 31. Temperature Contours with UFL and LFL – Methyl Acetate

	Calculated Adiabatic Flame Temperature			
	LFL - AIR *K*	UFL - AIR *K*	LFL - PURE O2 *K*	UFL - PURE O2 *K*
Alkanes				
Ethane	1530	1392	1490	1141
Propane	1477	1331	1438	1279
Butane	1448	1224	1411	1283
Pentane	1511	1032	1472	1332
Hexane	1627	1033	1562	1015
Carboxylic Acids				
Formic Acid	1096	1372	1185	1479
Acetic Acid	1403	1397	1370	1501
Propionic Acid	1467	950	1366	915
Butyric Acid	1580	898	1391	873
Valeric Acid	1484	940	1320	919
Hexanoic Acid	1476	939	1514	975
Acetates				
Methyl Acetate	1534	1042	1528	927

Table 1. Calculated Adiabatic Flame Temperatures at the UFL and LFL in Air and O_2

Hydrogen Safety Review Panel: Shaping Safety Awareness

Steven C. Weiner[1], **Bruce Kinzey**
Pacific Northwest National Laboratory
Washington, DC

Edward G. Skolnik
Energetics, Inc.
Washington, DC

ABSTRACT

An essential element for realizing the "hydrogen economy" is safe operation in all of its aspects – from hydrogen production through storage, distribution and use; from research, development and demonstration to commercialization. As such, safety is given paramount importance in all facets of the research, development and demonstration (RD&D) program of the U.S. Department of Energy's (DOE) Hydrogen, Fuel Cells and Infrastructure Technologies (HFCIT) Program Office.

The DOE Hydrogen Safety Review Panel was formed to bring a broad cross-section of expertise from the industrial, government and academic sectors to help ensure the success of the program as a whole. The Panel provides guidance to DOE on safety-related issues and needs, reviews individual DOE-supported projects and their safety plans, and explores ways to bring learnings to broadly benefit the DOE program.

This paper introduces the Hydrogen Safety Review Panel, explores its role in the DOE program, describes how it approaches safety review site visits as one of its functions and shares specific themes and examples of those interactions and lessons learned.

1. INTRODUCTION

Hydrogen has been used in the manufacturing, chemicals, and aerospace industries for nearly a century and has enjoyed a notably high safety record. Experience with and knowledge of safe production, shipping, and use of hydrogen is therefore quite extensive. About 9.5 million tons of hydrogen are currently produced and used annually in the U.S.[2]

One reason underlying the admirable safety record achieved is the use of highly trained staff and well controlled facilities in all aspects of the industry. Hydrogen producers control the material through the point of delivery and frequently beyond, using technology that the company also controls. For example, over-the-road tube trailer and

---

[1] Email: sc.weiner@pnl.gov

[2] *Hydrogen Production Overview*, National Hydrogen Association Fact Sheet Series, August 2004: http://www.hydrogenus.com/H2-Production-Overview-Fact-Sheet.pdf.

cylinder delivery trucks are operated by a gas supplier, or liquid supply is delivered by a merchant fleet from a supplier production site. Hydrogen's use inside the destination facility is controlled and monitored, again employing trained staff, so that the entire life cycle of the hydrogen from production to ultimate use is carefully managed.

The introduction of hydrogen as a commercial energy carrier in the hands of the general public promises a "whole new ballgame," however, with a host of new safety issues. Not the least of these is that the general public receives little if any official training on fuel use, at least with current fuels and fueling technologies. Another issue is that regulatory control over these technologies and products in use by the general public is somewhat limited, for example, once they have left the refueling station. Vehicles may be operated, modified, maintained (or not maintained) in a manner never intended. The mobile environment is one of admittedly harsh duty, including temperature extremes, vibration, shocks, collisions, frequent cycling, and possible long term exposure to the external environment. The expectation that the public would treat future technologies and products differently than they do currently seems unrealistic; therefore future technologies and products are going to have to increasingly rely on technological solutions, rather than regulatory and/or training solutions, to ensure safety through the most challenging of applications.

The hydrogen economy remains decades away, due in part to the physical inertia of our massive energy infrastructure and the large construction effort and economic investment that will be required to alter it. Despite the anticipated long lead time however, safety with hydrogen is as important now if not more so than when the hydrogen economy becomes a reality. The public's trust can be a fragile commodity and any catastrophic accidents may further delay or even preclude hydrogen's intended implementation.

The ability to produce, store, and use hydrogen in a manner perceived as safe by the general public will be essential to its success as a commercial energy carrier. Furthermore, when accidents do occur, as they frequently do with conventional fuels and vehicles, the demonstrated, reliable ability to respond with means that quickly contain and control damage and prevent catastrophe will be equally critical for the hydrogen economy to proceed.

2. THE HYDROGEN SAFETY REVIEW PANEL

The U.S. Department of Energy formed the Hydrogen Safety Review Panel to help the Hydrogen, Fuel Cells and Infrastructure Technologies Program Office develop and implement the practices and procedures that will ensure safety in the operation, handling, and use of hydrogen and hydrogen systems for all DOE projects. Emphasizing safety in its own program and utilizing these practices and lessons learned, the Program Office helps promote the safe use of hydrogen throughout the emerging hydrogen economy.

The Panel, formed in December 2003, brings a broad cross-section of experience upon which to build for the success of the DOE program as a whole. The experience resides in industrial hydrogen production/supply, government R&D and use, industrial liability and facility insurance, risk analysis, accident investigation and fire protection and is backed up with technical support from the national laboratories. The Panel members are noted in Table 1; Edward Skolnik, Energetics, Inc., and Steven Weiner,

Pacific Northwest National Laboratory, provide technical and management support to the Panel whose activities are focused on several key objectives:

- Provide expertise and guidance to the DOE and assist with identifying areas of additional research necessary to fill safety information gaps
- Integrate safety procedures into all DOE project-funding procurements to ensure that all projects incorporate hydrogen safety requirements.
- Publish a handbook of "Best Management Practices for Safety" (2007)

All new DOE-funded hydrogen projects are now required to submit a safety plan as an early project deliverable report. Safety plans, utilizing a DOE and Panel-developed guidance as a resource,[3] focus on the identification and analysis of safety vulnerabilities, mitigating the associated risks, ensuring an effective communications plan and other information. The Panel reviews each safety plan and makes any needed recommendations for improvement. Some subset of projects is then selected for more in-depth review.

Addison Bain, Chair	NASA (ret.)
Carol Bailey	Sentech
Harold Beeson	NASA White Sands
William Doerr	FM Global Research
Don Frikken	Becht Engineering
James Hansel	Air Products and Chemicals
Richard Kallman	City of Santa Fe Springs, CA
Harold Phillippi	ExxonMobil Research and Engineering
Jesse Schneider	Daimler Chrysler
R. Rhoads Stephenson	Jet Propulsion Laboratory (ret.)
Robert Zalosh	Worcester Polytechnic Institute

Table 1. U.S. Department of Energy Hydrogen Safety Review Panel

Through utilization of the Hydrogen Safety Review Panel, DOE hopes to capture and capitalize on some of the established safety experience gained from the long histories with hydrogen in the industrial and aerospace arenas. This paper explores some of the lessons learned and safety discussion from the safety review site visits conducted during the first year of the Panel's existence.

3. SAFETY REVIEW SITE VISITS

One important effort of the Panel is conducting site visits to perform safety reviews of DOE-funded projects that cover R&D in hydrogen production, storage and fuel cells, as well as demonstration projects that include hydrogen refueling stations. Projects are performed by companies of varying size, DOE national laboratories, and academic

[3] Guidance for Safety Aspects of Proposed Hydrogen Projects, U.S. Department of Energy, August 2004: http://www.eere.energy.gov/hydrogenandfuelcells/pdfs/safety_guidance.pdf.

institutions and typically involve a number of collaborators. Site visits are an opportunity for the contractor/project team and the Panel review team to:

- Share and discuss new insights that bear on safety.
- Address project-specific safety issues.
- Identify project-specific findings that can have a broader benefit in the DOE program

Typically the review team consists of 3-4 members from the Panel and its support staff.

3.1 A Protocol Helps Shape the Site Visit and Discussion

Site visits are intended to focus on engagement and discussion between the review team and the contractor/project team rather than as an audit or investigative exercise. Use of a site visit protocol helps achieve that intended purpose.

Once the site visit has been scheduled, the contractor project manager (or principal investigator) provides the relevant program review presentation, annual progress report, safety plan and other literature, as applicable, to the safety review team. Background material on the Hydrogen Safety Review Panel and the two-page protocol document is provided to the contractor. An agenda is established. The contractor is asked to provide the following as part of the agenda:

- Overview of organization-wide safety policies and procedures and how these are implemented/managed for a given facility down to the project and staff levels
- Project overview
- Laboratory/facility tour of relevance to the project that includes demonstrations, if possible

Topic questions for discussion, having been submitted to the contractor/project team in advance of the site visit, are discussed. The topic questions are tailored for each site visit based in part on the provided literature and input is sought from the HFCIT staff teams on safety issues they wish to have addressed; typically all reviews will address the following items:

- What safety planning approaches and methodologies are being applied?
    - Identification and analysis of safety vulnerabilities
    - Risk mitigation plans: standard operating plans, management of change, equipment maintenance and integrity, measuring and monitoring safety performance, training
    - Communication plans for safety reviews and "incidents" reporting
- What codes and standards/regulations apply to this system, how are they being applied and what is the extent/nature of interaction with local codes and standards officials?
- How are hydrogen safety issues being addressed?
    - Hydrogen leak prevention, detection, containment/exhaust

- o Hydrogen pressure
- o Ignition sources: identification and control
- o Materials: material properties, interaction/compatibility with hydrogen and hydrogen-containing materials

A Q&A session helps address specific questions from members of the review team. The contractor/project team is also encouraged to raise any safety issues and questions which they wish to discuss with the review team. Lessons learned from the project that relate to hydrogen safety are discussed. The safety review team also seeks input from the contractor/project team on the protocol/format used for the site visit meeting and ideas for improving the value of other such site visit meetings.

3.2 Reports Document the Discussion, Learnings and Actions

At the conclusion of the safety review site visit, two reports are prepared by the review team: (1) Site Visit Preliminary Report and (2) Safety Evaluation Report.

The Site Visit Preliminary Report (1-2 pages) is submitted by the review team to the HFCIT Program Office within 14 days of the site visit to ensure the timely dissemination of the meeting outcomes. This report briefly summarizes major areas of discussion, recommendations, lessons learned, etc. Safety-related action items and tentative recommendations from the meeting are reported even though actions, analysis and response may not be completed until after this report is submitted.

The Safety Evaluation Report is the comprehensive report of the safety review that is submitted to the Program Office on behalf of the Panel, describing the conduct of the site visit meeting, safety issues of concern, recommendations, lessons learned, etc. consistent with the purposes noted above. The draft report is sent to the contractor for comments, corrections and clarifications. The contractor's response to specific observations, questions and recommendations is included in the final report. This important feature ensures that the review team is providing DOE with the perspective of all site visit meeting participants. Regardless, the review team and the Panel retain the responsibility for making project-related safety recommendations as it feels appropriate.

The contractor is also asked to provide as much detail as is warranted on "lessons learned" to broadly benefit the DOE program. The Safety Evaluation Report is submitted to HFCIT and a copy is provided to the contractor.

4. EMERGING THEMES AND LEARNINGS FROM SAFETY REVIEWS

One excellent method for enabling the dissemination of hydrogen safety information among the hydrogen community and existing and potential hydrogen suppliers and users is the use of "lessons learned" and recommendations stemming from safety reviews. Over the course of the initial set of safety reviews conducted by the Panel, a few important themes have already emerged for safety planning, practices and implementation in the DOE-funded program. Issues, learnings and observations are categorized as follows:

- Design and assembly of equipment
- Operating procedures
- Equipment maintenance
- Hydrogen storage materials
- Other issues

Although not all-inclusive, the following sections reflect the meeting discussions of safety aspects of projects during the site visits.

4.1 Design and Assembly of Equipment

It goes without saying that if one designs with safety in mind, it will save time, money, and perhaps a failed system or project later on. Thus, it is always advisable to talk to safety representatives/officers both internal to the institution and in the community (e.g., fire marshals, local safety officials, etc.). The need for external involvement usually is proportional to the size of the hydrogen system, the amount of hydrogen being produced, stored, and/or used and the location of the equipment relative to other facilities and buildings. In addition, designing with versatility in mind allows one to expand and/or modify equipment more easily. Designing space rather than trying to fit into an existing laboratory or facility can reduce the associated risk.

For example, the design and siting of hydrogen storage/handling facilities may present several options. A project that uses hydrogen (perhaps a fuel cell tester, a developer of a new hydrogen compression system, or a developer of hydrogen storage systems) may go through large quantities of hydrogen in short periods of time. If the facility is not designed for handling, moving and distributing the hydrogen, one may wind up storing hydrogen in an area where safety is compromised. One might need to consider alternative supply options, e.g. an electrolyzer versus cylinder racks or tube trailers. For large quantities of hydrogen, the electrolyzer itself could have siting problems in certain circumstances.

Ideally, outdoor storage in cylinders or a tube trailer is advised, but sometimes this cannot be done. Wherever hydrogen will be stored, the questions should be asked: "What if a full tank of hydrogen or a manifold connecting several open, full tanks of hydrogen should rupture, releasing all of the hydrogen into a laboratory, storage room, or other contained space? Will the ventilation system be sufficient to keep the hydrogen concentration in that space sufficiently low to prevent a fire or explosion? If not, is the risk acceptable?

Hydrogen molecules, being small and light, have the ability to diffuse very quickly and to leak in situations where no other substance will leak. Thus, systems should be built with robust valves and a minimum number of mechanical connections. If the system is going to see large temperature/pressure cycles, compression fittings should not be used.

Although buried hydrogen pipelines are in commercial practice, one must consider that buried lines can corrode from the outside if not protected in certain types of soil. Buried lines also present the obvious challenge of leak detection.

4.2 Operating Procedures

The need for sufficient and accurate documentation has been a recurring theme of the safety reviews. Such documentation includes standard operating procedures (SOP), maintenance schedules, training schedules, warning placards, emergency call information signs, protocols, etc. While different companies or laboratories may each have their own practices and methods for documentation, the overall need for a means to disseminate knowledge to key personnel for system operations, to new employees, and to emergency responders is of utmost importance. Electronic, interactive company-wide websites are one good way of disseminating procedural and training information. Availability of hard copies of SOPs, warning placards, emergency phone numbers, and diagrams of operating systems, are needed to provide emergency call information and for first responders.

Industrial organizations typically establish policies for managing deviations from normal operations. Management of change (MOC) procedures may apply to facilities, e.g., replacing or adding a new piece of equipment, as well as abnormal situations for which there are deviations from standard operating procedures. Documented MOC policies and procedures can be tailored to apply to small laboratory-scale equipment and operations as well as larger demonstration-scale facilities. Trained and knowledgeable staff will be in a position to make informed judgments as to whether a deviation is significant enough to require application of the appropriate MOC procedures.

Numerous organizations are doing developmental work on fuel cells and/or hydrogen storage systems and require significant amounts of compressed hydrogen onsite. For many, the movement of hydrogen cylinders within a facility and changing hydrogen cylinders is an important consideration. One company, which must constantly move large quantities of hydrogen, designed and built an ergonomic hydrogen cart with which one can transfer two hydrogen cylinders at once on a hand-cart designed with a large platform to ensure tank stability, and with a configuration to take pressure off the workman's back. They use long, flexible-hose tubing to ensure quick and safe replacement of empty cylinders with full ones. Other hydrogen cylinder-related issues include (1) a logical (but not always followed) caveat of always moving a cylinder with the cylinder cap in place; (2) the tendency for hydrogen cylinders to leak slightly through the lead seal on the tank valve. Sound protocols should cover all aspects of cylinder transport and use.

Another emerging theme, and one that is not always obvious, is to ensure that labels on purchased materials accurately reflect content. Some larger facilities will have a central storage area from which materials are delivered. It is possible that the storage area personnel deliver the wrong material. Checking the labels should be a necessary step that is reflected in SOPs or protocols before using materials.

4.3 Equipment Maintenance

Proper and timely equipment/sensor maintenance is key to ensuring this aspect of a good risk mitigation plan. Large pieces of equipment may have service contracts and be handled accordingly, but that may not be the case routinely. An electronic spreadsheet can assist in scheduling and tracking equipment maintenance and sensor calibration requirements that might be noted in SOPs.

A strong valve maintenance and replacement program is also an important consideration. Hydrogen leaks can occur either out of the valve or through the valve. The former could be detected by a sensor near the valve, but the latter, especially if it is small, may be more difficult to detect, and result in a hydrogen presence where unexpected.

4.4 Hydrogen Storage Materials

A significant amount of hydrogen research is now centered on the storage of hydrogen in complex metal hydrides. Some of these storage materials are dangerous in their own right. Sodium alanate ($NaAlH_4$) is usually considered the typical material of this type, and is used for much of the testing. Hydrogen is recovered from sodium alanate in a two-step process:

$$3NaAlH_4 \rightarrow Na_3AlH_6 + 2Al + 3H_2$$

$$Na_3AlH_6 \rightarrow 3NaH + Al + 3/2H_2$$

Sodium alanate itself, the intermediate product Na_3AlH_6, and the byproduct NaH are all very reactive in the presence of either oxygen or moisture. For example, they are pyrophoric in moist air. The fact that there is likely to be some free hydrogen present as well exacerbates the issue. Great care is needed throughout the manufacturing and testing of these compounds to keep them in an inert atmosphere.

The disposal of these materials is an important consideration in metal hydride handling safety. Spent (oxidized) materials may contain pockets of active alanate or an active intermediate may be present, and the mix might be shock-sensitive. Thus, it is recommended that even supposedly fully oxidized alanates be treated as hazardous materials.

4.5 Other Issues

Although not directly involving hydrogen, oxygen depletion leading to asphyxiation is a serious issue discussed during several safety reviews. For example, laboratory work is often performed in a glove box, under a nitrogen or argon atmosphere. If the glove box develops a large leak, it may not be obvious when the inert gas leaks into the room itself, depleting oxygen. Asphyxiation by oxygen depletion in a facility where an inert gas is being used or is being piped through is an often overlooked danger that costs several lives each year. Methods of prevention include oxygen level sensors in the room itself, flow rate sensors on the inert gas feed, and adequate ventilation in the room to overwhelm any large inert gas leak.

Performing safety-based site visits has also revealed that even though large demonstration-sized hydrogen projects carry the larger risks, even small, laboratory-scale hydrogen production projects can evolve sufficient hydrogen to be a potential hazard. Personnel should calculate the worst-case scenarios for possible hydrogen buildup and ventilate the area adequately.

Two different facilities reported that some electrochemically-based sensors may mistake hydrogen for carbon monoxide, suggesting they not be used in a hydrogen-containing application.

The safety review site visit can be an effective mechanism for engaging organizations and project teams to bring learnings to the DOE program and help shape safety awareness. Even the importance of accepted safety practices as standard as the use of safety glasses in research laboratories can be reinforced. Site visits have already proven their value in tapping the expertise of the Hydrogen Safety Review Panel and providing broad benefit to the DOE program.

5. ACKNOWLEDGEMENT

The authors wish to thank the U.S. Department of Energy's Hydrogen, Fuel Cells and Infrastructure Technologies Program Office: Mr. Steven Chalk, Program Manager, and Mr. Patrick Davis, Manager, Safety, Codes and Standards, for their support of this work.

Hazard Potential Analysis for Freeway Transportation of Toxic Substances

Kuo-Shuh Fan
Department of Safety, Health and Environmental Engineering
National Kaohsiung First University of Science and Technology
Kaohsiung, 824, Taiwan
ksfan@ccms.nkfust.edu.tw

Shih-Chien Huang
National S&T Center for Hazards Mitigation
Taipei, Taiwan

Jenq-Renn Chen*
Department of Safety, Health and Environmental Engineering
National Kaohsiung First University of Science and Technology
Kaohsiung, 824, Taiwan
jrc@ccms.nkfust.edu.tw

ABSTRACT

A simple and manageable methodology that is capable of assessing the global risk of road transportation of hazardous materials is proposed. A transportation risk index (TRI) is first defined as the product of impact radius, transport frequency, incident probability, and release probability. The TRI has the unit of impact radius in km per year per million of vehicle kilometer. Its calculation is more amiable for the risk analysis of large area with multiple chemicals and transportation routes compared with the typical societal risk. Upon the completion of calculation of the TRI for every transport routes in the selected area, the total risk is obtained by summing overlapped TRI in the GIS platform. A contour is then generated, also performed directly in the GIS platform, and then map directly on the GIS map to give the final risk potential map. The National Freeway 1 in Taiwan with a total length of 370 km is used as an example to illustrate the methodology. The freeway is divided into 37 sections based on intersections. The historical incident related data are collected for each section. The highest risk area as analyzed by the method is found to match reasonably well with the incident records of past 10 years.

1. INTRODUCTION

The development of various industries has greatly increased the need for transporting toxic chemicals and hazardous materials. Although rail transportation has been known for its low incident rate [1], the road transportation by trucks or tank trucks remains the major choice for its flexibility and simplicity. However, the high incident rate of road transportation has rendered the great risk on the community and the environment. In 2003 along, there were eight incidents related to the transportation of hazardous materials in Taiwan's freeway and public road. It is thus crucial to plan and prepare for the response of these transportation incidents.

Most existing methods of transportation risk analysis [1,2] focused on the routing of particular hazardous materials. Societal risk, given by the F-N curve, is normally calculated for a given route. A route is then selected based on the lowest societal risk. This approach is useful for route selection of transporting hazardous materials from one facility to another facility. However, for regulatory authorities such as the Environmental Bureaus, the Fire Brigade, the Transportation Bureau, etc., the lowest societal risk for a particular facility does not necessarily reflect the actual risk as other facilities may also take the same route. Furthermore, the great efforts to calculate the societal risk also prevent the regulatory authorities to thoroughly perform these calculations and find the overall risk for each route in their jurisdiction. It will be useful to have a simplified approach such that the overall risk, rather than individual risk, can be calculated and alternative routes may be recommended or emergency plan can be prepared accordingly.

In this work, we developed a simple and manageable methodology that is capable of assessing the global risk of road transportation of hazardous materials. The National Freeway 1 in Taiwan with a total length of 370 km is used as an example to illustrate the methodology. The National Freeway 1 is the major freeway route for the connecting major cities from north to south Taiwan. It is also the major route for hazardous materials transportation. The freeway is divided into 37 sections based on intersections. The historical incident related data are collected for each section. The highest risk areas as analyzed by the method are compared to the incident records of past 10 years.

2. METHODOLOGY

2.1 The Risk Potential Map Approach

Normally, the transportation risk analysis (TRA) for a particular route of a known chemical consists of the following steps [2]: (a) identifying the possible release scenarios (b) assessing the occurrence frequency of each release scenario (c) evaluating the possible outcome cases and its occurrence frequency of each release scenario (d) determining the impact area to each outcome case (e) estimating the number of fatalities for the population in the impact area. Each step requires significant data input such as the weather condition, population density, incident rate, dispersion modeling, etc.

To assess the overall risk associated with all transportation routes in a particular area, it is almost impossible to follow exactly the above TRA approach. Also, it is necessary to consider the overlapping of the risks from different chemicals and different routes. For a given location, it is trivial to add the F-N curves for different risk from different TRA. For a whole regime, the direct summation of F-N curves would be a formidable task. It is always desired to have a simple map projected with risk associated with all transportation of hazardous materials. Recently, we have developed a methodology to create the global risk potential map for the analysis of risk from fixed facilities in a large area [4]. In this approach, a risk index (RI) is proposed and defined as:

$$RI = \text{release probability (1/yr)} \times \text{impact radius (km)}$$

The risk index is used to replace the exhaustive F-N curve for the risk of a given chemical in a given facility. The impact radius is calculated from dispersion model while the generic reliability data is used for the release probability. For simplicity, only toxic effects are considered. The impact area was then assigned with the corresponding risk index value. Risk from multiple chemicals or facilities for a given location are then calculated by adding the overlapped risk index value. The risk index associated with the affected area is plotted in a geological information system (GIS). A contour of the risk index is eventually generated on the GIS and the area with the highest risk can then be easily identified. The authorities of jurisdiction of the area can then take proper action plans for the high risk facilities. This risk potential map is found to match well with the locations of past incidents.

In this investigation, it is aimed to assess the feasibility of applying the developed technique from fixed facilities to the transportation risk analysis.

2.2 Transportation Risk Index

Similar to the risk index for fixed facilities, we defined a transportation risk index (TRI) as follows:

$$TRI = \text{impact radius (km)} \times \text{transport frequency (1/yr)} \times \text{incident probability (1/}10^6\text{ vehicle-km)} \times \text{release probability (-)}$$

Where the transport frequency is defined as the number of vehicle transport per year. The incident probability is the vehicle incident rate given by number of incident per 10^6 km of vehicle transport (MKV). It should be noted that normally only the incident rate for all vehicles is available and is used for the incident rate of truck and tank truck with hazardous materials. The release probability is the probability of the chemical released should a transportation incident occurred. The impact radius is calculated from dispersion model while the generic reliability data is used for the release probability. For simplicity, only toxic effects are considered in this work although fire and explosion can also be considered in a similar way. The TRI has the unit of impact radius in km per year per million of vehicle kilometer. Although the TRI is not as informative as the F-N curve, its calculation is more amiable for the risk analysis of large area with multiple chemicals and multiple transportation routes.

Upon the completion of calculation of the TRI for every transport routes in the selected area, the total risk is obtained by summing overlapped TRI, normally done directly in the GIS platform. A contour is then generated, also performed directly in the GIS platform, and then map directly on the GIS map to give the final risk potential map.

3. RESULTS

3.1 Data Input

In this section, results will be presented for the National Freeway 1 in Taiwan based on the selection of six toxic chemicals: benzene, acrylonitrile, chlorine, ethylene oxide, butadiene and vinyl chloride. Table 1 gives important properties of these chemicals. These chemicals are listed as toxic chemicals and regulated by Taiwan EPA [5]. According to the Toxic Chemical Substance Control Acts of Taiwan EPA [5], permits are required for the selection of transportation route of the listed toxic chemicals. Written notices to the EPA are also needed for each truck or tank truck carrying the listed toxic chemicals. The system thus allows precise information for the calculation of transport frequency. For non-listed chemicals or hazardous materials, the transport frequency is usually not available. Nevertheless, the Taiwan EPA intends to increase of the number of listed chemical every year to cover all major chemical in production or use. Table 2 shows the transport frequency of these six chemicals for the year of 2001 in National Freeway 1. It is apparent that majority of these transport centered in Kaohsiung area in south Taiwan where major chemical and petrochemical industries located.

Table1. Properties of the six toxic chemicals.

	Benzene	Acrylonitrile	Chlorine	Ethylene Oxide	Butadiene	Vinyl Chloride
Boiling Point(°C)	80.1	77	-34	10.7	-4.5	-14
Vapor Pressure (mmHg)	74.6 (20°C)	88 (20°C)	5.83×10^3 (25°C)	1314 (25°C)	245 kPa (20°C)	2660 (25°C)
Vapor Density (air = 1)	2.8	1.9	2.5	1.5	1.9	2.2
IDLH (ppm)	500	85	10	800	2000	50
TLV-TWA (ppm)	0.1	2	0.5	1	2	5

Table 2 Transport frequency of these six chemicals for the year of 2001 in different segment of National Freeway 1.

Segment of National Freeway 1	Transport Frequency (1/yr) Year 2001					
	Benzene	Acrylonitrile	Chlorine	Ethylene oxide	Butadiene	Vinyl chloride
Keelung~Hsichuh	0	0	0	0	0	0
Hsichuh~Taishan	0	0	0	0	0	0
Taishan~Yangmei	0	0	727	0	0	4120
Yangmei~Tsaochiao	1006	1920	727	204	0	11080
Tsaochiao~Houli	1006	1920	727	204	0	11080
Houli~Yuanlin	1006	1920	727	272	0	11080
Yuanlin~Tounan	1006	1920	727	272	0	11080
Tounan~Hsinying	1006	5472	727	272	750	11080
Hsinying~Hsinshih	1006	5472	727	272	750	11080
Hsinshih~Kangshan	1006	13072	727	272	7973	11080
Kangshan~Kaohsiung	3147	14272	727	272	11423	11080

Figure 1 gives the incident rate for the National Freeway 1 in Taiwan. This information is ready available from the National Highway Police Bureau [6]. However, only the incident rate for all vehicles is available and is applied for the incident rate of truck and tank truck with hazardous materials.

For the release probability, a fixed value of 0.011 is assumed [7].

For impact radius, a fixed hole size of 101.6 mm (4-in) on the bottom of the tank is assumed. The sizes of the tank for different chemicals are, according to typical tank truck sizes in use in Taiwan, 25 m^3 for benzene and acrylonitrile, 16.5 m^3 for chlorine, 13 m^3 for ethylene oxide, 20 m^3 for butadiene and vinyl chloride. The dispersion is performed using the ALOHA program, developed by US EPA and NOAA [8]. IDLH is chosen as the level of concern. Wind speed of 1.5 m/s, atmospheric stability class F, air temperature of 35°C, and relative humidity 95% were used for the dispersion calculation. The maximum plume length is then used as the impact radius.

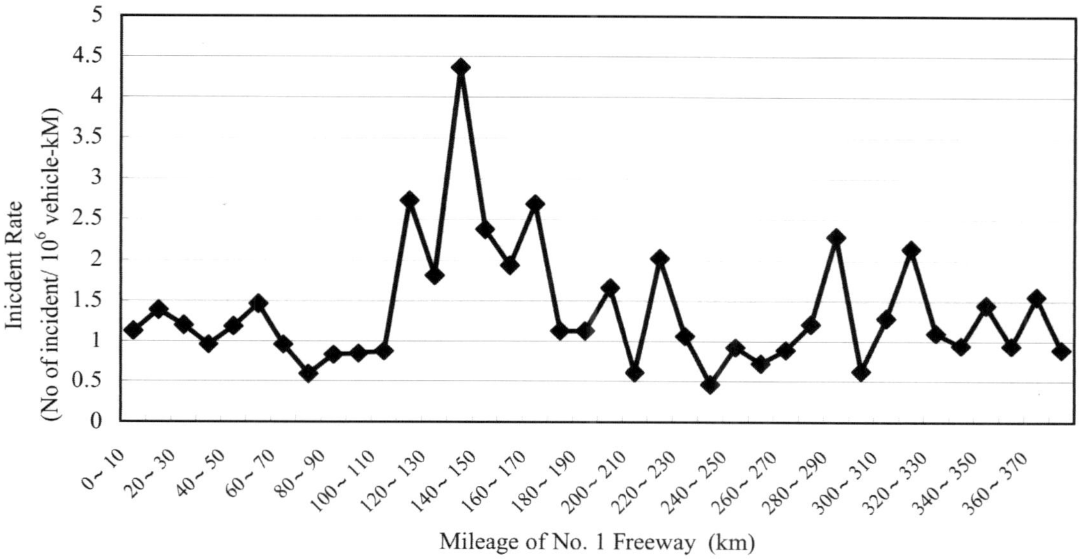

Figure 1 The incident rate for the National Freeway 1 in Taiwan.

3.2 Risk Potential Map

Figure 2 shows the individual risk potential map for the six chemicals. For benzene, the highest risk sections were found to near Taichung in the central Taiwan and Kaohsiung in south Taiwan. For acrylonitrile, the highest risk sections were found to be in Tainan and Kaohsiung, both in south Taiwan. For chlorine, vinyl chloride and ethylene oxide, the highest risk sections were in Maoli, and Taichung in central Taiwan and Tainan in south Taiwan. For butadiene, the highest risk sections were in Kaohsiung and Tainan in south Taiwan. Figure 3 shows the combined risk potential for the six toxic chemicals. The areas where the most risk impacted are Maioli, followed by Taichung, Tainan, and Kaohsiung. For individual chemical, the major contributing factors for the TRI are the transport frequency and the incident rate. For combined risk, the impact radius, determined by the level of concern and thus related to the toxic properties of the chemical, was also a dominant factor in the TRI. This is reflected in the fact that chlorine and vinyl chloride contributed to majority of the risk.

3.3 Discussions

Figure 3 also includes a map showing the location of past incidents involving transportation of all hazardous materials on the National Freeway 1. Table 3 gives the complete list of these incidents. It is not appropriate to compare directly the risk potential map with incident record as the present risk potential map focused on only six listed chemicals. It is unlikely that the limited incident data will reflect properly the actual risk. Nevertheless, two out of three incidents relating to the six listed toxic chemicals in Table 3 were in Maoli and Tainan Counties, which were the areas with the highest combined risk potential. Further analysis of Table 3 reveals that the highest number of incidents were in Taipei County (5), followed by Kaohsiung County (4), Hsinchu County (3), Tainan County (3), Maoli County (2), Yunlin County (2), Taichung County (1), Chiayi County (1), and Taoyuan County(1). This is still compared reasonably well with risk potential map in Figure 3 where the most risk impacted area are Maioli, followed by Taichung, Tainan, and Kaohsiung. It is believed that the risk potential map will match the incident data even better if more chemicals were included in the analysis.

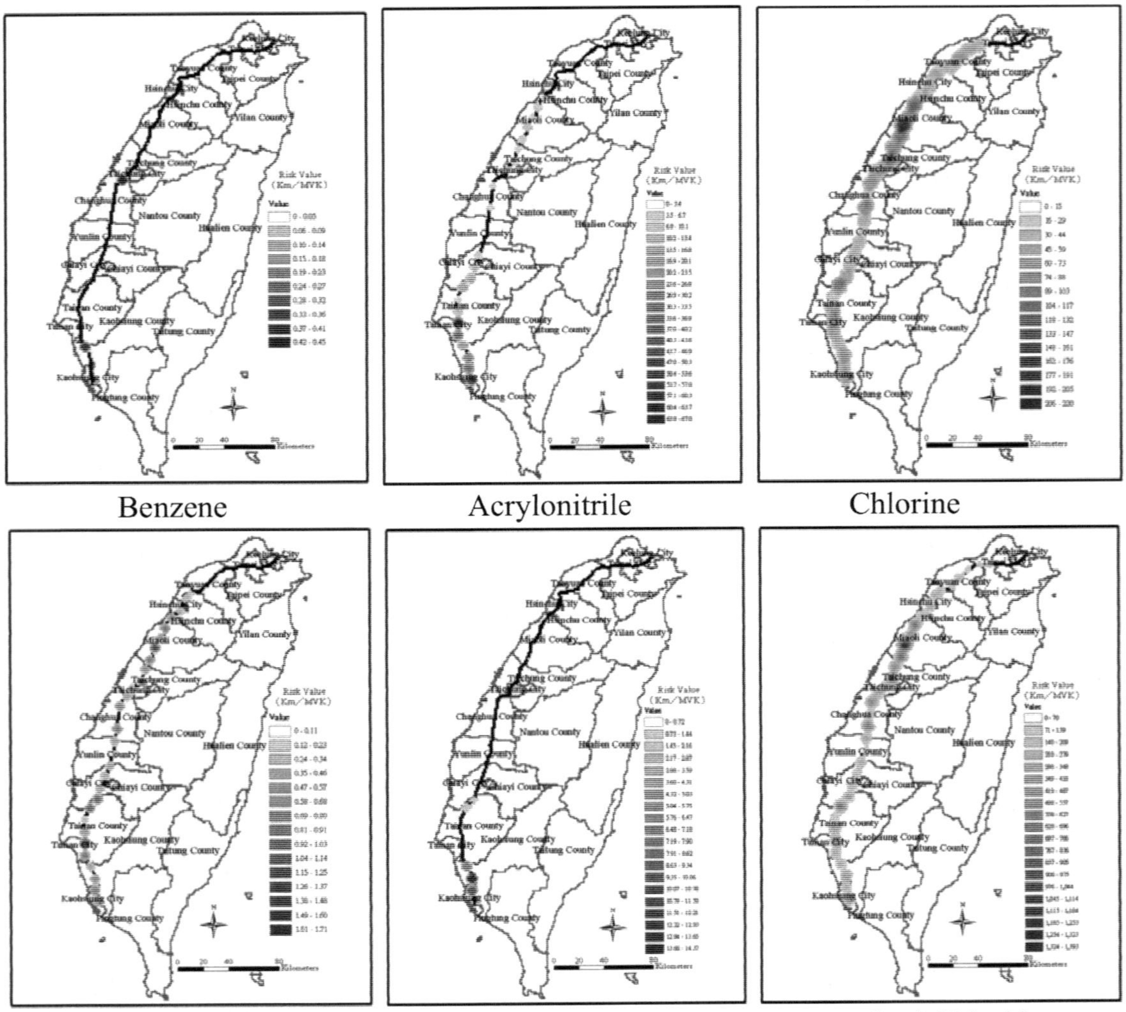

Benzene | Acrylonitrile | Chlorine

Ethylene Oxide | Butadiene | Vinyl Chloride

Figure 2 Risk potential for individual chemicals.

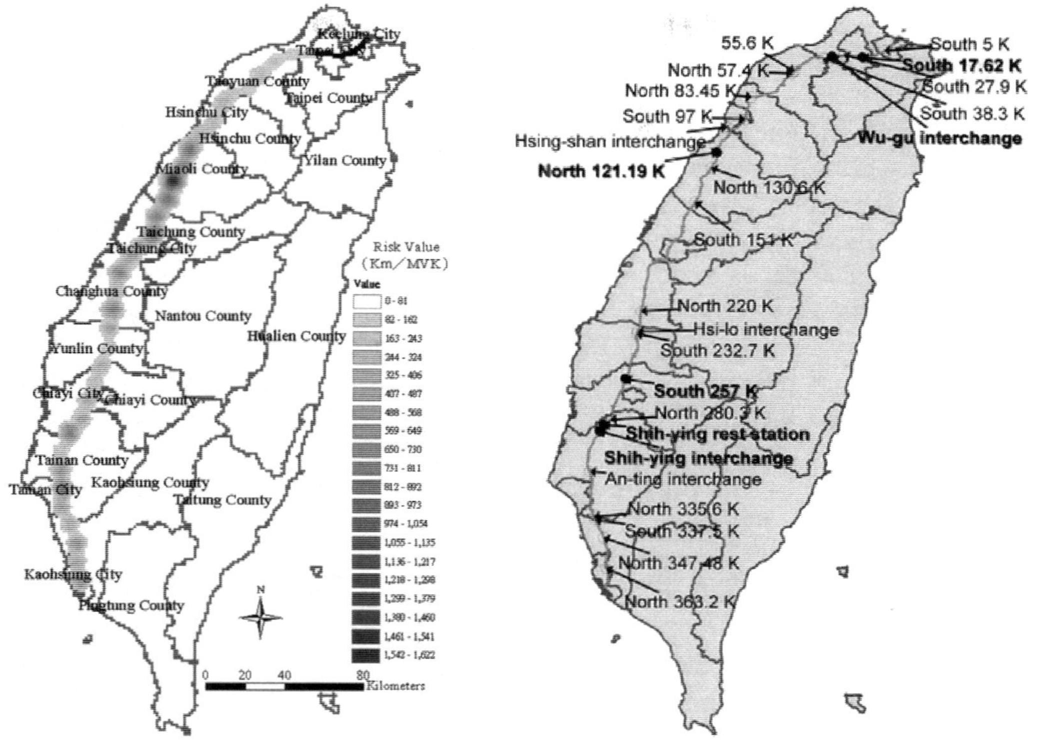

Figure 3 Combined risk potential map (left) and incident records (right) for National Freeway 1 transportation of toxic chemicals.

Table 3 Complete list of incident of transportation of all hazardous materials on National Freeway 1 up to the year of 2003.

Date	Location	Damage situation	Incident materials
1993/10/15	Wu gu interchange, Taipei County	Overturn and release	Formaldehyde
1994/01/13	South 17 K, Taipei County	Overturn and release	Epichlorohydrin
1994/01/15	South 232K, Yunlin County	Overturn, release and fire	Toluene
1994/05/16	North 121K, Maoli County	Overturn and release	*Acrylonitrile*
1997/06/04	North 130K, Maoli County	Release and fire	Xylene
1998/01/21	North 347K, Kaohsiung County	Release and fire	Styrene
1998/04/10	North 335K, Kaohsiung County	Overturn and release	Hexane
1998/04/27	North 363K, Kaohsiung County	Overturn and release	Styrene
1998/10/30	South 5K, Taipei County	Traffic incident, no release	Ethylene glycol
2000/02/27	His-Lo interchange, Yunlin County	Overturn, no release	Octyl acrylate
2001/05/04	Shin-Ying rest station, Tainan County	Traffic incident, no release	Ethylene dichloride
2001/05/15	An-Ting interchange, Tainan County	Overturn and release	Fuel Oil
2001/05/29	Hsing-Shan interchange, Hsinchu Cty	Overturn and release	Ethylene glycol
2002/06/13	South 337K, Kaohsiung County	Overturn and release	Ethylene glycol monobutyl ether
2002/07/25	Shin-Ying rest station, Tainan County	Fire of tires without release	*Butadiene*
2003/04/22	South 27K, Taipei County	Overturn and release	Ethylene glycol
2003/04/25	South 257 K, Chiayi County	Overturn and release	*Chlorine*
2003/08/06	North 83K, Hsinchu County	Traffic incident, no release	Sodium hydroxide
2003/08/12	North 57K, Taoyuan County	Overturn and release	Hydrogen peroxide
2003/10/11	South 38K, Taipei County	Fire on cylinders	Acetylene
2003/10/21	South 151K, Taichung County	Release from cylinders, no fire	LPG
2003/11/07	South 97K, Hsinchu County	Traffic incident, no release	Fuel oil

4. CONCLUSIONS

A simple and manageable methodology that is capable of assessing the global risk of road transportation of hazardous materials is developed. A transportation risk index (TRI) is defined as the product of impact radius, transport frequency, incident probability, and release probability. The impact radius is calculated from dispersion model while the generic reliability data is used for the release probability. For simplicity, only toxic effects are considered in this work although fire and explosion can also be considered in a similar way. The TRI has the unit of impact radius in km per year per million of vehicle kilometer. Its calculation is more amiable for the risk analysis of large area with multiple chemicals and transportation routes compared with the typical societal risk. Upon the completion of calculation of the TRI for every transport routes in the selected area, the total risk is obtained by summing overlapped TRI in the GIS platform. A contour is then generated, also performed directly in the GIS platform, and then map directly on the GIS map to give the final risk potential map.

The National Freeway 1 in Taiwan with a total length of 370 km is used as an example to illustrate the methodology. The resulted risk potential map is compared reasonably well with the incident records. It is believed that the risk potential map will match the incident data even better if more chemicals were included in the analysis.

5. ACKNOWLEDGEMENTS

This work was supported by the National Science Council, Taiwan, and the National S&T Center for Disaster Reduction. All supports are greatly acknowledged.

6. REFERENCES

1. Bubbico, R., Di Cave S., and Mazzarotta, B. Transportation Risk Analysis for Road and Rail Transport: a GIS approach. *J. Loss Prev. Process Ind.*, 17 (2004) 483–488.
2. Center for Chemical Process Safety (CCPS). Guidelines for Chemical Transportation Risk Analysis, New York, American Institute of Chemical Engineers, 1995.
3. Bubbico, R., Di Cave, S., and Mazzarotta, B., Risk Analysis for Road and Rail Transport of Hazardous Materials: A Simplified Approach, *J. Loss Prev. Process Ind.*, 17 (2004) 477–482.
4. Chen, J R, Liang, S C, Fan, K S, Hsu, P C, and Su, J L., GIS-Based Global Risk Potential Map for the Panning of Chemical Emergencies, First International Conference on Urban Disaster Reduction, Kobe, Japan, January 2005.
5. Toxic Chemical Substances Control Act, Environmental Protection Administration, Taiwan, December 22, 1999. http://cemnt.epa.gov.tw/eng/
6. National Highway Police Bureau, Taiwan, http://www.hpb.gov.tw/advert_04.htm.
7. Center for Chemical Process Safety (CCPS), Guidelines for Process Equipment Reliability Data, with Data Tables, New York, American Institute of Chemical Engineers, 1989.
8. EPA, ALOHA (Areal Locations of Hazardous Atmospheres) code, http://www.epa.gov/ceppo/cameo/aloha.htm, 2004

Achieving Effective Alarm System Performance: Results of ASM® Consortium Benchmarking against the EEMUA Guide for Alarm Systems

Dal Vernon C. Reising
Honeywell Laboratories
Minneapolis, MN
dalvernon.reising@honeywell.com

Tim Montgomery
ChevronTexaco Energy Technology Company
Richmond, CA,
tim.montgomery@chevrontexaco.com

ABSTRACT

The Abnormal Situation Management®[1] Consortium has recently completed a series of studies related to effective alarm management practices for the refining and petrochemicals industry. These studies related directly to the alarm system performance guidelines published in the Engineering Equipment and Materials User Association's (EEMUA) Publication No. 191. Results from 37 unique operator consoles indicate that the EEMUA recommendation for average alarm rate during normal operations (i.e., less than one alarm per 10 minutes), while not universally demonstrated, is achievable today. Our study found that about one-third of the consoles surveyed were able to achieve this recommended alarm rate guideline for normal operations and about one-quarter more consoles were achieving the EEMUA "manageable" level of 1 to 2 alarms per 10 minute period. However, the EEMUA recommendation for peak alarm rates following a major plant upset (i.e., not more than 10 alarms in the first 10 minutes) appears to be a challenge, given today's practices and technology. Only 2 of the 37 consoles came close to achieving the alarm rate guideline for upset conditions. This suggests that to achieve alarm system guidelines for upset conditions, more advanced site practices and alarm-handling technology (e.g., dynamic or mode-based alarming) are required. In studying the relationships between the observed alarm rate performance and other metrics collected, along with anecdotal information gathered (a subset of which is included here), we conclude that there is no

[1] ®ASM and Abnormal Situation Management are U.S. registered trademarks of Honeywell, Inc.

"silver bullet" for achieving the EEMUA alarm system performance recommendations. Rather, a metrics-focused continuous improvement program that addresses key lifecycle management issues is most appropriate.

1. INTRODUCTION

In an informal survey of industry practice that was conducted at the 2001 Honeywell Users Group meeting, only 30% of the 20 refining and petrochemical companies surveyed reported "Doing Quite a Bit" for alarm management, whereas 50% of those companies reported "Just Starting", and the remaining 20% reported "Doing Nothing or Little". Taken together with results of a Health & Safety Executive (HSE) survey (HSE, 2000) that showed an approximate average alarm rate of 5 alarms per 10 minute period, the conclusion one might draw is that very few operating sites would be achieving the Engineering Equipment and Materials User Association (EEMUA) recommendations for average alarm rate in its Publication No. 191 document (EEMUA, 1999). This document defined some aggressive alarm rate recommendations that need to be confirmed as being attainable. These recommendations included:

- An average of less than 1 alarm per 10 minute period for steady-state operations, and
- Less than 10 alarms per 10 minute period following a plant upset

Following the publication the EEMUA alarm rate recommendations, there were questions on whether these recommendations were appropriate, particularly for the recommendation of not more than 10 alarms per 10 minutes following an upset condition. Reising, Downs, and Bayn (2004), as part of another ASM study, concluded this recommendation appears to be supported by human cognitive capabilities (more accurately, the limits associated with human cognitive capabilities). Two different modeling analyses were conducted, and both analyses suggest that EEMUA recommendation for upset conditions is a very reasonable guideline, with respect to human performance limitations. Based on video-recorded observations of operators responding to upset conditions in high-fidelity simulator training, they were able to code various response behaviors, the frequency of those behaviors, the average time for those behaviors, and derive an averaged probabilistic time estimate needed to respond to an alarm at approximately 49 seconds. This result, taken together with past research on human mental workload that suggests not taxing people for more than 80% of the time available (Parks & Bouchek, 1989), suggests that – with today's alarm management practices and alarm system tools -- an operator requires approximately 1 minute per alarm, on average, during alarm burst conditions.

The Abnormal Situation Management (ASM) Consortium recently completed a series of studies to benchmark the performance of alarm systems in ASM members' facilities, primarily throughout North America. The objectives for this work were to:

(i) assess whether members were achieving the EEMUA recommendations for alarm system performance as well as to

(ii) relate that performance to contextual metrics such as plant complexity, degree of alarm rationalization, nature of plant operation, and so on.

In the ASM study, alarm system metrics (e.g., alarm rate average, alarm rate distribution), system configuration metrics (e.g., total number of control valves, total number of configured points, etc.), alarm system maintenance metrics (e.g., percent of total alarms from Top 10 Worst Actors, average number of standing alarms) as well as general information (e.g., process type, DCS type, frequency of monitoring average alarm rate, extent of alarm rationalization) were all collected to achieve these two objectives.

We present the results of this data collection effort below, along with anecdotal information about successful alarm management initiatives, and conclude the paper with suggestions on how to best focus alarm management resources.

2. STUDY APPROACH

The ASM benchmarking study was conducted in two phases. In Phase I, the participating ASM members agreed upon the set of metrics, duration of the data collection periods and committed to providing at least 3 months of data to Honeywell for each console in the study. A console was defined as representing a portion or area of a plant that was typically operated by a single console operator under normal plant conditions, even if that console had multiple workstations (i.e., keyboards).

Honeywell collated and analyzed the information and presented the anonymous results back to the Consortium members. In addition to the alarm rate performance collected for each console, three broad areas of metrics were also collected, so as to better relate the alarm rate performance data to other potential factors. Each of these broad areas had specific metrics; the three areas were:

- General descriptive data (e.g., type of unit, type of DCS, extent to which the console was rationalized, extent to which worst actors were tracked, etc.)
- Alarm system configuration data (e.g., total number of configured alarms, alarm priority distribution, total number of points, etc.), and
- Alarm system maintenance data (e.g., average percent of the total alarms that the top 10 worst actors make up, average number of standing alarms, average number of disabled alarms, etc.).

In Phase II, participating members once again agreed on the set of metrics and were asked to provide additional console data and provide anecdotal information about and lessons learned from their alarm management initiatives and practices; this latter information then could possibly be used to examine potential critical elements that might explain the varying alarm performance results.

While the majority of the metrics collected in each phase remained identical, a few metrics were revised in Phase II. For example, some of the general descriptive metrics in Phase II were revised in attempt to get more objective values for those metrics. The data sets for consoles participating in Phase I were also re-categorized by the original contributors on these revised general descriptive metrics so that both the Phase I and Phase II data sets could be combined for more thorough analyses.

The combined phases collected a total of 90 sets of monthly data from 37 unique consoles. There were 22 unique consoles from Phase I -- each with three months of data submitted -- and 15 new, unique consoles from Phase II -- each with just one month of data submitted. Three consoles appeared in both phases, which account for the 40 entries in the following result figures.

3. STUDY RESULTS

The data collected from the two phases of the ASM study are combined in the following discussion of results. First, the benchmarking results against the alarm rate recommendation for normal operations are presented. Second, the results of peak alarm rate for non-normal operations are discussed. Third, this section concludes with a discussion of a multivariate regression analysis that investigated possible relations between the average alarm rate and other influencing metrics.

3.1 Average Alarm Rate for Steady-State Operations

Figure 1 presents the monthly overall average alarm rate for the 40 console data sets. Each bar represents the lowest, average and highest monthly average alarm rate for a single console that submitted 3 months of data. Single points represent consoles in the second phase that were only required to submit one month of data. On the right hand side of the chart is a scale representing the 5 alarm performance categories described in Table 1 below.

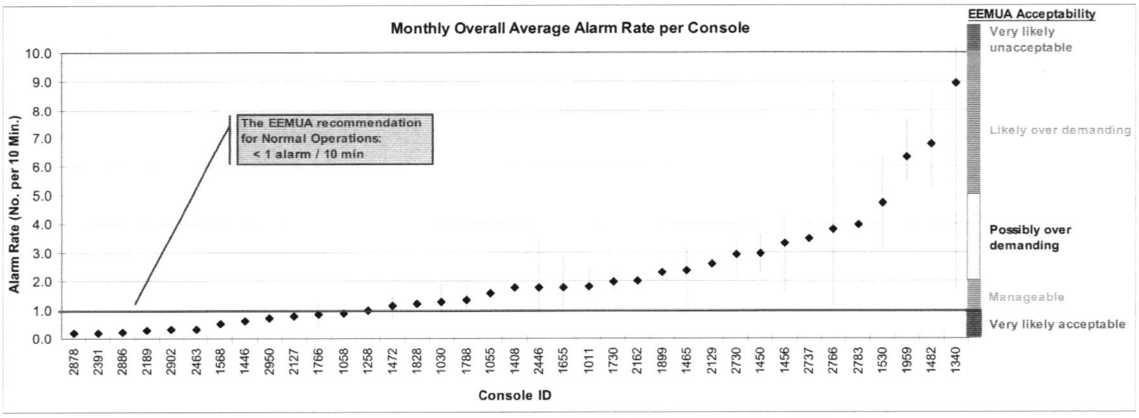

Figure 1: Monthly overall average alarm rate per console. Consoles that submitted three months of data show the low-average-high monthly values with bars present.[2]

The overall average alarm rate for each month was calculated for the entire month of data for a given console and included all modes of plant operation (normal, planned off normal, abnormal and emergency), not just steady-state operations. Participants stated that removing or filtering out non-steady-state modes would be prohibitive, given the state of the art in alarm system tools circa 2001-2003. Because the EEMUA acceptability

[2] The last three digits of the console identification number were randomly generated and randomly assigned to the various consoles. The first digit of the identification number (either 1 or 2) specifies which phase of the ASM benchmarking study data was reported. Three consoles participated in both phases (446, 730, and 766).

categories were meant to categorize alarm rates for normal plant operation, the consoles in the ASM study that had non-normal operation during the month will be biased on the high side against the true steady-state average alarm rates.

Average Alarm Rate in Steady-state Operation, *per 10 minute period*	Acceptability Categorization
More than 10 alarms	Very likely to be unacceptable
More than 5 but less than 10	Likely to be over-demanding
More than 2 but less than 5	Possibly over-demanding (see **Note 1**)
1 or more but less than 2	Manageable
Less than 1 alarm	Very likely to be acceptable

Note 1: *This range was categorized by ASM study team in absence of range categorization in EEMUA Publication No. 191*

Table 1: Acceptability categories for average alarm rate in steady-state operations, as outlined in EEMUA Publication No. 191 (adapted from Figure 42, p. 105)

The monthly overall average alarm rate across both phases of the ASM study was 2.3 alarms per 10-minute period, compared to the British Health & Safety Executive study of 5 alarms per 10-minute window that led to the EEMUA guidelines document (HSE, 2000). The median monthly average alarm rate was 1.77 alarms per 10 minute period, meaning that over half of the consoles were at "Manageable" or "Very likely to be acceptable" average alarm rate levels.

There are two caveats to be considered when interpreting the results presented in Figure 1. First, as was stated above, because these data points may contain modes of operation other than steady-state, these might be considered "conservative" estimates of the actual steady-state alarm rates for the consoles participating in the ASM study. Second, while the reported average alarm rates might be conservative estimates for the participating consoles, these same rates probably do not reflect the general performance across all consoles of the ASM member companies, or presumably the industry as a whole. The ASM members acknowledged that the data collection process did not represent the full range of performance at their faculties and that, typically, better performing consoles were better represented because alarm rate performance data was more readily available from these consoles.

Regardless, based on the average alarm rates reported in Figure 1, one may conclude that the EEMUA recommendation is an achievable target for which to strive.

3.2 Averaged Peak Alarm Rate for Non-Normal Operations

In addition to characterizing the acceptability of various average alarm rate values, EEMUA (1999) also characterizes the acceptability of various peak alarm rate values following major upsets. Table 2 presents this characterization below. The ASM study team, in an effort to understand better how consoles were performing against these acceptability categories, asked participants to report the percent of the time for a month spent in additional categories (see the legend in Figure 3). Two new categories were: (i) 1-10 alarms/10 minute period, and (ii) > 100 alarms/10 minute period. In addition, the EEMUA category of 20-100 alarms/10 minute period was broken into three separate categories: 21- 30 alarms/10 minute period, 31-50 alarms/10 minute period, and 51-100 alarms/10 minute period.

Number of alarms displayed in 10 minutes following a major plant upset	Acceptability Categorization
More than 100 alarms	Definitely excessive and very likely to lead to operator abandoning use of the system
20-100	Hard to cope with
10-20	Possibly hard to cope with (see **Note 2**)
Under 10	Should be manageable – but may be difficult if several of the alarms require a complex operator response

Note 2: *This range was categorized by ASM study team in absence of range categorization in EEMUA Publication No. 191.*

Table 2: Acceptability categories for peak alarm rate following a major upset, as outlined in EEMUA Publication No. 191 (adapted from Figure 45, p. 107)

Figure 2 presents the averaged peak alarm rate distribution for each console to examine how much of the time each console was experiencing a specific alarm rate (see the legend in Figure 2). This contribution chart in Figure 2 was generated by averaging values across each console's set of three month data, if applicable, for the Peak Alarm Rate Distribution intervals (e.g., "% with < 1", "% with 1-10", and so on). The "% of 10 minute periods" for the various alarm count intervals are plotted on a Log scale to highlight those periods in which there were peak alarm rates and increased operator workload.

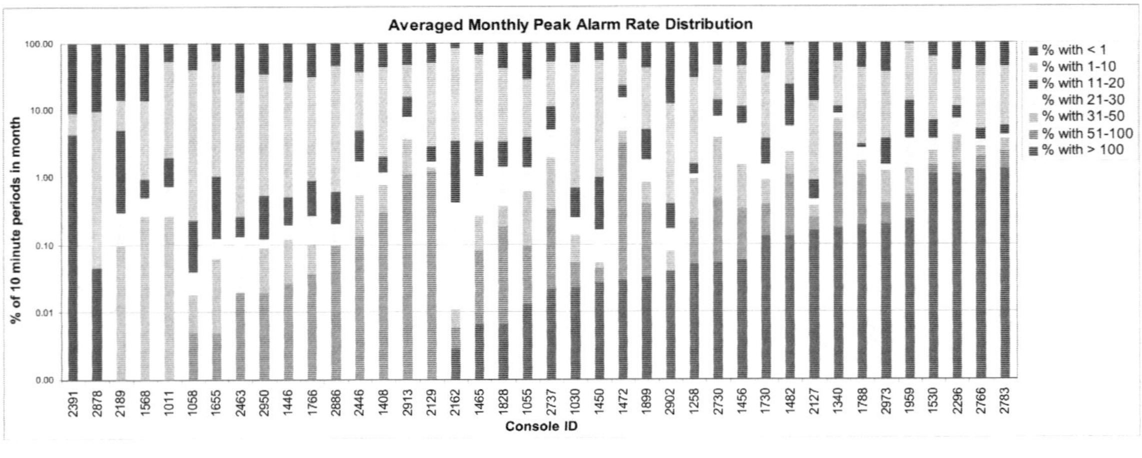

Figure 2: Monthly average peak alarm rate distribution per console. Bars for consoles that submitted three months of data show the averaged value -- across the three months reported -- for each peak alarm rate category.[3]

[3] Care is needed in interpreting the Log-plotted results. As a general rule, bar lengths can not be compared between consoles (e.g., Console 2878 has 9.5% for 1-10 alarms and the associated, green bar looks as long as Console 1730's 1-10 alarms bar, which is actually representing 34.0%).

Figure 2 shows that no console was able to meet the peak alarm rate recommendation of less than 10 alarms per 10 minute period for the duration of their data collection timeframe. Only two (2) consoles were able to come close to this recommendation; these two consoles had ten minute periods with 11-20 alarms but did not have periods in which the peak alarm rate exceeded 20 alarms. The remaining 35 unique consoles (95% of the unique consoles in the studies) had peak alarm rate periods with 31-50 alarms per 10 minute period, with 24 of those consoles (65% of the unique consoles in the studies) reporting at least one occurrence where the peak alarm rate periods exceeded 100 alarms per 10 minute period.

While the two best consoles (consoles 2391 and 2878) did not indicate 10 minute periods exceeded a peak alarm rate of 20 alarms during their respective data collection timeframe, it is not possible to assess whether these consoles' peak alarm rate performance was due to superior performance or due to the fact that they did not experience significant process upsets in that timeframe. To that end, our data collection methodology was insufficient insofar as we did not collect alarm rate data specifically tied to plant upsets. The only conclusion that can be drawn from the current data set is that this latter EEMUA recommendation alarm rate performance is a challenge with today's technology and practices.

3.3 Relating Average Alarm Rate for Normal Operations to Benchmarking Metrics

One of the observations made by ASM members about EEMUA Publication No. 191 is that the recommendations for alarm rate performance are devoid of context. That is to say, the recommendations for 1 alarm per 10 minute period during steady-state operations and not more than 10 alarms in the first 10 minutes following an upset are not accompanied with statements about expected level of automation to support those recommendations, the size or operations complexity of the unit being considered, the number of control loops or total number of points in the unit, and so on.

The ASM study did analyze whether there were strong linear correlations between the average alarm rate for normal operations and many of the other general, configuration, and maintenance metrics collected. Not surprising, few -- if any -- strong bivariate correlations existed between any one of the other metrics and average alarm rate. In light of this, a multivariate best subsets regression was performed to investigate the relation that multiple, potentially influential metrics might have on the average alarm rate for normal operations. This would allow the consideration of multiple relations with average alarm rate simultaneously, as opposed to the single comparisons allowed by the bivariate correlation analyses. Intuitively, such an analysis is more reasonable, given the multiple tactics being employed at sites to address alarm management issues.

While the ASM study did collect over 30 different metrics for each console, in addition to the average alarm rate and peak alarm rate data, the ASM study team identified fourteen potentially influential metrics for inclusion in the multivariate regression analysis, based on the results of the bivariate correlation analyses and their collective industry experience (Reising & Montgomery, 2005). The fourteen potential metrics included:

- the total number of configured alarms for the console
- the percent of configured emergency priority alarms out of all configured alarms on the console
- the percent of configured high priority alarms out of all configured alarms on the console
- the percent of configured low priority alarms out of all configured alarms on the console
- the percentage of points rationalized on the console
- the total number of control valves for the console
- the total number of configured alarms per control valve on the console
- the total number of analog inputs for the console
- the degree of modal alarming on the console (i.e., the number of strategies that dynamically change alarm settings based on either process conditions or operating mode)
- the control sophistication of the console (a qualitative rating based on the presence of multivariate predictive control on the unit and the number of emergency shutdown systems for the console)
- the frequency of monitoring the top ten worst actors on the console
- the frequency of monitoring the average alarm rate
- the frequency of monitoring the peak alarm rate distribution
- the average number of standing alarms on the console

An "optimal" set of variables for the best subsets regression resulted in the best model, although the selection of the optimal set was somewhat subjective, given that the R^2 only changed slightly after subsets of eight predictor variables. This resulting optimal model was selected as the lowest-order model with $R^2 > 53.0$, which was the point where R^2 improved only incrementally as additional predictor variables were considered. The plotted residuals for this model followed a normal distribution on a Normal Probability Plot, and there was no obvious pattern in a scatter plot of the residuals, both suggesting the assumptions of regression analysis were being met.

The resulting equation for the Best Subsets regression contained the following factors (presented in the relative order of weighting):

- The *degree of rationalization* had a negative coefficient, meaning that as more points on the console were rationalized, the average alarm rate decreased.
- The *frequency for monitoring average alarm rate* had a positive coefficient, meaning that the more frequently monitored consoles also had the higher average alarm rates. This was a surprising result and the lack of a negative correlation suggests that simply having a monitoring program is not sufficient. Rather, supporting work processes (e.g., a worst actors program) need to be in place to take advantage of monitoring programs.
- The *degree of control sophistication* had a negative coefficient, meaning as control sophistication increased, the average alarm rate decreased.
- The *frequency for monitoring Top 10 worst actors* had a negative coefficient, meaning that as the worst actors were more frequently identified, the average alarm rate decreased.

- The *average number of standing alarms* had a positive coefficient meaning that the average alarm rate increased with an increase in the average number of standing alarms.

The actual values for the constant term and coefficients from the Best Subsets regression equation are not presented because we do not believe these values can be used to "predict" an expected average alarm rate. Our purpose for the regression analysis was not for predictive purposes, but rather to explore the relations between various potential influencing factors and the average alarm rates observed. In addition, this model only accounted for approximately 50% of the variance in the average alarm rate data. The conclusion to be drawn from the regression results are that good alarm system performance requires a battery of techniques and practices, from alarm rationalization upfront, to ongoing maintenance activities such as addressing worst actors. And there are other influencing factors not captured in our data set that account for the other 50%. One of the limitations in our study, which might account for some portion of that other 50%, was that we did not ask participants to quantify:

- how proactive or reactive their operators were in running the plants.
- how well equipment was maintained or how hard the unit was being pushed against its limits.
- the quality of the alarm rationalization efforts or how long it had been since the rationalization was complete.

4. ANECDOTAL SUCCESS STORIES

In addition to collecting alarm system performance data, the ASM study also asked participants to provide anecdotal success stories around their alarm management initiatives, including how those initiatives impacted their average alarm rate performance. The following discussions highlight several points that will be called out in the Conclusion section of this paper.

4.1 Success through Alarm Rationalization

Two separate pairs of units of similar manufacturing processes were compared on alarm system performance at Site A. The two newer units had approximately 40% more tags than their comparable older units. The two newer units were considered fully rationalized with respect to alarm configuration while the older units were reported to have very little, if any, rationalization completed.

For the three-month data collection period, the older units were reasonably steady (i.e., no trips, just normal furnace swings), whereas the newer, rationalized units had a complete plant trip in the first month of the collection. When comparing the older comparable units to the new comparable units, Site A found:

- The newer units had an average alarm rate at approximately the EEMUA recommendation of 1 or less alarms per 10 minute period when considering steady-state operations
- The older units (with no trips) had up to 7 times the average alarm rate per minute when compared to the new units of the same type (with trips)

- The older units (with no trips) had up to 20 times the average alarm rate per minute when compared to the new units of the same type if the month with a trip for newer units is excluded (no trips)
- The older units had 2-3 times the number of alarm bursts when compared to the new units of the same type (with trips), where alarm burst is defined by the EEMUA definition of more than 10 alarms in a 10 minute period.

4.2 Success through Addressing Worst Actors

Site B, which was about to embark on an alarm rationalization process, was able to reduce its average alarm rate by nearly 40% by doing a "batch-wise" worst actors review[4] and addressing the most troublesome points through either elimination or repair. Most of the improvements ended up addressing problems introduced in the original engineering of the alarm system including status alarms (alarms indicating a non-critical status changes) and duplicate alarms (alarms for the same measurement or condition). The plant is now operating regularly within the EMMUA "Manageable" average alarm rate benchmark of <1 alarm per 5 minutes (see EEMUA, 1999, Figure 42, page 105 for these benchmarks). Further improvements are expected following the completion of the rationalization process.

Site C, which has over 10 operating consoles, had been doing alarm rationalizations for the last 10 years, but still had high average alarms rates relative to the EEMUA recommendations. This site was able to reduce its average alarm rate by about 75% other a 6 month period by implementing software tools and supporting work processes to routinely monitor alarm rates and address alarm worst actors. Approximately 30% of the consoles experienced much better performance over this period while another 30% experienced noticeably better performance. (The rest of the consoles remained about the same because they already had low alarm rates.) Following this improvement, the number of consoles achieving the EEMUA "Very likely to be acceptable" average alarm rate benchmark (<1 alarm/10 min) doubled to approximately 65% of the site's consoles, and the number of consoles achieving the EEMUA "Manageable" average alarm rate benchmark (<1 alarm/5 min) or better increased to 80+% (see EEMUA, 1999, Figure 42, page 105, for these benchmarks). The site, however, still suffers from high alarm rates following major plant upsets and is now using its alarm metrics monitoring infrastructure to learn how dynamic alarming could reduce/eliminate alarm flood situations.

4.2 Success through Alarm Rationalization and Addressing Worst Actors

Site D, which had a number of operator consoles, was able to reduce its average alarm rate by approximately 33% following an alarm rationalization project. This was despite the fact that the number of configured alarms actually increased by nearly 20% as a result of the alarm rationalization project.

In the following year, monthly worst actors monitoring led to another 33% reduction in the average alarm rate. In the second year following its initial alarm rationalization

---

[4] "Batch-wise" is meant to connote a review of one large block of historized alarm data at a time, in contrast to continuously reviewing the worst actors each week or month.

efforts, continued monthly worst actors monitoring helped the site achieve an additional ~20% average alarm rate reduction for an 86% reduction overall.

Part way through its third year of worst actors monitoring, Site D achieved another 8% average alarm rate reduction and was experiencing only ~6% of the alarms it experienced before initiating its alarm management improvement efforts. This achievement is particularly noteworthy because the site has added ~15% more configured alarms since the initial alarm rationalization work. These additional alarms are mainly due to additional facilities that have been installed at this site.

The site now consistently operates below the EMMUA "Manageable" average alarm rate benchmark of <1 alarm per 5 minutes on all of its consoles, and half of these consoles operate below the EMMUA "Very likely to be acceptable" benchmark of <1 alarm per 10 minutes (EEMUA, 1999, Figure 42, page 105).

5. CONCLUSIONS

Our objectives in the ASM study that benchmarked alarm system performance against the EEMUA recommendations (EEMUA, 1999) for was to (i) determine if these recommendations were being achieved, and (ii) relate these alarm system performance results to other characteristics and metrics of the consoles themselves. From our study results, we believe that the EEMUA recommendation for less than one alarm per 10 minute period for normal operations is achievable. However, the recommendation for not more than 10 alarms in the first 10 minutes following a major upset appears to be more difficult to achieve. Only two of the 37 consoles in this study came close to meeting that recommendation, which suggests that today's practices and technology are insufficient for achieving that level of performance. Our study results indicate that more sophisticated alarm handling techniques (Campbell Brown, 2002), such as dynamic alarming and/or alarm suppression, will have to be applied for alarm flood situations so that this recommendation can be achieved. In addition, the application of such advanced alarm handling techniques will require more explicit consideration of alarm flood situations during the alarm rationalization process.

While our study did not have strong bivariate correlations between average alarm rate and various other metrics, the multivariate regression suggests there are alarm management activities that will have an impact on average alarm rate performance, such as alarm rationalization, improved control strategies, and addressing worst actors. In addition, common sense tells us that good and superior alarm rate performance should be supported by ongoing maintenance (i.e., lifecycle management) practices, such as following sound alarm system design practices, routinely monitoring alarm system performance, and investigating alarm system performance following major upset conditions (beyond the root cause incident investigations often conducted for regulatory compliance).

There is no "silver bullet" or "one shot wonder" for good alarm management. The most successful sites will likely approach alarm management as an ongoing, continuous improvement activity, not unlike preventive maintenance or total quality management programs. However, based on our anecdotal reports above and the improvement roadmap recommended in the EEMUA guide (see "Roadmap," page xii), we would strongly recommend addressing alarm worst actors on a regular basis as the best action a site with limited resources could take to improve its alarm system performance.

6. ACKNOWLEDGEMENTS

This research was funded by the Abnormal Situation Management Consortium (www.asmconsortium.com). We would like to extend our sincerest thanks to the control engineers that provided alarm system performance data from the 37 consoles. In particular, we would like to recognize the following contributors: Rob Angerer, Pat Benavides, Robert Carroll, Darrell Carver, Jim Cawood, Mike Clark, Bruce Colgate, Tim DeMaere, Jamie Errington, Bridget Fitzpatrick, Detlev Imorde, Chuck Mikels, Andy Nichols, Perry Nordh, George Pohle, Sean Ward, and Jeff Zeiger. Special thanks to Wendy Foslien for performing the multivariate regression analysis and helping with its interpretation.

7. REFERENCES

1. Campbell Brown, D., *Horses for courses – a vision for alarm management*, IBC seminar on "Alarm Systems," London: Informa House, June 26 – 27, 2002
2. Engineering Equipment and Materials User Association, *Alarm systems: A guide to design, management, and procurement* (Pub. No. 191), London: EEMUA, 1999.
3. Health & Safety Executive, *Better Alarm Handling*. HSE Information Sheet – Chemicals Sheet No. 6, Sudbury, Suffolk, UK: HSE, 2000.
4. Parks, D. L., & Bouchek, Jr., G. P., Workload prediction, diagnosis, and continuing challenges, In G. R. McMillan, D. Beevis, E. Salas, M. H. Strub, R. Sutton, & L. Van Breda (Eds.), *Applications of human performance models to system design*, New York: Plenum Press, 1989.
5. Reising, D. C., Downs, J., & Bayn, D., Human performance models for response to alarm notifications in the process industries: An industrial case study. In *Proceedings of the Human Factors and Ergonomics Society 48th Annual Meeting* (pp. 1189-1193). Santa Monica, CA: Human Factors and Ergonomics Society, 2004.
6. Reising, D. C., & Montgomery, T., *ASM Consortium Technical Report: Final Report on Alarm System Performance (ASPM) Benchmarking Study – Phase 1 and 2 & Human Performance Modeling for Alarm Response*, Minneapolis, MN: Honeywell Controlled Document, 2005.

Dynamic Risk Assessment of Inherently Safe Chemical Processes: Accident Precursor Approach

Anjana Meel and Warren D. Seider
Department of Chemical and Biomolecular Engineering
University of Pennsylvania
Philadelphia, PA 19104-6393
anjanam@seas.upenn.edu

September 2004

The potential for the loss of thousands of human lives and economic losses, in addition to social and legal complications, have increased the desire to have *inherent safety and security* and *dynamic risk assessment and reliability* prominent in the planning, development, design, control, and operations phases of chemical plants.

This paper focuses on the usage of *game theory* to achieve inherently-safer chemical reactors by shifting the operating regimes from unstable and non-minimum phase behavior towards stable and minimum phase behavior at comparable profitability levels. Emphasis is placed on an exothermic, stirred-tank reactor involving the van der Vusse reactions, a fermentation reactor with substrate inhibition, and a styrene polymerization reactor. Designs are presented that account for the tradeoffs among profitability, controllability, safety, and flexibility. Potential operation changes, redesigns, and reconfigurations are explored to modify the designs using bifurcation and singularity theory. Inherently safer designs ensure a lower frequency of initiating events (undesired events) and provide better controllability and more robust performance.

The failure probabilities of various accident scenarios with different severity levels, given an initiating event, are estimated using the latest developments in probability theory; for example, *copulas* and *Bayesian analysis*, to ensure better predictions. Failure probabilities of safety systems, as opposed to individual plant components, are estimated initially using recently-developed, *information gap decision theory* under severe uncertainties and the CCPS guidelines for quantitative risk analysis. Accident precursor (near-miss and incident) data are used to modify dynamically the initial failure probabilities to obtain posterior failure probabilities with the help of Bayesian theory (both empirical and hierarchical), thereby, generating danger signals far in advance of potential accidents. The proposed technique also helps to assess human reliability under different scenarios and provides a quantitative technique for more effective designs of maintenance schedules and alarm systems.

Anjana Meel and Warren D. Seider
Department of Chemical and Biomolecular Engineering
University of Pennsylvania
Philadelphia, PA 19104-6393

20th Annual CCPS International Conference
RISK MANAGEMENT: THE PATH FORWARD
April 11 – 13, 2005
Atlanta, Georgia

Judging Effectiveness of Inherent Safety for Safety and Security of Chemical Facilities

David A. Moore, PE, CSP
President and CEO
AcuTech Consulting Group
Chemetica, Inc.
88 Kearny Street
San Francisco, CA 94108
dmoore@acutech-consulting.com

Introduction:

Inherent Safety is gaining popularity as a process risk management strategy. Industry's knowledge and use of the methodology is increasing since it was first documented in 1976 by Professor Kletz. But some regulators seem to perceive inherent safety as a panacea for safety and security risk reduction. Recent proposed safety and security regulations are focusing on the method as a first choice.

These developments provoke several concerns. Firstly, most of industry is not formally documenting how they use inherent safety as a strategy for risk management. Engineers tend to make orderly, inherently safer decisions by practice for the most part. This has been expected of industry as a matter of principle, and there is evidence it is being practiced but without a degree of measurement of their actions or the benefits. One of the suspected reasons for this is the lack of formal and agreed inherent safety analysis approaches, and the other is that these requirements simply haven't existed. No one is sure, therefore, of the degree of difficulty that requirements such are being proposed will cause.

Secondly, there seems to be an overemphasis of inherent safety as a singular strategy for safety and security assurances in many of the proposed regulations. Inherent safety has to be considered in light of other risk management approaches where one is not necessarily preferable over another. That decision should be made on a case by case basis rather than blanket requirements.

Thirdly, there is little guidance on how to judge effectiveness of inherent safety, particularly in a fair and equitable way to all parties. This could prove to be a major dilemma for both industry and regulators as they try to justify that 'enough' inherent safety has been applied to be considered 'in compliance' with inherent safety regulations. Experience has shown that regulators and industry have a difficult time interpreting inherent safety and agreeing on adequacy of efforts.

Lastly, blanket regulations such as this may reverse trends in developments of new technologies that indeed may benefit society and could be operated at risk levels acceptable to all.

Given that inherent safety is a rather subjective concept, it makes the matter a difficult one to understand, implement, and regulate. Companies should be knowledgeable of inherent safety, actively document attempts to use the concepts, and should prepare themselves for the challenges of regulation of the issue should it come to that requirement. AIChE should take the lead in research and in defining guidance for industry to follow.

Chemical Security and Inherent Safety

The newly appreciated concerns for terrorism have naturally highlighted the issue of the potential for attack on facilities handling hazardous materials. Out of this concern has sprung a potentially far-reaching proposed act titled the Chemical Security Act of 2001, S.1602. The Act was introduced on 10/31/2001 by Senators Corzine (D; NJ), Jeffords (D; VT), Boxer (D; CA), and Clinton (D; NY).

Inherent safety is a safety strategy, but it is being mentioned in several chemical security-related proposed bills. These regulations have some significant momentum and receive a lot of attention at the moment since the after effects of the terrorist attacks on the United States of September 11th, 2001, are generating many concerns about chemical plants being targets of terrorism. There are several issues of concern for regulations that attempt to make inherent safety a chemical security regulatory requirement:

1. The need to introduce inherent safety as a strategy at all facilities subject to such a regulation is questionable. This would result in a great deal of analysis to consider a single strategy has been applied, vs. setting security objectives and allowing the owner/operator to determine the relevant issues and vulnerabilities and make appropriate risk management decisions. It should be considered as a potential strategy rather than the first priority.
2. The previous issues of common methods for analysis and judging adequacy of inherent safety measures applicable to safety analysis would apply to security analysis.
3. We should not introduce regulatory requirements while under the perception of an urgent threat that are not well thought out and that use anything other than proven, assured benefit methods with a regulatory framework well defined.

Summary of the Chemical Security Act Proposals

The proposed Chemical Security Act bills generally state that there are significant opportunities to prevent theft from, and criminal attack on, chemical sources and reduce the harm that such acts would produce by:

- reducing usage and storage of chemicals by changing production methods and processes; and

- employing inherently safer technologies in the manufacture, transport, and use of chemicals;

Implications of the Proposed Regulations are Far-Reaching

This proposed regulations would have sweeping applicability and significant implications for design and operation of facilities handling hazardous materials. Any new initiatives such as this have to be rational, measured, cost-effective, and fully justified.

Ripe Atmosphere in Congress

Despite gains against terrorist threats in America, U.S. Homeland Security Director Tom Ridge has expressed several times recently that "the risk of another terror attack on U.S. soil was high" and urged the police and the public to ``be more vigilant." According to Ridge at a press conference on December 5, 2001, "Any time you make progress on one front, you have to be prepared for that enemy to use some of its resources and its assets elsewhere. We have to prepare for the potential use of those assets," Ridge told Reuters.

This atmosphere means that the chances are high that legislation such as the proposed Act will be promulgated at break-neck speed. With good intentions, there is a significant potential for possible overreaction and for the extension of these vast increases in

Implied Expectations

The EPA RMP regulations did not include explicit requirements for inherent safety, but there may be an expectation that inherent safety principles were used since its enactment to make major reductions in risk. No doubt the evaluation of data from EPA RMP submissions will focus on whether risk reduction was accomplished over the past 5 years from registrants under that program. Reduction in inventory is one such measure that may be implied to represent degree of inherent safety applied. If not, industry may face criticism caused by the high expectation of the public and misconceptions about the ease of inherent safety to be applied to all risks. The inference will then be that facilities are not doing what is necessary to reduce security risks.

Research on the Evaluation of Inherent Safety

Some methods have been proposed to provide a benchmark for inherent safety. Most of these involve indices or fuzzy logic. While these are excellent developments in the right direction, they are not fully validated or comprehensive enough to assure that the aforementioned issues are satisfied.

There is a need for metrics and rules for how to evaluate inherent safety before regulations can be effective. Without a fair and legitimate way to measure the total risk balance created by changes in the name of inherent safety it will be subjective and possibly unfair.

Complex process systems, particularly with a long history of safe performance, cannot suddenly be dictated that a system is inherently safer without a great deal of individualized risk-risk tradeoff evaluation. Inherent safety is not fully understood, so regulating it and forcing change against typical engineering practices (with a strong empirical basis of success) is not recommended

There have been many experts recognize that this may be creating many other problems by overly relying on one strategy vs. a holistic approach. Facilities should be given that flexibility all the while bounded by appropriate layers of safety to reduce risk to an acceptable level.

Public Perception of Inherent Safety

Often inherent safety is seen as 'obvious' and 'common sense' when in reality the big picture may not be that simple. Risk-risk tradeoffs can have unfortunate results if not properly evaluated. Priorities to inherent safety may mean compromises elsewhere. Efforts to reduce risks often neglect the possibility that measures to reduce the "target risk" may introduce or enhance "countervailing risks." [1]

An example of a format for evaluating risk-risk tradeoffs is shown below:

Table 1. Risk Tradeoff Analysis Classification System

Compared to the Target Risk, the Countervailing Risk affects:	Compared to the Target Risk, the Countervailing Risk is:		
		Same Type	Different Type
	Same Population	*Risk Offset*	*Risk Substitution*
	Different Population	*Risk Transfer*	*Risk Transformation*

An important point is that we need to consider risk management interventions, not a single risk reduction strategy alone. Like medications, any intervention can have side effects. Instead industry and government should condone advocate a proactive, holistic approach rather than heuristic, piecemeal reactions to emergencies-of-the-month.

Confusing Terminology as an Obstacle to Progress

---

[1] Harvard Center for Risk Analysis, "Risk/Risk Tradeoffs in Pesticide Regulation: Evaluating the Public Health Effects of a Ban on Organophosphate and Carbamate Pesticides", George M. Gray and James K. Hammitt, Harvard Center for Risk Analysis And Department of Health Policy and Management, Harvard School of Public Health 718 Huntington Ave, Boston, MA 02115, August 6, 1999

A contributing problem is the failure of the industry to have an agreed, uniform terminology. Several terms are floating that are meant to describe the same concept, which makes for several important issues. For example, several terms used are:

Inherently Safer Technology (IST) – this implies that there is a particular 'technology' that is to be introduced vs. the existing technology behind a plant operation. It limits options for risk management if there is an expectation of a magic bullet technology vs layers of safety that can be used in concert

Inherently Safer Systems (ISS) – this implies there is a physical or engineered 'system' that is being applied. It has similar issue to IST above.

Inherently Safer Design (ISD) – this implies that the opportunity is limited to design, and not to operational inherently safer strategies.

Alternative approaches – This term has been used in some recent regulation, most likely to disguise the concept of inherent safety. This is the broadest term, but is vague and is not universally agreed upon as the same as inherent safety.

Obviously the industry would benefit from an effort to document and stress uniform terminology.

Inherent Safety Challenges

Inherent Safety (IS) is emerging as a key process risk management issue. Process safety professionals have embraced the concepts voluntarily for years and it is an established method for addressing process risks. AIChE/CCPS published a concept book on the topic in 1996[i] and there have been numerous other publications and presentations.

IS is now being considered by regulators and environmentalists as the first process safety strategy industry should be using for reducing process risk from such events as reactive chemistry, terrorism, acute and chronic environmental releases, and process safety in general. Their expectation is that inherent safety can reduce hazards to de minimis levels.

The ultimate goal is to see all companies applying inherently safer principles throughout the design and operating lifecycle of projects and for IS to be effectively used to reduce process risks. But the formal implementation of IS has not been universal. Several key obstacles have caused IS to be underutilized, namely (a) the perception that IS is technically and economically impractical for all but new processes, (b) the lack of an inherent safety infrastructure – a frame work for evaluating IS, technologies supporting IS approaches, methodologies that permit IS to be integrated into technical, economic and EHS design considerations, and (c) the lack of industry guidance on the how to specifically conduct an IS study, particularly for existing plants.

Therein lies a dichotomy – the expectations of the regulators on the use of IS are high, and the actual use by industry is mixed. Regulators see implementing IS as an effective way of minimizing or eliminating the hazards of offsite impact from process incidents. However, in actual practice this has proven to be problematic because IS, at this stage in its development, is more of a conceptual methodology rather than a codified procedure with a well established and understood framework for evaluation and implementation. Furthermore, IS cannot be regarded as the sole design criteria as it must be integrated with other considerations. Despite these limitations there has been a proposal for regulating IS - the Corzine bill "Chemical Security Act of 2002" - and there is also one example of an implemented IS requirement which is part of the Contra Costa County, California, Industrial Safety Ordinance (ISO).

The issues that limit IS from being widely adopted by industry were discussed in a workshop that was held at the 17th Annual CCPS International Conference & Workshop on Risk, Reliability and Security in Jacksonville, Florida, on October 11, 2002.. Speakers from the USEPA, AIChE, Contra Costa County, and industry presented their experiences on the issue.

Barriers Identified For Implementing IS:

In summary of the workshop discussion, the audience agreed that there were barriers for effectively implementing IS, and issues and challenges for any regulation of IS. Understanding these issues is helpful to finding solutions to achieve more widespread use of the concept. Some of the constraints were reported to be as follows:

A. Constraints Related to the Adoption and Implementation of IS by Industry:

1. Feasibility for Existing Facilities vs. New Facilities – A key dilemma is that the majority of the applications for IS are retrospective with the existing industrial installed base whereas the feasibility of applying IS to the fullest at least cost generally diminishes as the facility is built. This leaves many companies where new processes (and particularly new technologies) are rarely implemented resulting in few occasions to practice the methods. It also poses a challenge to find ways to employ IS concepts cost-effectively while considering existing facility constraints.

"Although a process or plant can be modified to increase IS at any time in its life cycle, the potential for major improvements is greatest at the earliest stages of process development. At these early stages, the process engineer has maximum degrees of freedom in the plant and process specification. The engineer is free to consider basic process alternatives such as fundamental technology and chemistry and the location of the plant. Imperial Chemical Industries (ICI) describes six stages of hazard studies, including three during the process design phase and three during construction, startup and routine plant operation. The identification of inherently safer process alternatives is most effectively accomplished between the first and second process design hazard studies (Preston and Turney 1991 – needs ref). At this stage the conceptual plant design meets the general rule for an optimization process - that a true optimum can be found only if all of the parameters are allowed to vary simultaneously (Gygax 1988)" (CCPS, "Guidelines for Engineering Design for Process Safety, 1993)" [ii].

There may be a perception issue in industry that IS is only for new facilities and that once operating there are no feasible IS opportunities. Instead, IS may be applicable throughout the facility lifecycle. IS concepts need to be creatively applied to take a broader view.

"Marshall categorizes safety approaches as "strategic" and "tactical." Inherent and passive approaches are strategic, must be implemented early in the facility development, and have broad and wide ranging impacts on the process design. Tactical approaches, including the active and procedural categories, can be implemented late in the design process and are characterized by repetition and high costs associated with maintenance." (Hendershot, 1997[iii])

2. Costs for Implementing IS Changes – Closely related to the issue of perception is the issue that costs to implement IS may be extraordinary and IS may require changes not otherwise part of the facility's business plan. Costs are a primary concern, particularly to modify existing facilities. Costs have to be evaluated on a holistic basis, in which case it has been proven that the lifecycle costs and risks may be reduced after implementing IS. The CCPS/CWRT recent book "Making EHS an Integral Part of Process Design" is an attempt to provide a broader evaluation framework. There is some evidence that if Environment, Health, & Safety (EHS) input is done concurrently up front there could be savings in life cycle costs. The book also discusses total cost and benefit assessment (a CWRT developed tool) that allows one to include EHS costs in monetary terms and combine these with standard financial analysis.

Companies may be unclear on the value of IS or may be unable to easily prove that IS is beneficial to employ. Methods to prove the value of IS and to quantitatively measure whether a given process is 'as inherently safe as is practicable' are generally unavailable or unproven. Case studies showing the economic and other benefits are not available for a wide array of industrial situations.

3. Knowledge of IS - It is not clear that all parties involved in the design and operation of process facilities are knowledgeable of how IS can be applied, what constitutes IS, and how to make inherent safety judgments. Education is required to both help in facilitating the process and in inspiring users to practice it.

4. Agreed Upon Methods and Tools for Systematically Conducting IS Reviews Are Not Available – With the exception of the INSET project tools (Mansfield, et al, 1996)[iv] or checklists, there are few inherently safer analytical methodologies available. The Mary Kay O'Connor Process Safety Center has ongoing research on the development of an inherent safety index based on fuzzy logic theory. (Gentile, et.al.,2001[v])

Constraints Related to the Regulation of IS:

Constraints to the regulation of IS include the concerns above plus:

1. Criteria for Making Compliance Decisions – An obstacle to clear cut regulation is the lack of consensus on appropriate IS metrics. Assuming that the regulation is performance-based, there must be metrics for consistent regulation. These criteria are very hard to define with a broad conceptual topic such as IS for the wide variety of chemical processes to be regulated. This dilemma was recently described by the Mary Kay O'Connor Process Safety Center – "Regulation to improve inherent safety faces several difficulties. There is not presently a way to measure inherent safety. Process plant complexity essentially prevents any prescriptive rules that would be widely applicable. It would seem that legislation could explicitly require facilities to evaluate inherently safer design options as part of their process hazard analysis. But inherent safety would be almost impossible to enforce beyond evaluation because there are unavoidable technical and economic issues." (Mannan, et.al, 2003[vi])

2. Need To Consider Risk Rather Than Only Hazard - There is general industry disdain to the idea of imposing a requirement for 'change for the sake of change', i.e., requiring that every hazardous situation be made inherently safer. Industry is interested in referencing a measure of acceptable risk which limits the need for additional risk reduction since beyond that resources may be better spent on other matters.

In the United Kingdom, the Health and Safety Executive, in their well-known report on the tolerability of risk, have proposed an upper level of the risk to life from an industrial accident risk that should not be exceeded, and a lower level that is "broadly acceptable". In between the risk should be made as low as reasonably practicable (ALARP), using cost-benefit analysis (Kletz, 2000[vii]). Some consideration such as ALARP with criteria for feasibility of IS implementation is required.

3. Criteria To Measure Performance – Inherent safety regulations would have to show measurable benefit. If there was an reduction or increase in the number of incidents it could be incorrect to infer whether IS was the leading factor or whether other measures were involved. It is, therefore, difficult to measure the effectiveness of IS regulations.

The USEPA representatives at the workshop reported that the EPA intends to include IS in their analysis of the effectiveness of the Risk Management Plan (RMP) regulation (USEPA, 1996)[viii] when they review the next submittals of registrations and hazard assessments. This is likely to be challenging given the state of implementation of IS and EPA's own admission on their expectation for inherent safety in the Risk Management Planning regulation. When EPA promulgated the RMP rule, some commenters asked EPA to require facilities to conduct "technology options analyses" to identify inherently safer approaches. EPA declined to do so, stating that "PHA teams regularly suggest viable, effective (and inherently safer) alternatives for risk reduction, which may include features such as inventory reduction, material substitution, and process control changes. These changes are made as opportunities arise, without regulation or adoption of completely new and unproven process technologies. EPA does not believe that a requirement that sources conduct searches or analyses of alternative processing technologies for new or existing processes will produce additional benefits beyond those accruing to the rule already. (FR, 1996[ix])

4. IS Means Different Things To Different Audiences - One person's opinion of IS is not another person's necessarily, and the risk reduced to one may impose risk on another. These risk tradeoffs need to be considered and pose a significant constraint to objective regulation.

Contra Costa County, California, has the only chemical process safety inherent safety regulation in the United States at this time. Lessons learned from the implementation of inherent safety requirements for their Industrial Safety Ordinance were reported to include the following: (Moore, 2002)[x].

- Companies found IS to be difficult if not impractical to accomplish for existing processes without major unplanned investments;
- Companies subject to the regulations believed that risk should be considered, not only hazard reduction, to determine the need for IS;
- There were different perspectives on what was reasonable and feasible when determining the need for implementing IS;
- The regulatory guidance provided to ensure that IS was being considered consistently and fully needed to be complete to avoid confusion;
- The public and regulators often mistrust industry if wholesale technology changes and other first principles of IS aren't applied to achieve major risk reduction vs. less substantial inherently safer changes;
- Guidance/training is needed for a team to know how to apply IS effectively;

Recommendations

The workshop concluded that, industry and regulators have a major opportunity and a major challenge ahead to develop effective programs to encourage the broad adoption of IS. AIChE should be a clearinghouse for success stories by sponsoring Best Practice sharing sessions and publishing papers on the topic. AIChE should work with the American Chemistry Council and American Chemical Society to develop technical approaches for IS and to promote use throughout their memberships.

Endnotes

[i] Bollinger, R. E., D. G. Clark, A. M. Dowell, R. M. Ewbank, D. C. Hendershot, W. K. Lutz, S. I. Meszaros, D. E. Park, and E. D. Wixom (1996). *Inherently Safer Chemical Processes: A Life Cycle Approach*, ed. D. A. Crowl. New York: American Institute of Chemical Engineers.

[ii] Center for Chemical Process Safety (CCPS) (1993). *Guidelines for Engineering Design for Process Safety*. New York: American Institute of Chemical Engineers.

[iii] Hendershot, Dennis C., "Chemistry - The Key To Inherently Safer Manufacturing Processes", Presented Before the Division of Environmental Chemistry, American Chemical Society, Washington, D. C. August 21-25, 1994, Paper No. ENVR-135.

[iv] Mansfield, D., Y. Malmen, and E. Suokas (1996). "The Development of an Integrated Toolkit for Inherent SHE." *International Conference and Workshop on Process Safety Management and Inherently Safer Processes*, October 8-11, 1996, Orlando, FL, 103-117. New York: American Institute of Chemical Engineers.

[v] Gentile, M. and M.S. Mannan, "Development of an Inherent Safety Index Using Fuzzy Logic,"*Proceedings of the 4th Annual Mary Kay O'Connor Process Safety Center Symposium - Beyond Regulatory Compliance: Making Safety Second Nature*, College Station, Texas, October 30-31, 2001, pp. 510-526.

[vi] Mannan, M.S., "Challenges in Implementing Inherent Safety Principles in New and Existing Chemical Processes," White Paper, Mary Kay O'Connor Process Safety Center, College Station, Texas, August 2002.

[vii] Kletz, Trevor, "Green Intention, Red Result", IChemE, Symposium Series No. 147., 2000

[viii] Environmental Protection Agency (EPA) (1996). "Accidental Release Prevention Requirements: Risk Management Programs Under Clean Air Act Section 112 (r) (7)." *Federal Register* 61, 120 (June 20), 31668-730.

[ix] 61 Fed. Reg. 31699 (June 20, 1996).

[x] Moore, David A., "Experiences in the Regulation of Inherent Safety", Mary Kay O'Connor Process Safety Center, Texas A&M University System, 2002 Annual Symposium, Beyond Regulatory Compliance, Making Safety Second Nature, October 29 - 30, 2002, College Station, Texas

The Safety Quality Factor – Tuning LOPA in Risk Analysis

J. Gort, MSc MoSHE
TNO Quality of Life
Hoofddorp, The Netherlands
j.gort@arbeid.tno.nl

R.J.A. Kersten, ir. & M. N. Boers, MSc
TNO Defence, Security & Safety
Rijswijk, The Netherlands

ABSTRACT

Over the years safety researchers have integrated human factors components in accident investigation tools or have developed models in which organizational and technical elements are coupled together. These approaches are useful for either the organization that wishes to learn more from accidents or for analysts who wants to know if installations are prone for human error. Management of industrial companies however likes to know firstly its actual safety performance and secondly in what area of safety, money should be spent. Some recent research indicates that it is possible to couple the two worlds of technical and organizational safety and to give an answer to that management question. In the method described, a LOPA (Layer of Protection Analysis) is used to calculate the probability of accident scenarios. Besides that, an analysis of the safety culture, OHS (Occupational Safety & Health) management system and the latent failures within the organization using the Tripod methodology, leads to a Safety Quality Factor (SQF). The SQF is used in the LOPA to incorporate the organization and culture aspects in the analysis. In this paper it is further explored how a SQF can be determined, what the implications of such a SQF can be for the scenarios and what it means for management to use it.

1. INTRODUCTION

TNO has performed a research project to integrate a technical safety approach with an organizational safety approach. The technical study used the Layer of Protection Analysis (LOPA) as a tool to be developed further for a combined risk assessment with the organizational and cultural part of safety. More especially, we integrated the organizational safety analysis with the more technical and (semi)quantitative safety-analysis. This leads to new insight in the most effective way for companies to decide on which safety measures had to be implemented. Eventually, it leads to a decision making tool for management. This tool helps companies to make integral and rational decisions around safety and business.

This paper describes the possibilities we found in this research to combine technical safety-analysis and organizational analysis. We describe the individual instruments and the way we combine them. At the congress we will present the results of the case-study and the way companies can integrate these analysis in their business.

2. METHOD AND USED INSTRUMENTS

2.1. Introduction

The central tool used for the technical risk analyses of the chemical process under study was the Layer Of Protection Analysis (LOPA) . This method is a semi quantitative tool for analysing and assessing risks and is based on the identification of potential accident scenarios. A particular scenario will only take place if certain undesired initiating events (e.g. overpressure, operator failure) occur and if protective layers fail or are absent. There are various protection layers in a chemical process, all of which help to prevent an accident scenario to develop.

New in this instrument is the Safety Quality Factor (SQF). The SQF is based on qualitative analysis of the safety organization and the maintenance organization. The SQF influences the probability of occurrence of the accident scenarios by reducing or increasing the frequency of initiating events and/or one or more protection layers. To determine the SQF an extensive organization survey is performed. For this organizational analysis we combine three instruments:

1. Quick scan OHSAS 18001
2. Tripod Delta survey
3. Quick scan safety culture.

These three methods analyse different parts of a organization:

1. the OHSAS audit focuses on the safety management system, the structural measures an organization has taken to understand and control the OHS (Occupational Safety & Health) risks involved with the working process;

2. the Tripod® Delta survey analyses the level of control the management has on the working environment which can be seen as a measure of the effectiveness of the safety management system; this is described by Latent Failures or Basic Risk Factors; and

3. the Quick scan safety culture is developed to analyse the assumptions which are underneath the existing problems with behavioural safety; for this analysis the ideas about safety culture that management has are compared with evidence of the existing safety culture found at the shop floor.

The results of these combined tools lead to insights of the good and weaker parts of the management influence on safety and give priorities for improvements. Besides that it leads to an overall judgement of the safety management organization, expressed by the Safety Quality Factor.

The overall analysis leads to knowledge of quantitative risks, the technical safety level of the company and the quality of the safety organization. Calculating the investments for the different measures and the results that will be achieved with them, leads to the situation that management have insight in the cost-effectiveness of several options of investment. A management-services tool helps the company to make more integral and more rational decisions around safety and business and to monitor the safety improvement by the use of safety performance indicators. This tool helps companies to improve safety in an effective and efficient way.

In the next two paragraphs we describe the different analysing instruments in more detail.

2.2. Layer of protection analysis (LOPA)

LOPA was developed in the 1980's to facilitate decision-making and answering key questions using a rational, objective and risk-based approach. The key questions are e.g. How safe is safe enough? How many protection layers are necessary? How much risk reduction will each layer provide? LOPA evaluates the individual protection layers as proposed or as present in a given process for their effectiveness and subsequently compared against risk tolerance criteria. In this way the method provides a tool to compare required safety levels with safety measures.

LOPA is a simplified form of risk assessment as typically "order of magnitude" categories for initiating event frequencies, consequence severity and the likelihood of failure of independent protection layers (IPLs) are taken into account. Using this information, the risk of a scenario is assessed. The method thus falls in between qualitative methods like HAZOP, What-if or FMEA and a quantitative method like QRA. LOPA is thus not a fully quantitative risk assessment approach. It is a simplified method to assess the value of protection layers for a well-defined accident scenario. LOPA does not suggest which IPLs to add or which design to choose. It merely assists in judging between alternatives for risk mitigation

2.2.1 The layers of protection

Typically, the following types of IPLs are defined:
1. Process design,
2. Basic process control system (BPCS),
3. Critical alarms and human interventions,
4. Safety instrumented functions (SIF), SIS or ESD,
5. Physical protection (relief devices),
6. Post release physical protection (walls, dikes),
7. Plant emergency response,
8. Community emergency response.

Layers 1 to 4 are preventive measures. These measures are aimed at the prevention of a Loss Of Containment (LOC). In terms of risk such a measure is considered to reduce the *probability* of an LOC. Layers 5 to 8 are mitigation measures. These measures are aimed at minimising the consequences. In terms of risk, a mitigating measure is considered to reduce the *effect* or *consequence*. The characteristics of the safety layers are that each safety layer is totally independent of other safety layers.

To prevent the occurrence of a scenario (or the consequences of an initiating event) only one layer of protection is required. The essence of LOPA is based on the fact that both preventive and mitigating measures have a *probability of failure on demand* (PFD). The primary purpose of LOPA is to determine if there are sufficient layers of protection against the scenario to balance the safety measures with the required safety levels. If insufficient safety measures are present, additional layers may be added or existing layers may be improved. Or, the other way around, if risks are excessively reduced, layers can or may be abandoned (safety critical equipment or instrumentation). Because for a scenario to occur, *all* IPLs should fail.

2.2.2. Scenario and safety gaps

Scenario development is an essential step in the LOPA analysis. Scenarios are developed by a team of experts on the particular installation or process, often making use of existing safety studies like e.g. Hazard and Operability Analysis (HAZOP's) or Failure Mode and Effect Analysis (FMEA's) The result of such an analysis is a list of accident scenario's, with associated frequencies of occurrence. One LOPA scenario is a single cause – consequence chain.

To find out if a particular scenario frequency is acceptable, it is needed that a company sets up acceptability criteria. These criteria are called the LOPA Target Factors (LTF). LTFs can e.g. be created using a risk matrix. For the considered consequence categories (typically five or so) a maximum allowable frequency is defined. Frequencies of LOPA scenarios that exceed the LTF are not acceptable. The difference between the scenario frequency and the LTF is called the Safety Gap. A production process is safe enough when there is no safety gap (left).

2.3. Human factor in LOPA

There are studies of integrating the human factor into the LOPA . For instance Baybutt (2002) proposes the use of a human factors issues list to provide LOPA analysts with various types of human failures which may occur and affect the initiating events, IPLs and enabling conditions. He also suggests to use credits in the calculations to correct for measures taken to prevent the human failures. The LOPA-HF he introduces is a badly needed tool to implement possible human failures into the scenarios of high risk installations. However, the main question is still what will be the probability of a human failure to occur? What are the influences and possible control mechanisms of management on the human factors? This has been the focus of our research.

2.4 Organizational instruments

Each method used for the organizational analysis has a different focus. By factfinding on different organizational levels we triangulate our findings as a validation step. The three different instruments for the organizational analysis are described in the following paragraphs.

2.4.1. The OSHAS 18001 quick scan

The OSHAS 18001 quick scan is mainly an audit method based on the OHSAS 18001. OSHAS 18001 contains requirements for safety management systems based on the ISO 9000 series of quality management. In our research we assess:
- The level of description or the actual presence of the management system, written procedures, instructions etc.
- To what degree the written management system is implemented
- The way management evaluate the quality of the system
- The way the company improves the management system and the actual safety and health of the working place.

The results of the quick scan are translated in a figure per element of the OHSAS 18001. We use the following scoring system for each element: 1= Not present, 2= Intention available, 3= Procedures available, 4= Procedures implemented, 5= Monitoring and controlling in place.

2.4.2. Tripod Delta

Tripod is a model developed for Shell company by the University of Leiden and Manchester (Groeneweg 1998) and is used to measure latent failures in the working environment that can cause human error. Accidents are often attributed to human error, but accident analysis points out that contributing causes of accidents can be traced back to elements of the working environment and eventually management decisions. These controllable parameters are called Latent Failures or Basis Risk Factors (BRF's). The BRF's are: Design (DE), Hardware (HW), Maintenance management (MM), Housekeeping (HK), Error enforcing conditions (EC), Procedures (PR), Training (TR), Communication (CO), Incompatible goals (IG), Organization (OR), Defences (DE).

The Tripod® Delta survey is a checklist-based survey and evaluates actual performance of an organization. The results of the survey is a quantitative profile supported by a qualitative interpretation of the findings. This gives management information on how to eliminate the structural weaknesses in the organization, how to consolidate the strong points and how to set priorities in action planning.

2.4.3. The Safety Culture Quick Scan

The Quick Scan Safety Culture is a new TNO instrument that has been tested in a few case studies. The quick scan focuses on perceived and actual values and believes about safety of different actor groups within an organization, such as the management, operators, (technological) engineers and maintenance employees. By comparing the values and believes of these groups it is possible to determine the real status of the safety culture.

We start with a management meeting in which 10 pre-selected safety management problems are ranked. Pre-selection is based on literature data and on our experience from earlier research. In the next steps we look for evidence in the company for the top three of perceived problems by document analysis, interviews and observations. Also the employees fill in a questionnaire about the safety culture.

The data of the safety culture questionnaire are statistically analysed. Subsequently these results are verified in interviews. At the end we know which perceived problems are truly present in the company, and also if there are any other major management problems with safety culture.

2.5 The safety quality factor

Many activities with respect to safety are working in an indirect way. Training, for example, will affect the IPL 'critical alarms, human intervention'. It may also reduce the frequency of certain initiating events (e.g. operator failure). However, the training activity itself will never prevent a scenario from occurring. Likewise, maintenance will affect the failure frequency of many IPLs, but the activity itself won't stop a scenario from developing. The quality of such safety (related) activities can be determined by the organizational review methods described above. By calculating the outcome and combining the results of the different surveys, it is possible to determine the quality of organizational safety. This is semi-quantified in a so-called Safety Quality Factor (SQF). This SQF can be included in the assessment of the failure frequency of an IPL or the frequency of initiating events as part of the determined scenarios.

2.6 The management tool

All results are discussed in a meeting with the management of the company. Goal of this first meeting is to commit management to the findings and to set priorities. Central in the discussion with management are the safety gaps and how to direct the investment to fill these. In a second meeting we focus on these priorities, trying to find solutions, resulting in concrete actions. We also formulate performance indicators to be included in a Balanced Score Card of the company.

3. RESULTS OF THE CASE STUDY

The OHSAS and the Tripod survey reveiled that the problem areas in the company were about the use of hardware and equipment, maintenance management of the plant and housekeeping. For instance we found that planned maintenance work was postponed because the costs were considered too high, safety procedures and the inspection system were insufficiently implemented in several places, and that safety meetings did not resulted in obvious improvements in the working area. Especcially different actor groups of the company differed in their opinion about the maintenance management and use of hardware and equipment. Overall was management more positive in their reflection on the quality of the basis risk factors which in itself can be riskful.

The following Tripod profile has been made from the case company (see figure 3).

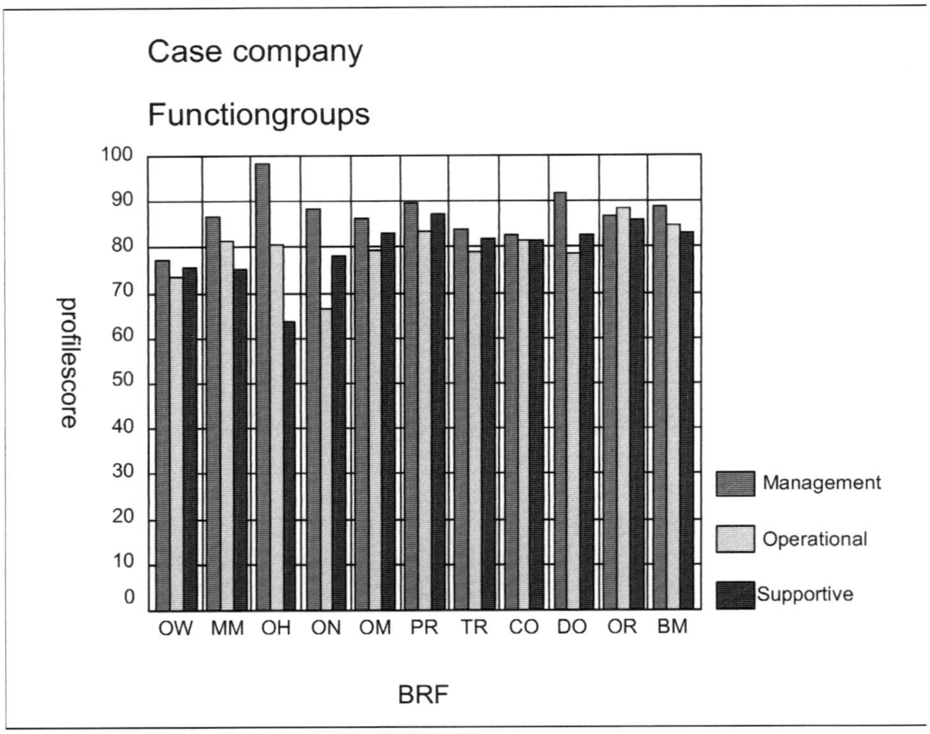

Figure 1: Profile of the case company on the Tripod questionnaire (Abbr. In Dutch)

With the Quick scan safety culture we found that middle managers set a good example in safe behaviour. However supervision and correcting employees is not implemented consistently, because management does not know the actual safety performance. There is less knowledge of daily safety and safe behaviour at the shop floor. Also, in some instances production was reported to be more important than safety, but we think this is more the exception than the rule in this company.

4. Calculating the Safety Quality Factor

The Safety Quality Factor (SQF) should (at least) give an indication of the quality of organizational safety within a company and of the degree in which it influences the independent protection layers or initiating events of the identified scenario's. At this stage we have not been searching for a real quantitative method but we have used the general outcomes of the performed surveys to calculate the quality of the organizational safety. The safety quality factor is based on the OHSAS quick scan and the Tripod survey. Because the Safety culture quick scan has been predominantly a qualitative assessment we have used this instrument as a validation for the semi-quantitative outcome. This leads to a semi-quantitative verdict of the influence this has on the IPL's.

4.1 Calculating the results

The overall score of the case company on the quick scan **OHSAS** audit has been 3,9 on a scale from 1 to 5. Because we are interested in a score which is comparable to the Tripod score (1-10) the OHSAS score has been transferred to this scale: Score[1-10] = (Score[1-5] – 1)/4*9 + 1. This gives 7,5 as a result. To determine the relevance of this score, we consider the following scale division for

an OHSAS-audit based on experiences with the use of the OHSAS audit as a norm for safety management systems[1]:
- 10% of the companies score between 9 – 10
- 60% of the companies score between 5 – 9
- 30% of the companies score between 1 – 5.
This leads to a score division of: Good: 8 – 10 (best 25% of the companies); Pass: 6,3 – 8; Moderate: 4,3 – 6,3; Bad: 1 – 4,3. So the score of the case company is in the scale 'pass'.

The overall score of the case company on the **Tripod**-survey was 81 on a scale of 1 to 100. A comparison with the results of companies that underwent the same survey reveals the following[2]:
- 1% of the companies score between 90-100
- 66% of the companies score between 70 – 90
- 33% of the companies score between 50 – 70.
This leads to a score division of: Good: 82,7 – 100 (best 25% of the companies); Pass: 75,2 – 82,7; Moderate: 65,2 – 75,2; Bad: 50 – 65,2. So the score of the case company is in the scale 'pass'.

4.2. Determining the SQF

The SQF is determined by combining the three techniques. Because organizational failures are often very important contributors to the causes of accidents the four point scale is divided in two negative, one neutral and one positive result. This is to make sure that any upgrade of the quantitative calculation on a risk assessment is only allowed when organizational safety is really good. The SQF is determined in the following manner.

Table 1 Conclusion of organizational safety of the case company

Survey	Bad (--)	Moderate (-)	Pass (0)	Good (+)	Final conclusion
- OHSAS 18001			X		The SQF is considered to be '0'.
- Tripod meting			X		
- Quick scan safety culture	Some ineffective safety cultural findings				

4.3 Use of the SQF in the LOPA

The Safety Quality Factor can be used as a directive to choose the right order of magnitude in human failure. In fact it can be seen as 'fine tuning' the frequencies used for the calculations of the scenarios (a.o. initiating event and probability of failure on demand frequencies). A typical range for PFD of an IPL is between 1 (no effect) and 0.01. For the human factor the range is normally smaller: between 1 (no effect) and 0.1.

An appropriate method for determination of the frequencies is desirable, especially when both an IPL and the initiating event are based on human action. Table 2 presents how the SQF can be used

[1] This scale divison is based on several quick scan OHSAS audits performed between 1998 - 2003.
[2] This scale division is based on more than 100 Tripod Delta surveys performed between 1996 and 2003 (information of Tripod International BV).

to determine frequencies, where 'good' represents the upper limit of the range and 'bad' the lower limit of the range.

Table 2 Effect of Safety Quality Factor (SQF) on frequencies with a range of 1 to 0.1.

SQF	Bad	Moderate	Pass	Good
Score	--	-	0	+
Correction factor	10	5	2	1
Index number	-1	-0.7	-0.3	0

The correction factors can be used in the calculation of scenario's which are influenced strongly by human handling. For instance if a runaway can be induced by putting a wrong substance into a process and this scenario is mainly covered by procedures than the quality of the safety organization should at least be adequate. The SQF can even be cumulative in a scenario where there are several human interventions which can fail. If the example scenario of putting in a wrong substance in the process is furthermore covered by operator control and inspection and maintenance, the company culture will have a strong impact in the scenario frequency. The effect of (improvement of) the safety culture will become visible in the amount and size of the safety gaps.

5. DISCUSSION

The Safety Quality Factor is used to correct the results of the LOPA analysis. This may result in higher frequencies of failure of protection layers or initiating events resulting in higher scenario frequencies. In this case the company studied had a safety quality factor of '0', which means that for any range of numbers to be used in the calculations of the scenario's the middle or mean number has to be used.

The SQF seems to be useful in determining the influence of the organizational safety in manifestation of scenarios. It is quite a rough correction factor to use and can be further developed into specific human factors areas like e.g. procedures, communication, maintenance and competence. By thoroughly analyzing the OHSAS, Tripod and safety culture analysis on the topics which are involved with scenarios it may be possible to determine the correction factor per issue. For instance a company with good procedures and bad maintenance needs more correction on the calculation of scenarios where maintenance is involved. This will be subject of our coming study.

Apart from the calculation of the SQF and the integration of its result in the LOPA analysis, we drew some general conclusions based on the three different analyses. In this case we concluded that in this company many safety measures are present, in a general sense. However, when looking into more detail we found that the company lacks precision and is nonchalant in implementing and controlling procedures and regulations. This leads to several occurrences of non-safe behaviour or dangerous situations. With respect to the three organizational safety methods it can be concluded that it is possible to give both a (semi-) quantitative and a qualitative judgement on the organizational safety of a company. A combination of three organizational analysis methods discloses additional facts and evidence and it leads to more opportunities to improve safety performance. At the same time the three different methods all revealed weak spots in the organization with different focus, and can therefore be considered good internal validation tools. In the end the case company has got a lot of relevant information about the process safety and the quality of the safety organization.

The next step will be development of a management tool to increase rational decision-making based on the LOPA-analysis and the organizational safety findings. In such a tool the cost-effectiveness of proposed safety measures (aimed at closing safety gaps) will have to be assessed. Also, safety performance indicators, to be used in a Balanced Score Card, will be developed.

References

- Sonnemans PJM, Körvers PMW, Brombacher AC, Van Beek PC and Reinders JEA. *Accidents, often the result of an 'Uncontrolled Business process'- a study in the (Dutch) Chemical Industry.* Quality and Reliability Engineering International 2003 (in press).
- Baybutt P, *Layers of Protection analysis for human factors (LOPA-HF).* Proceedings of the Symposium on Juman Error in Occupational Safety, American Society of Safety Engineers, pps 163 – 175, March 13-14, 2003, Atlanta, Georgia.
- le AR, Guldenmund FW, Smit K & Bellamy L. *Modification of technical risk assessment with management weighting factors.* Safety and reliability - ESREL 1998.
- llamy L. *I-RISK, integrating management and technical systems.* OECD workshop on audits and inspections 2001.
- rvers PMW, Schaafsma J, Sonnemans PJM. *Invest in safety or invest in production: a dilemma?* Making process safety pay, the business case. Center for Chemical Process Safety, International Conference and Workshop. 2001
- Go J, Zwetsloot G.I.J.M., Lemkowitz S, Steijger N. and Moonen C. *Ensuring safety during manpower changes in the chemical process industry.* Symposium on Human error, London. 2003.
- CCPS of the American Institute of Chemical Engineers, *Layer of Protection Analysis, simplified process risk management,* 2001.
- *OHSAS 18001, Occupational health & safety management systems - specification.* ICS code 13.100. NEN, 2000.
- Groeneweg, J. *Controlling the controllable. Preventing business upsets.* fifth edition. Center for Safety research, Unit of Cognitive Pychology. Global Safety Group, 2002. ISBN 90-6695-140-0.

Paper (152b)

FERC Consequence Analysis Model for LNG Spillage onto Water: Effects of Cargo Tank Release Assessment Variables

Yuanhua Qiao[1], Harry H. West[1], M. Sam Mannan[1], David W. Johnson[2] and John B. Cornwell[2]

[1]Mary Kay O'Connor Process Safety Center
Chemical Engineering Department
Texas A&M University System
College Station, TX 77843-3122, USA

[2]Quest Consultants Inc.
908 26th Avenue N.W.
Norman, Oklahoma 73069-8069, USA

Prepared for Presentation at
American Institute of Chemical Engineers
2005 Spring National Meeting
20th Annual CCPS International Conference
Atlanta, Georgia
April 10-13, 2005

February 15, 2005

UNPUBLISHED

FERC Consequence Analysis Model for LNG Spillage onto Water: Effects of Cargo Tank Release Assessment Variables

Yuanhua Qiao[1], Harry H. West[1], M. Sam Mannan[1*], David W. Johnson[2] and John B. Cornwell[2]

[1]Mary Kay O'Connor Process Safety Center
Chemical Engineering Department
Texas A&M University System
College Station, TX 77843-3122, USA

[2]Quest Consultants Inc.
908 26th Avenue N.W.
Norman, Oklahoma 73069-8069, USA

ABSTRACT

Models and published experiment data for predicting the consequence of LNG spills onto water are presented in this paper. LNG release process is modeled based on the steady state Bernoulli equation. The evaporation and dispersion processes are modeled by FREC model. The spillage consequences are dependent upon the tank conditions, release scenarios and the environmental conditions. Sensitivities of the spillage consequence to those release assessment variables are assessed in this paper.

INTRODUCTION

Liquefied natural gas (LNG) is playing an increasingly important role in the natural gas industry and energy markets. Take the U.S market as example, industry analysts predict that LNG imports could increase to 5% of total U. S. gas supply by 2007 (1). Marine transportation of LNG has been carried out with a very good safety record since 1959 (2, 3). However, the risks associated with LNG have been debated for decades. The events of September 11, 2001 have raised concerns about the potential for terrorist attacks on LNG tankers. That combination of recent interest in expanding or building new facilities to receive LNG carrier, along with increased awareness and concern about potential terrorist action, has raised questions about the potential consequences of incidents involving LNG marine transportation.

LNG HAZARDS

Natural gas is combustible, so an uncontrolled release of LNG poses a serious hazard of explosion or fire. The major LNG hazards include:

Pool Fires

If LNG spills near an ignition source, the evaporating gas in a combustible gas-air concentration will burn above the LNG pool (Methane, the main component of LNG,

* Corresponding author. Tel.: + 1-979-862-3985; fax: + 1-979-458-1493.
E-mail address: mannan@tamu.edu (M. Sam Mannan).

burns in gas-to-air ratios between 5% and 15%). The resulting pool fire would spread as the LNG pool expanded away from its source and continued evaporating.

Jet fires

When compressed liquefied gases are released from storage tanks or pipelines, and an ignition source is available in the flammable range, a jet fire may occur. Jet fire could occur during tanker unloading or transfer operations when pressure is increased by pumping operations. Since LNG carrier is not operated at high pressure, there is no scenario that includes a jet fire during a cargo tank incident.

Flammable Vapor Clouds

If LNG spills but does not immediately ignite, the evaporating natural gas will form a vapor cloud that may drift some distance from the spill site. If the cloud subsequently encounters an ignition source, those portions of the cloud with a combustible gas-air concentration will burn, and the vapor cloud fire would gradually burn its way back to the LNG spill where the vapors originated and would continue to burn as a pool fire.

Rapid Phase Transition or Flameless Explosion

The phenomenon of rapid vapor formation with concomitant loud "bangs" has been observed when LNG was released on water. This non-flaming physical-chemical interaction is referred to as "rapid phase transition" or "flameless explosion". It is believed that the rapid phase transition in the process of LNG spillage onto water will not propagate into a significantly larger damage scenario (4).

BLEVEs and Confined Space Explosions

If significant confinement of the vapor cloud occurs after an accidental LNG release, damaging overpressures (explosion) may occur. BLEVE scenarios are not realistic for LNG containment systems, but may be possible with associated process equipment used for refrigerant systems or BTU control systems.

BACKGROUND

Since the 1970s, large amount of theoretical and experimental research has been conducted on the evaporation of LNG spills on water, the dispersion and combustion of the evolved vapor clouds, and the establishment of pool fires at the spill site. Experimental data on LNG spread evaporation process on water have been obtained by the Bureau of Mines (5, 6) and other agencies or institutes. Those experimental results, such as the relationship between maximum spread and the quantity of LNG spilled on water_have been reviewed in reference (7).

Model Validation against Available Experimental Test Data

Qualitative data began to emerge from the Lake Charles experimental project in the 1950s. Beginning in 1970, the Shell Pipeline Company in Houston conducted about 300 laboratory tests consisting of spilling from $1*10^{-3}$ ft^3 (30 ml) to $0.21 ft^3$ (6 liters) of cryogens on water (8). In 1968 and 1969, the U. S. Bureau of Mines Safety Research Center at Pittsburgh conducted LNG spill tests up to about 16.6 ft^3 (0.47 m^3) on a quiescent pond (5,6). Throughout the 1980s and 1990s, several other LNG spill experiments were conducted. The largest scale LNG vapor dispersion test to date was about 1412 ft^3 ($40m^3$).

However, the consequence models validated with these small scale experimental data are being used for spill scenarios of 3,302,125 gallons ($12{,}500m^3$), which is the scenario modeled in Federal Energy Regulatory Commission (FERC) report (9), and even up to a full cargo tank volume of 6,604,250 gallons ($25{,}000m^3$).

LNG Source Term Calculations

Almost all authors use the orifice model, but variations exist in assumed initial conditions and orifice coefficient.

Recently Fay (10) presented two release rate models for LNG cargo tank ruptures, one for holes above the seawater level and the other for holes below seawater level. Further detail analysis in his paper is based only on the above seawater level model.

In May of 2004, ABS Consulting Inc. (3) performed the work on consequence assessment methods for incidents involving releases from LNG carriers for FERC. The simple orifice model was used in their report to model the rate of LNG release from the tank. FERC (9) then updated the ABS report in June of 2004, with some modifications made to the consequence modeling methods, such as using a discharge coefficient of 0.65 to the calculation of outflow from the ship.

DNV (11) used an expert team Delphi type method to assess hole sizes in the hull and LNG cargo tank. They concluded that 0.82 ft (0.25m) would be a credible hole size for a "puncture" type event, 2.46 ft (0.75m) is the maximum credible hole from accidental operational events, and 4.92 ft (1.5m) is the maximum credible hole size from terrorist event.

LNG Vaporization Rate on Water

LNG vapor generation is calculated based on the heat transfer from water into the pool. In Otterman's model (2), which is the most widely accepted LNG evaporation model, the vaporization rate of $0.04 lb/ft^2/s$ (that is about $0.18\ kg/m^2/s$) is based upon the experimental data from Bureau of Mines. Sometimes vaporization rates are reported as thickness regression rates, with a typical commonly used value of 1 inch.

While in Opschoor's model (12), the evaporation rate of $0.01\ lb\ ft^{-2}\ s^{-1}$ ($0.05\ kg\ m^{-2}\ s^{-1}$) was derived from convective heat flux equation, but it is not widely used. FERC recommended using the value of $0.037\ lb\ ft^{-2}\ s^{-1}$ ($0.17\ kg\ m^{-2}\ s^{-1}$). The evaporation data was then employed in the mass conservation equation for LNG evaporation process, which is combined with the pool spread models to get the maximum radius of LNG pool and the time needed for all LNG to be evaporated.

LNG Pool Spread on Water

Early spread models are based on the steady state Bernoulli equation and approximate axi-symmetric spread on water (13). With this approach, spread is driven strictly by gravity, and the rate is given as a function of pool height only. Raj and Kalelkar derived a different spreading relationship by equating gravitational force and inertial resisting force (14). This approach also defines spread as driven strictly by gravity. Otterman derived the spread model based on experimental data of oil spills, and concluded that those three methodologies yield almost identical predictions for the maximum pool radius (2).

Quest applied the modified Webber model for spread based on a balance of fluid inertia and gravity force. In addition, Quest attempted to account for the effects of waves on spread, which resulted in a smaller pool radius (15).

For the model of LNG spread on unconfined water pool, usually the method of Webber is applied based on review of the literature and example implementation of the Shaw and Webber spread methodologies. FERC also recommends the Webber model (9).

LNG Pool Fires

If pool fire occurs soon after the LNG is released onto water, the evaporation rate will be the result of both heat from the water and heat from the fire, with the resulting increase in the evaporation rate. The value of 0.075lb/ ft^2/s (0.34 $kg/m^2/s$) was addressed in Fay's model (10). FERC uses a value of 0.058 $lb/ft^2/s$ (0.28 $kg/m^2/s$) for the evaporation rate of LNG under pool fire conditions (9).

For the potential LNG pool fire hazard assessment, the most widely used model is based on the solid flame model. The fire is represented as gray-body emitter, and the geometry is approximated as an upright cylinder that is tilted as a result of wind. Geometric view factors are applied to calculate the thermal radiation to a given receptor. Quest (15) and FERC (9) both employ the solid flame model.

The point source model was employed by Fay (10). In this model, the energy generated by combustion is assumed to emanate from a single point at the center of the pool fire.

LNG Flash Fire

The hazard assessment of flash fires is primarily a matter of applying a vapor dispersion model to estimate the maximum distance to the Low Flammability Level (LFL). No thermal radiant target is considered, rather the flash fire is assumed to burn back to the source causing damage to all vulnerable things in its path.

METHODOLOGY

Release scenario

In the absence of models that account for the true structure of an LNG carrier and the physics of a release of cryogenic LNG, the simple orifice model is generally employed to calculate the LNG release rate.

This model represents flow from a hole in the side of a single hull cargo tank that allows the LNG to flow directly from the tank into the water:

$$Q = C_d \pi \rho_l R^2 \sqrt{2gH} \quad \text{(E1)}$$

Where Q = mass flow rate

C_d = discharge or orifice coefficient

ρ_l = density of LNG

R = radius of hull breach

H = static head above hull breach

It is worth to note that static head consists of both the liquid height and the ullage pressure. Many analysts forget to include the ullage pressure.

From (E1), it can be seen that the mass flow rate is the direct function of the square root of the static head, H. Therefore, the LNG liquid height above the opening, the ullage tank pressure must be defined for each rupture scenario. Since H will decrease as the cargo tank empties, the LNG flow rate will decline with time.

The assumed location of the rupture opening is also a factor since it will determine the liquid height. Most analysts assume a full cargo tank and the rupture is at the waterline. The vapor pressure of the LNG is dependent on the LNG composition, and can vary from about atmospheric pressure to about 2psig.

The hole size assumption in the LNG cargo tank release is the governing factor since it is a linear function of the assumed area of the ruptures. In FERC/ABS reports (9, 3), 1.64 ft (0.5m), 3.28 ft (1m), and 16.4 ft (5m) diameter hole sizes were selected for comparison.

The tank volume is not a direct factor in (E1), however it does have an effect on the release scenario, since the depletion time will change with the different tank volume, even though the initial release rate will be the same given the same liquid height above hull breach.

The discharge coefficient C_d with a value 1.0 is assumed in ABS report (3). After several comments suggesting modification of the orifice coefficient, FERC updated ABS with recommendation to use 0.65 as the discharge orifice coefficient (9). FERC recommended data would be used in this paper.

FERC Scenario for Cargo Tank Vapor Dispersion

The base scenario modeled by FERC is:

<u>LNG property:</u>
LNG composition: Methane (FERC implicit)
LNG density: 422.5kg/m^3 (26.38 lb/ft^3)
<u>Release scenario:</u>
Volume of vessel: 45,567 ft^3 (13,889 m^3) (CANARY input)
Percent of vessel volume filled with liquid: 90% (CANARY input)

Total spill quantity: 4.4 × 10^5 ft^3 (12,500 m^3) (FERC input)
Hole diameter: 3.3 ft (1 m)
Initial liquid height above hole: 43 ft (13 m)
Environmental condition:
Air temperature: 71 °F (22 °C)
Relative humidity: 70%
Wind speed: 6.7 mph (3.0 m/s) and 4.5 mph (2.0 m/s)
Pasquill-Turner stability class: D and F
Surface roughness: 0.03 ft (0.01 m)
Averaging time: 1 sec (i.e., peak concentrations are used)
Heat transfer parameters:
Film boiling heat flux to pool: 26,900 BTU/hr/ft^2 (85 kW/m^2)
Evaporation mass flux: 0.034 lb/ft^2/s (0.17 kg/m^2/s)

FERC Results for LNG Pool Spread & Vapor Dispersion from Cargo tank accident
The spilled LNG, if not ignited immediately, will evaporate, mix with air and move downwind. In FERC report, the method of Webber was employed to model the pool spread process. DEGADIS was selected to model the dispersion process resulting from large releases of LNG. Based on above models, FERC got consequence result shown in table 1 for above base accident scenario.

PARAMETERS	FERC RESULTS
Initial spill rate	7,600 lb/s (3,400 kg/s)
Total spill duration	51 min
Maximum pool radius	420 ft (130 m)
Wind speed and stability class	6.7 mph (3.0 m/s) and D stability
Downwind distance to LFL	6,500 ft (2,000 m)
Time LFL reaches maximum distance	16 min
Wind speed and stability class	4.5 mph (2.0 m/s) and F stability
Downwind distance to LFL	11,000 ft (3,300 m)
Time LFL reaches maximum distance	29 min

Table 1. FERC result for LNG release base scenario

Sensitivity of Scenario Assumptions to the LNG Hazard Assessment

FERC models will be employed in the following to examine the sensitivity of the consequence analysis results to different variables, such as LNG property, spill/spread scenario, and weather condition. All the parameters will keep the same as FERC scenario as listed above, except for the studied parameter for sensitivity analysis in each table. Based on FERC model, WinFERC, the model programmed in Fortran, is compiled to calculate the pool spreading process. WinFERC will supply the rate of vapor evolution from the pool and the pool radius as a function of time. This is then used with the DEGADIS model to assess the dispersion process. WinFERC interface is shown figure 1.

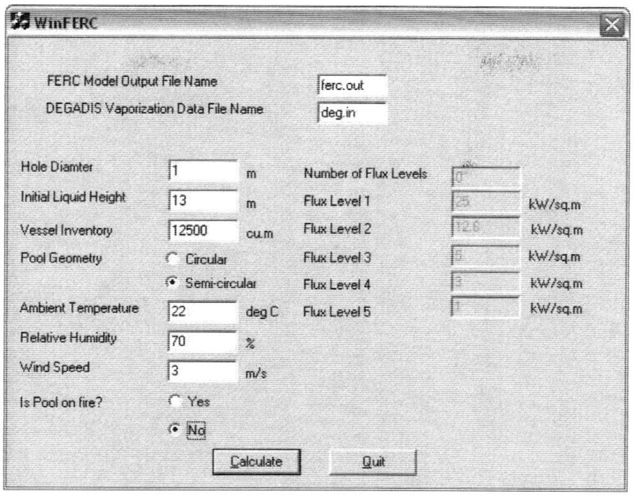

Figure 1. WinFERC model interface

Orifice Size

The following results in Figures 2 & 3 illustrate the effects of orifice size on the release results, including the time to empty vessel and pool size. The spillage rate is in proportion to the square of orifice radius, so the time to empty vessel is decreased dramatically with the increase of hole diameter as shown in figure 2. The pool size was increased at first with the increase of hole diameter, and the pool radius is a linear function of the hole diameter when the hole size is less than 5m, but the pool size would not change after hole diameter is larger than 5m. Figure 4 & 5 also show the effect of orifice size to the LNG vapor dispersion process. The distance to reach LNG LFL is changed by the orifice size as the same way of pool radius vs. orifice size.

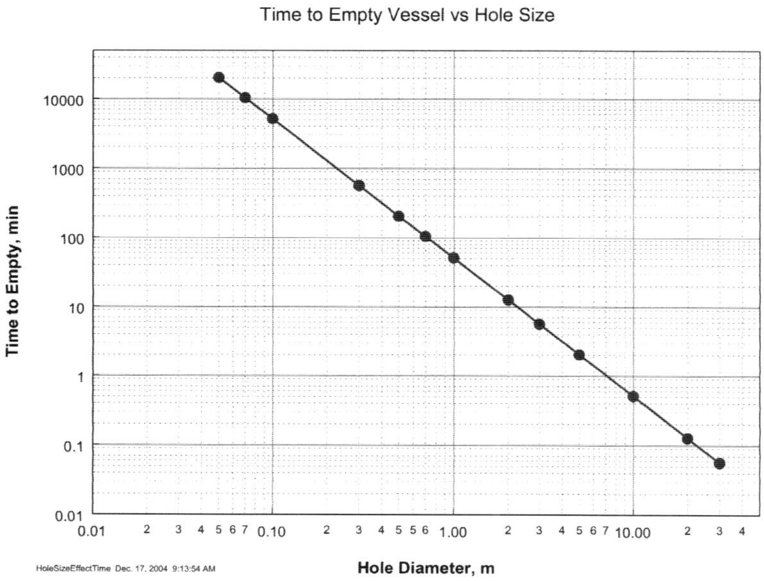

Figure 2. Time to empty vessel vs. hole diameter

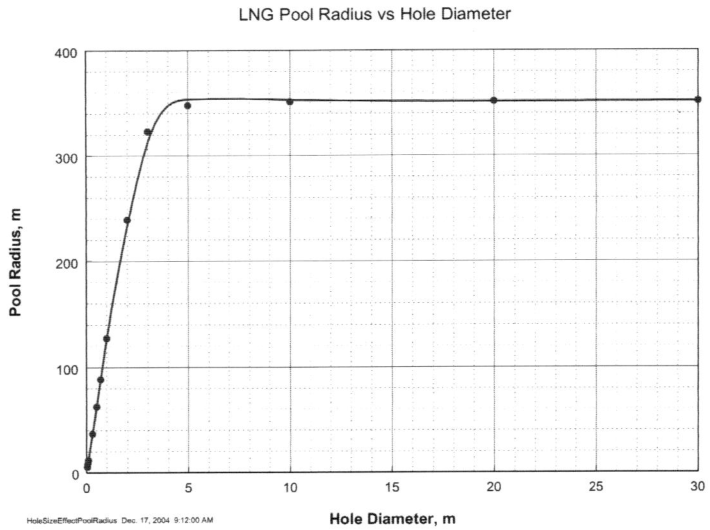

Figure 3. LNG pool radius vs. hole diameter

Wind Stability Class & Wind Speed

Figure 4 & 5 illustrate that the LFL downwind distance is increased with the wind stability class changed from D to F and the wind speed decreased from 5m/s to 2m/s. The wind stability class depends on the amount of incoming solar radiation as well as other factors, and it is an indicator of atmosphere turbulence level. When the wind stability class transforms from neutral class D to moderately stable class F, the atmosphere turbulence tends to be weak, so that the LFL downwind distance would be increased by this change. The decrease of wind speed also diminishes the turbulence level, so the LFL downwind distance will be increased. The difference in Figure 4 & 5 presents the combined effects of wind stability and speed on the LNG vapor dispersion process.

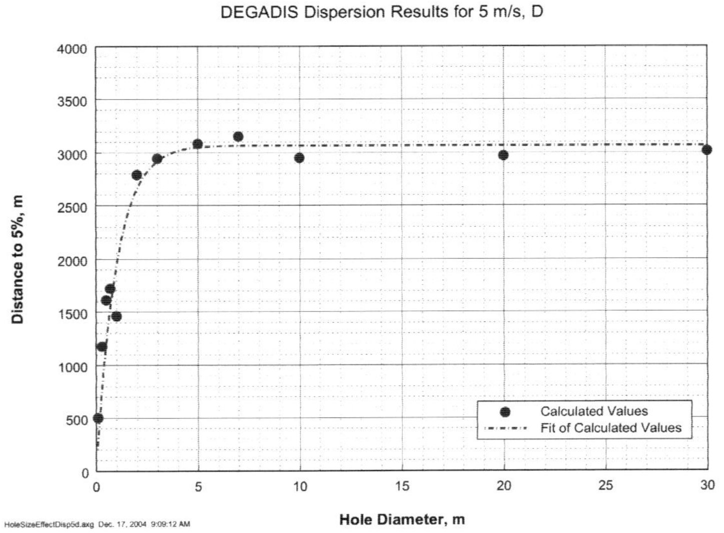

Figure 4. LNG vapor dispersion results at 5m/s wind speed & D stability

DEGADIS Dispersion Results for 2 m/s, F

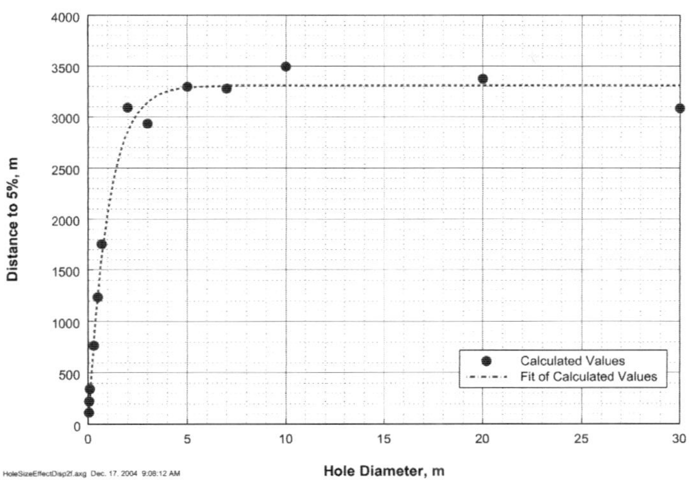

Figure 5. LNG vapor dispersion results at 2m/s wind speed & F stability

Cargo Tank Ullage Pressure

Usually the cargo tank ullage pressure in LNG tank is less than 2 psig. For sensitivity assessment purpose, the maximum ullage pressure considered here is 4 psig. According to Equation (E1), the spill rate is proportional to the square root of static head above hull breach, thus the time to empty vessel would be decreased with the increase of ullage pressure, as shown in figure 6. Figure 7 illustrates that the pool radius is enlarged with the increase of ullage pressure. The vapor dispersion process is also varied by the change of ullage pressure. As shown in Figure 8, when the hole diameter is 1m, the LFL downwind distance is increased with the ullage pressure increased from 0 psig to 2 psig, but LFL downwind distance will not be affected when ullage pressure is higher than 2 psig. While for LNG released from 5m-diameter hole, the LFL downwind distance remains constant with the increase of ullage pressure.

Time to Empty Vessel vs. Ullage Pressure

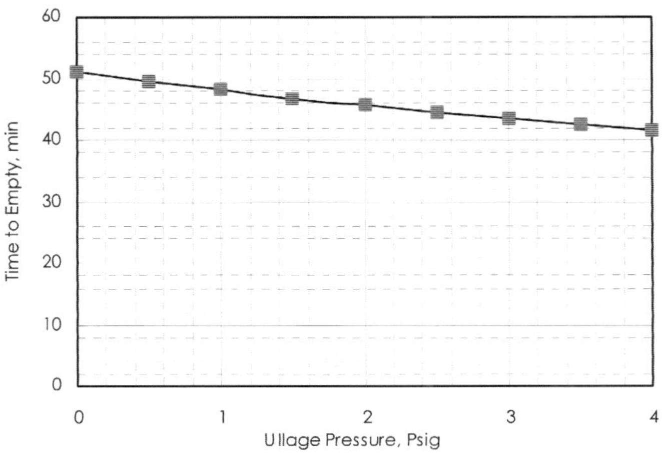

Figure 6. Time to empty vessel for spilled LNG vs. Ullage pressure

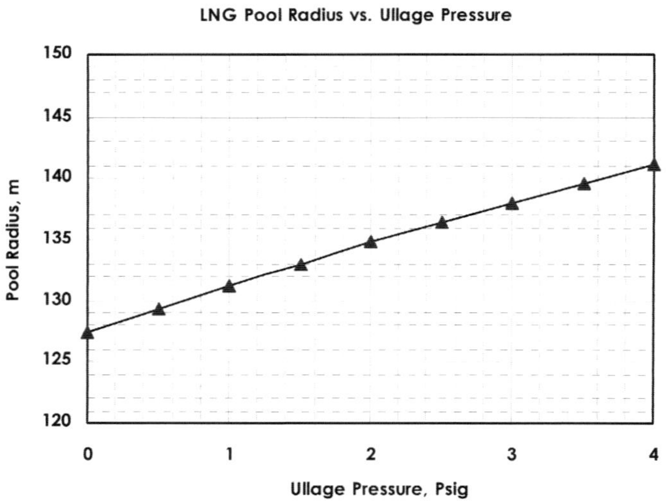

Figure 7. LNG pool radius vs. Ullage pressure

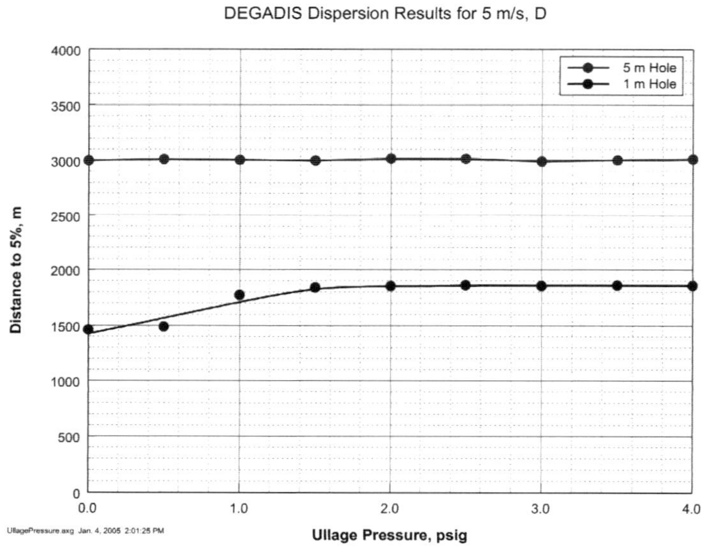

Figure 8. LNG vapor dispersion vs. Ullage pressure

Surface Roughness

Figure 9 illustrates the effect of surface roughness on the dispersion of flammable vapor clouds. The roughness data in FERC mode is 0.01m, which is corresponding to surface of wooded or urban area. The LFL downwind distance is decreased with the increase of surface roughness. The increment in surface roughness increases the degree of atmospheric turbulence near grade, which causes a decrease in the downwind extent of vapor clouds that form and disperse near grade. Such kind of influence is similar to that of atmospheric stability.

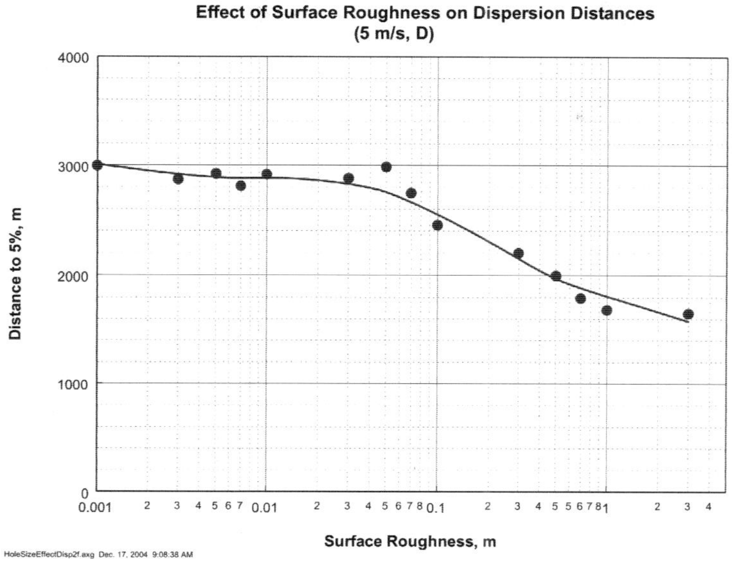

Figure 9. Effect of Surface roughness on dispersion distances

CONCLUSION

When the LNG cargo tank is ruptured, the fluid flows out of the hold onto the surface of water, in an amount and at a rate depending on the tank size, dimension and location of the rupture. The spilled fluid spreads on the water surface, eventually evaporating entirely by heat coming from water, mixing with air and dispersing downwind.

FERC model was employed in this paper to analyze the LNG spillage consequence. The model results illustrate that the changes of orifice size and ullage pressure will change not only the spill duration and pool size, but also the dispersion process. The variation of tank configuration takes effect on the spill process by changing the spill rate and duration. The wind stability, wind speed and surface roughness affect the vapor dispersion process.

Further work needs to be done about the effects of LNG composition on the consequence of LNG spillage onto water. The potential fire scenarios resulted from the spillage of LNG onto water will be explored in detailed in next work.

REFERENCES

1. http://www.energy.ca.gov/reports/2003-07-17_700-03-005.PDF
2. Otterman, B., *Analysis of large LNG spills on water. 1. Liquid spread and evaporation*, Cryogenics, 15 (8): 455-460, 1975
3. ABS Consulting, Inc., *Consequence assessment methods for incidents involving releases from liquefied natural gas carriers*, May 2004
4. West, H. Harry, Qiao, Y. and Mannan, Sam M., *LNG-water rapid phase transition: A literature review*, under submission

5. Burgess, D., Murphy, J. and Zabetakis, M., *Hazards associated with the spillage of liquefied natural gas on water*, U.S. Bureau of Mines, R#17448, U. S. Department of the Interior, November 1970
6. Burgess, D. and Biordi, J., *Hazards of spillage of LNG into water*, U.S. Bureau of Mines, PMSRC Report No. 4177, 1972
7. Gideon, D., and Putnam, A., *Dispersion hazard from spills of LNG on land and on water*, Cryogenics, 17 (1): 9-15, 1977
8. Holt, R. and Muenker, A., *Flameless vapor explosions in LNG/Hydrocarbon spill*, Report No. GRU.1 GPR.72, Esso Research & Engineering, Linden, NJ, 1972
9. FERC, *Notice of availability of staff's responses to comments on the consequence assessment methods for incidents involving releases from liquefied natural gas carriers*, Docket No. AD04-6-000, June 18, 2004
10. Fay, J., *Model of spills and fires from LNG and oil tankers*, Journal of Hazardous Materials, 96 (2-3): 171-188, 2003
11. Pitblado, R., Baik, J, Hughes, G., Ferro, C. and Shaw, S., *Consequences of LNG marine incidents,* presented at the CCPS Conference, Orlando, June 2004
12. Opschoor, G, Spreading and evaporation of LNG-spills and burning LNG-spills on water, Journal of hazardous materials, 3 (3): 249-266 1980
13. Enger, T. and Hartman, D., *Explosive boiling of liquefied gases on water*, Presented at the Conference on LNG Importation and Terminal Safety, Boston, 1972
14. Raj, P. and Kalelkar, A., *Fire hazard presented by a spreading burning pool of liquefied natural gas on water*, presented at the Combustion institute (USA) Western section meeting, 1973
15. Quest Consultants, Inc., *Modeling LNG spills in Boston Harbor*, Letter from John Cornwell of Quest to DOT, October 2,3, 2001; and letter from John Cornwell of Quest to Clifford Tomaszewski of the Office of Natural Gas & Petroleum Import & Export Activities, Nov. 17, 2003

Center for Chemical Process Safety - 20th Annual International Conference
Atlanta, Georgia
April 2005

Facility Siting Analysis – *Results Presented Using a Company's Risk Protocol*

Cheryl A. Grounds, P.E., CSP
Principal Engineer
Baker Engineering and Risk Consultants
3330 Oakwell Court, Ste. 100
San Antonio, TX 78218-3024
(727) 771-7853
cgrounds@bakerrisk.com

ABSTRACT

It has been 5 or 10 years since industry conducted its initial facility siting analyses. It could be time to revisit those studies, if not because the facility has changed, then because there might be room for improvement on the first efforts. Many of the initial studies focused primarily on explosion impacts. The results may or may not have been phrased in terms of risk. The recommendations were likely difficult to integrate with other plant improvement projects. And now resources for such an analysis are even tighter than they were then. This paper will utilize a facility siting case study to highlight analysis improvements including inclusion of explosion, fire, and toxic risks, consideration of toxic ingress into buildings, phrasing results of all scenarios in risk terminology, and presenting results and recommendations in the language of a company's risk management process. The method used is efficient and provides meaningful results that are easily translated into actions.

Introduction

It has been a number of years since many companies conducted their Facility Siting Analyses in response to the OSHA standard "Process Safety Management of Highly Hazardous Chemicals," 29 CFR 1910.119 mention of facility siting. API RP 752[(1)] provided a method that many companies used as the basis for their Facility

Siting Analysis. The challenge with using this approach is that, although the method mentions that explosions, fire, and toxic releases should be considered, it focused primarily on explosions. Also, the risk-based approach described was difficult for some companies to use since their companies had not embraced the concept of detailed risk analysis.

Years have passed now, plant process changes have been made, new buildings have been added, staffing levels and occupancy locations have changed, and risk analysis usage has increased. It might be time now to revisit those original facility siting studies. And given the opportunity, perhaps a greater focus can be paid to fire and toxic release, while utilizing more of a risk-based approach.

A case study will be used to illustrate a current, more comprehensive approach to a facility siting study that addresses the risk of explosions, fires, and toxic releases on occupied buildings and presents the results in a company's risk framework. A few select scenarios will be highlighted from the comprehensive, full report.

A facility siting study was conducted for an agricultural products production plant as a part of the company's ongoing program to address the OSHA standard. The study estimated potential risks to occupied buildings within the plant from postulated explosions, fires, and toxic releases. Occupant vulnerability and accident scenario likelihood were estimated for each occupied building and each hazard, respectively. Risk analysis was then performed using the company's Corporate Risk Matrix. Recommendation options were developed to eliminate or reduce the higher risks.

The methodology used was in keeping with the three-stage approach to identifying and managing risk to building occupants recommended in API 752. The first stage was to screen buildings within the facility by occupancy criteria. The second stage quantitatively evaluated the consequences to the occupied buildings. The third stage estimated the risk of specific explosion, fire, and toxic scenarios to the occupants of buildings, and identified buildings needing risk reduction measures to bring the risks within company risk tolerance criteria.

Plant staff and BakerRisk engineers surveyed the buildings at the site to generate building occupancy data. The determination of occupancy was made according to the Corporate occupancy criteria which included consideration of the following:

- *Emergency command, response equipment location, or safe havens*
- *Individual occupancy in terms of hours per day*
- *Total occupancy in terms of hours per week*

Based on these occupancy criteria, 34% of the identified buildings were considered occupied. The locations of these buildings are shown in Figure 1.

Scenarios were evaluated for explosion, fire and toxic hazards at the facility. The scenarios selected were based on a review of plant and process information including discussions with operations and engineering personnel. The locations of the postulated scenarios are shown on the facility map in Figure 1. The explosion scenarios are prefaced by the letter "E", fire scenarios by the letter "F", and toxic scenarios by the letter "T".

Figure 1. Building and Scenario Locations

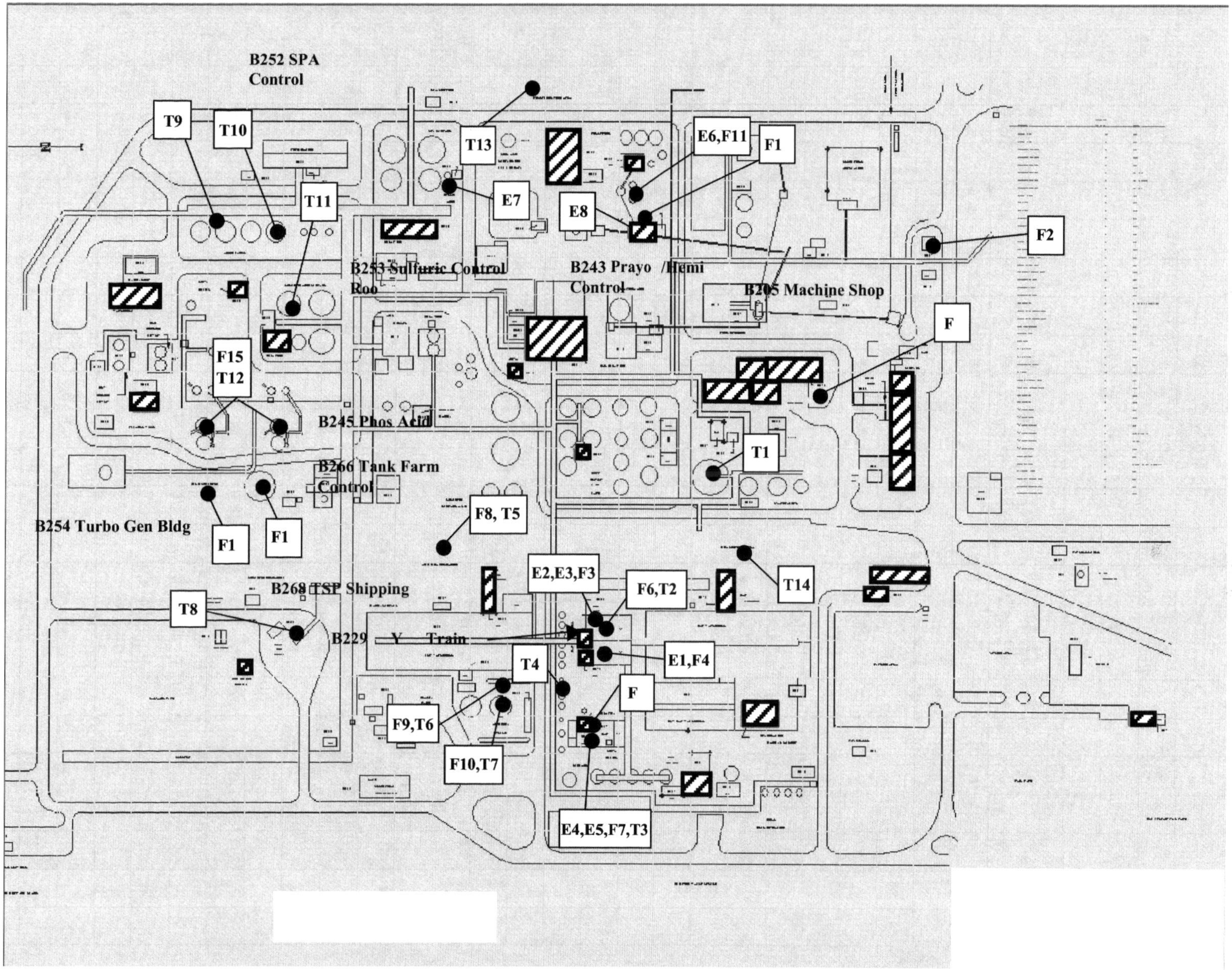

Explosion Scenarios

Explosion scenarios selected included natural gas explosions and an explosion of an ammonium nitrate process vessel. The natural gas explosion scenarios were from process units, some of which were process buildings containing control rooms. The ammonium nitrate case, E7, will be highlighted in this presentation.

Fire Scenarios

Fire scenarios included potential jet fires, pool fires, and flash fires. Fire scenarios included sulfur pit fires, natural gas jet fires, and a gasoline tank fire. All relevant classes of fire hazards were evaluated for all scenarios.

Toxic Scenarios

The toxic scenarios included releases of five toxic chemicals handled at the facility. This presentation will focus on two ammonia releases, cases T2 and T6.

Explosion Consequences

BakerRisk used its facility assessment code(2), $\text{SafeSite}_{3G}{}^{TM}$, to predict blast loads and evaluate building damage from explosions. SafeSite_{3G} employs a release and dispersion model integrated with the most current blast load and building damage prediction models. Scenarios are assessed in SafeSite_{3G} by first calculating the flammable mass in a vapor cloud through dispersion modeling, and then predicting blast loads using the Baker-Strehlow-Tang methodology.(3, 4, 5) The flame speed of the predicted vapor cloud explosions were based, in part, on the degree of equipment congestion and confinement in the areas near the scenarios.

Some of the explosion scenarios involved explosions inside of process buildings that also contained control rooms. The rising pressure of an explosion inside a building will act against the wall panels, which will eventually release from the framework and vent the explosion. BakerRisk estimated the response of the internal control room structural components to pressures developed during the build-up to venting of the building.

Other explosions originated outside of structures. For scenario E7, the ammonium nitrate (AN) process vessel explosion, an equivalent TNT charge was used.

The contours of blast overpressure and impulse were generated from all of the postulated explosion scenarios and are shown in Figures 2 and 3. Building damage was predicted with models based on experimental explosion testing and Single-Degree-of-Freedom structural analysis. The vulnerability of building occupants was based on the methods developed from bombing and accident data.[6] Estimated damage to building structural components was determined and an overall building damage level (BDL) rating was identified. A description of the building damage levels is provided in Table 1. Occupant vulnerability was estimated for the buildings impacted by explosions based on the BDL of the explosion scenarios. The vulnerability is a function of building damage level and construction type.

Table 1. Building Damage Levels

Building Damage Level (BDL)	Summary of Potential Damage
4 - Building Collapse	Primary and secondary structural members will fail or sustain major damage resulting in building collapse
3 - Major Building Damage	Walls facing the blast will fail while other walls have compromised structural integrity. This may cause eventual collapse of the building. Repair of this building is not practical
2b - Heavy Building Damage	There is widespread building damage. Walls facing the blast will sustain major damage while other walls and the roof sustain moderate damage. Building repair may not be practical in some cases.
2a -Moderate Building Damage	There is localized building damage. Walls facing the blast will sustain moderate damage while other walls and the roof sustain minor to moderate damage. The building can be repaired and reused. Window breakage and fallen overhead items are hazards.
1 - Minor Building Damage	Walls sustain the onset of visible damage (denting or bending of metal panels, cracking of CMU, window breakage). Repairs are necessary for cosmetic reasons only. Window breakage and fallen overhead items are hazards.

A single occupied building was predicted to collapse (BDL 4) for a detonation of the AN process vessel, scenario E7. Seven buildings were expected to receive no more than moderate damage (BDL 2a) and a vulnerability of less than 0.1% from the postulated scenarios. None of the other buildings were impacted by the explosion scenarios.

Fire Consequences

Fire consequences modeled included jet fire, pool fire, and flash fire. Where direct flame impingement was possible, there was the potential to engulf a building in a pool fire, or the building could be exposed to prolonged fire radiation, the structure was evaluated to determine the likelihood of fire penetration.

Fire consequences on building occupants were determined through consideration of the factors that influence an occupant's ability to successfully exit a building, including:

- The location and severity of the fire
- The fire resistance of the building
- The number, location, and arrangement of the building exits

Six buildings at the facility were impacted by the postulated fire scenarios.

Toxic Consequences

Toxic impacts on exposed buildings were determined using SafeSite$_{3G}$. The mitigation effect of the building can be appreciable — the indoor toxic concentration may be substantially less than the concentration outdoors. It is necessary to estimate the indoor concentrations based on the outdoor modeling results. The degree of protection given by a building depends on the ventilation rate and the duration of the release. The fraction of the time that the wind is in the direction of concern was included in the estimated likelihood of that scenario.

The toxic vulnerability values were based on the indoor concentration and their relationship to IDLH concentrations. Additionally, building administrative and engineering control strategies (such as provision of supplied air systems) were taken into account.

All occupied buildings were estimated to be impacted by the two ammonia cases of concern.

Risk Assessment

The risks of the potential explosion, fire, and toxic scenarios were qualitatively assessed using the Corporate Risk Matrix. This risk matrix plots scenario likelihood with severity of scenario consequences, as shown in Table 2. Both likelihood and consequences are expressed in terms of order-of-magnitude bands. The intersection of the likelihood and consequence is the risk level, categorized as levels A, B, C, D, or E. This Corporation correlated the risk levels with the timing and type of response required.

Table 2. Risk Matrix

		Severity of Consequence				
		1S Negligible	2S Low	3S Medium	4S Major	5S Catastrophic
Likelihood or Frequency	5L Probable	D	B	B	A	A
	4L High	D	C	B	B	A
	3L Medium	D	D	C	B	B
	2L Low	E	D	D	C	B
	1L Remote	E	E	D	D	C
	0L Extremely Unlikely	E	E	E	D	D

The scenario likelihoods were developed using event trees. This method considers factors such as the frequency of the initiating event, the probability of immediate or delayed ignition, the wind blowing the release toward a target, and other relevant data. The consequences to building occupants were identified in terms of building occupancy vulnerability calculated for each scenario. As would be expected, the process unit design and process safety measures taken at the plant make the scenarios a low likelihood (0L, 1L). The scenarios chosen were severe cases resulting in a high consequence (4S, 5S). Combining the consequence, "5S" and the likelihood, "1L" for scenario E7 we have been considering, we estimated the risk to be level "C". Table 3 lists the percentage of building/scenario combinations considered that resulted in each risk matrix risk level.

Table 3. Risk Assessment Results – Building/Scenario Combinations at Each Risk Level

	Risk Level				
Scenario Type	A	B	C	D	E
Explosion			1%		99%
Fire				2%	98%
Toxic			7%	18%	75%

The analysis showed the risk for the facility was primarily due to the toxic exposure threats. Action Level C risks were associated with the ammonia feed to the plants. A number of other toxic scenarios impacted those buildings immediately local to the release at an Action Level D. The ammonium nitrate process vessel explosion impact on a single building was also a C level risk. All fire risks were a level D or lower.

<u>Risk Reduction Measures</u>

Recommendations were suggested for all buildings with higher risk matrix action levels. Recommendations included items such as the following:

- Measures to reduce potentially released volumes
- Measures to reduce the likelihood of a release
- Relocation of building occupants
- Structurally upgrading a building

<u>Conclusion</u>

This case study illustrates the benefits of re-evaluating facility siting analyses. In the last 10 years, operating plants have undoubtedly changed occupancy levels and process unit design. This prompts the need for re-evaluation of outdated facility siting analyses. Given this need, companies should take the opportunity to consider fire and toxic hazards, and not focus solely on explosions.

In today's economic climate, a cost-effective analysis is required and is feasible using a screening analysis approach with the more efficient tools currently available. As seen in this case study, we considered a large number of buildings and scenarios in a screening approach, which resulted in a relatively small number of risks to address.

By adopting a risk-based approach, the analysis conclusions can be presented in categories of relative risk. This clearly identifies which facility siting hazards pose the greatest risk, and hence, where it may be most appropriate to use resources to reduce that risk. In this case, the site easily translated the suggested risk mitigation measures into projects with definite timelines based on the risk level categories. This facility siting analysis method considered and compared explosion, fire, and toxic risks and enabled resources to be utilized to reduce the greatest risks first.

1. "Management of Hazards Associated with Location of Process Plant Buildings," API 752, American Petroleum Institute, May 1995.
2. SafeSite$_{3G}$™, Version 1.1, Baker Engineering & Risk Consultants, December 2002.
3. Baker, Q. A., Tang, M. J., Scheier, E. A., and Silva, G. J., "Vapor Cloud Explosion Analysis," 28th Annual AIChE Loss Prevention Symposium, Atlanta, GA., April 19, 1994
4. Baker, Q.A.; Doolittle, C.; Fitzgerald, G.A.; Tang, M.J., "Recent Developments in the Baker-Strehlow VCE Analysis Methodology", 31st Annual Loss Prevention Symposium, American Institute of Chemical Engineers (AIChE), March 1997.
5. Tang, M.J.; Baker, Q.A., "A New Set of Blast Curves from Vapor Cloud Explosions", 33rd Annual Loss Prevention Symposium, American Institute of Chemical Engineers (AIChE), March 1999.
6. Charles J. Oswald, P.E., Ph.D., Quentin A. Baker, P.E., "Vulnerability Model for Occupants of Blast Damaged Buildings", 34th Annual Loss Prevention Symposium, March 6-8, 2000.

Methods for determining risks from major accident hazard installations.

Stuart Reston, Tom Maddison, Andrew Fowler, Mark Lawton.
Chemical Industries Division, Hazardous Installations Directorate, Health and Safety Executive, St Anne's House, Bootle, Merseyside, L20 3RA.
(stuart.reston@hse.gsi.gov.uk)

Abstract

This paper describes some methods, tools and techniques that can be used for assessing risks surrounding on-shore major hazards subject to the UK's COMAH 1999 regulations and demonstrating risks have been reduced to a level that is as low as is reasonably practicable (ALARP). The paper describes semi-quantitative techniques based on lines of defence principles such as LOPA and TRAM. It then introduces more quantitative methods developed by the UK's Health and Safety Executive (HSE) such as the approximate risk integral screening tool (ARI) and quick FN. These specifically look at the off-site risk from installations and enable judgements to be made on the impact of risk reduction measures. A tool developed by HSE for determining populations around installations for use in risk calculations is introduced.

Further tools developed by HSE to help duty-holders and regulators judge whether further risk reduction measures are appropriate are described. These include databases on risk reduction measures and their costs, and guidance on what should be included in weighing up costs and benefits. A brief example is given. The work concludes that no one method fits all and that a suite of risk assessment tools and techniques is needed to help make decisions on whether risks are ALARP for the diverse range of installation included under the COMAH regulations.

1. Introduction

The European Union's Seveso II Directive (96/82/EC) requires member states to take all measures necessary to prevent, control and mitigate major accidents from on shore gas, petroleum and chemical installations. In the UK, the Control of Major Accident Hazard Regulations (COMAH) implements this directive. (HMSO 1999) Regulation 4 of these regulations requires operators to "take all measures necessary (AMN) to prevent, control and mitigate major accidents". This is interpreted as the equivalent of reducing risks "as low as reasonably practicable" (ALARP).

Maddison (2003) describes the ALARP concept and proposes how under the COMAH 1999 regulations, a duty holder can make a demonstration that risks have been reduced ALARP. One of the key features of this is that the duty holder's risk analysis and depth of ALARP demonstration should be proportionate to the risk.

This paper considers methods that are commonly used in major hazard risk assessment and describes several other methods and tools that the UK's Health and Safety Executive (HSE) has developed specifically for risk assessment of on-shore major hazards.

2. Methods to determine risks from hazardous installations.

Three general types of risk assessment provide three different levels of depth; qualitative (Q), semi-quantitative (SQ) and quantitative risk assessment (QRA). All three require hazard identification but the extent of the detail needed for the frequency and consequence assessment and therefore the risk analysis is dependent on the level of risk to individuals and to groups of people. (HSE 1999)

2.1 Qualitative risk assessment

Where the risks are small and well known, and the site is not located in the vicinity of possibly incompatible development, a simple description of the types of major accident, their consequences and their likelihoods (e.g. based on published information), and a review of compliance with relevant standards is sufficient. Examples would include low-pressure gasholders, chlorine drum installations in rural areas, ambient storage facilities for low volatility substances, and warehouses containing low toxicity solid products.

2.2 Semi-quantitative risk assessment.

More complex installations will need site-specific hazard analysis with estimates of the consequences and frequencies of a representative sample of events. Accident pathways are often best represented by the 'bow-tie' diagram where the initiating events, preventive measures, release, mitigating measures and consequences are described in a single diagram. Such diagrams aid identification of safety critical systems.

Examples would include most pressurised flammable gas storage sites; some bulk liquefied toxic gas user sites, and pesticides warehouses. As the complexity of the operations and the possible risk control methods increase then more quantified site-specific analysis is necessary. Some examples of semi-quantitative risk assessment methods that have been utilised in safety reports include risk matrices and lines of defence.

HSE's Hazardous Installations Directorate commissioned guidance on the appropriate use of risk matrixes in COMAH safety reports and a synopsis of this was published in 2001. (Middleton and Franks 2001) It was noted that Risk Matrices can bridge the gap between purely qualitative and fully quantitative approaches. They enable combinations of likelihood and consequences of major accidents to be combined in a single diagram. [Figure 1] When all significant events are plotted on the same diagram safety critical events can be recognised and some indication of cumulative risks can be obtained.

Figure 1. An example of a risk matrix

Outcome Likelihood	Single fatality	2-10 fatalities	11-50 fatalities	50-100 fatalities	100+ fatalities
Likely $>10^{-2}$ per year	Intolerable	Intolerable	Intolerable	Intolerable	Intolerable
Unlikely 10^{-2}- 10^{-4}	Intolerable if $>10^{-3}$	Intolerable if $>10^{-3}$	Intolerable	Intolerable	Intolerable
Very Unlikely 10^{-4}- 10^{-6}	Tolerable	Tolerable	Tolerable	Tolerable	Intolerable
Remote 10^{-6}- 10^{-8}	Acceptable	Acceptable	Tolerable	Tolerable	Tolerable

There are a number of techniques that invoke the concept of Lines of Defence (LODs) to help with assessing the adequacy of protection measures. One, developed by HSE, is known as TRAM (Technical Risk Audit Method) and is described by Naylor *et al.* (2000) It is a technique for the approximate estimation of both risk and associated risk reduction for given process sites/plant. TRAM is realized as a computer modeling package whose algorithm requires fault sequence frequency data and end consequence data to be input, from which a computed estimate is derived for the number of LODs required to reduce risks of the scenario to a tolerable level. The auditor must provide input in the form of technical judgments as a result of inspection activity, expressed in terms of quantified "attributes" associated with the various LODs relevant to the scenario being studied. Where there are sufficient LODs in place, risks may be judged to have been reduced ALARP, if insufficient, further consideration would be required.

Another such technique is Layer of Protection Analysis (LOPA) that is a semi-quantitative methodology that is based on 'lines-of-defence' concepts. (CCPS 2001) Simple event sequences are described, starting with an initiator and resulting in a consequence. The consequence can be described in general terms as is usual in 'matrix' type representation systems (eg serious injuries, on site fatalities, off-site fatalities). Safety systems (including operator actions) are represented by IPLs (independent protection layers) that have to satisfy three tests: independence, effectiveness, and auditability. Initiating event frequency and conditional probabilities are given order of magnitude values.

The result is the frequency of an event sequence producing a certain level of consequence. These can be represented on the two axes of a matrix or graph. This can be used to determine the number of IPLs. Where a system has the appropriate number of IPLs the risk may be considered to be 'broadly acceptable' and no further action is required. Otherwise the urgency of the remedial action is proportionate to the number of additional IPLs required and the severity of the consequences. This is shown in figure 2 where the events are plotted on a matrix, the number of IPLs required is shown, and the shading represents the urgency of remedial action. The frequencies are given in cpm (chances per million per year). The result frequencies may be summed to give the probability of certain consequences arising from certain activities (e.g. the frequency of off-site fatality incidents from over-pressurisation of a reactor).

Figure 2. LOPA matrix showing appropriate numbers of IPLs (illustrative)

Freq cpm	Minor	Injury	Severe	Fatality	Disaster
1	1.5	1.5	2	2.5	3
0.5	1	1	1.5	2	2.5
>1k		0.5	1	1.5	2
>100			0.5	1	1.5
>10	No	Action		0.5	1
>1					0.5

There are no HSE published criteria for judging whether the 'risk from' something is considered to be as low as is reasonably practicable (ALARP) or broadly acceptable, and individual risk of fatality criteria cannot be directly compared with the result of the LOPA analysis. The use of LOD/LOPA in the context of COMAH was analysed on behalf of HSE by Amy Vectra Ltd. (HSE 2000) and identified some important limitations:

- Very large consequences are not considered.
- There is no consideration of costs and benefits.
- It cannot consider complex event sequences.
- Other than in a very simple way human factors are not represented.
- Safety measures that do not meet the definition of an IPL are ignored.

For most major accident hazard events at a COMAH installation, LOPA presents a simple methodology with order of magnitude accuracy, which complements hazard identification studies and matrix representation schemes. The more severe or complex cases need more detailed methodology (e.g. QRA). When LOPA suggests more IPLs are necessary but expense is an important factor, the case would need to be referred for cost/benefit analysis.

2.3 A Stepwise approach to demonstrating ALARP.

A stepwise approach is one that Duty Holders have found useful to help them make a demonstration that they have reduced risks ALARP. The steps are outlined below.

1) Identify controlled substances, their inventories and locations. Show the local environment including on and offsite populations that may be affected and other hazardous installations (including those at designated domino effect sites) that might be affected by major accidents or be initiators of a major

accident. Show that the required measures are, at least, to current authoritative good practice.

2) Identify all major accidents and develop a qualitative view on the significance of each one. In the light of the view on the significance of all the identified major accidents, choose a representative subset for detailed consideration.
3) Refine the prediction of the hazard range and its likelihood, for each event in the chosen representative subset.
4) Refine the prediction of the consequences, for each event in the chosen subset, including estimates of the number of fatalities, major and minor injuries to man and damage to the environment, and develop a view on the extent of lesser harms such as major and minor injuries to persons.
5) Show the consequences and the likelihood, for each event in the chosen subset, on a suitable matrix or fN plot (with suitable error bands).
6) Divide the area of the matrix (or plot) into 3 bands (broadly acceptable risk, tolerable if ALARP, intolerable risk) and calibrate these bands against criteria. Suitable numerical criteria for individual risk are set down in R2P2.
7) For intolerable risks immediate action should be taken to reduce risks. For those events in the broadly acceptable region a comparison with relevant standards should be appropriate. For those events in the TIFALARP region look at the major accidents in a particular consequence band and select the ones with the highest frequencies.
8) For these, determine what additional risk reduction measures, beyond relevant good practice standards, (software as well as hardware) may be implemented.
9) Implement these measures unless a reasoned argument is presented showing the costs to implement this scheme are grossly disproportionate. Arguments based upon `strength in depth` concepts such as Layers of Protection (LOPA) or `Lines of Defence` may be used when these have been developed sufficiently. Where the risks have been assessed as high, the use of formal cost benefit analysis may be required to test the cost-benefit balance for the prospective remedial measures.
10) Revisit step 7 for lesser consequences and continue until proportional demonstrations are made.

2.4 Quantified Risk Assessment.

Quantified risk assessment is nearly always time consuming and requires a high level of technical capability but the cost of performing a QRA should not be disproportionate to the value of solving the problem. Consistency in the application of QRA to major chemical hazards can be achieved by standardised methodology, and this can be very effective for comparative studies although care has to be exercised when interpreting the outcome in absolute terms. The methodologies are now well established and provide good insight into the relative merits of risk reducing options. QRA is well suited to complex problems, particularly where high consequence/low frequency events are significant and there is a choice of remedial measures (e.g. Alkylation Units at refineries, bulk toxic gas production facilities) or situations where a standard methodology can be established and easily applied (e.g. assessments for land use planning).

The risk to a population surrounding an installation needs to be measured objectively so as not to be subject to transient concerns and thus form a consistent basis for decision-making. One way of expressing this is by means of a set of frequency and consequence (usually fatality) pairs plotted in the form of a log-log graph, also showing the cumulative frequency and numbers or 'F-N curve'.

2.5 FN relationships:

It is normally accepted that the best available technology for studying risks from major hazard installations is full scope application of quantitative risk assessment. Such studies lead to a relationship between the number of fatalities (N) that can follow a major accident, ranging from 1 to some maximum value (N_{max}) and a frequency (f) at which that number of fatalities is estimated to occur. This relationship, or more usually the corresponding relationship involving F, the cumulative frequency of events having N or more fatalities, is usually presented as a graph with log-log axes (figure 3).

Figure 3 Example of an FN Plot

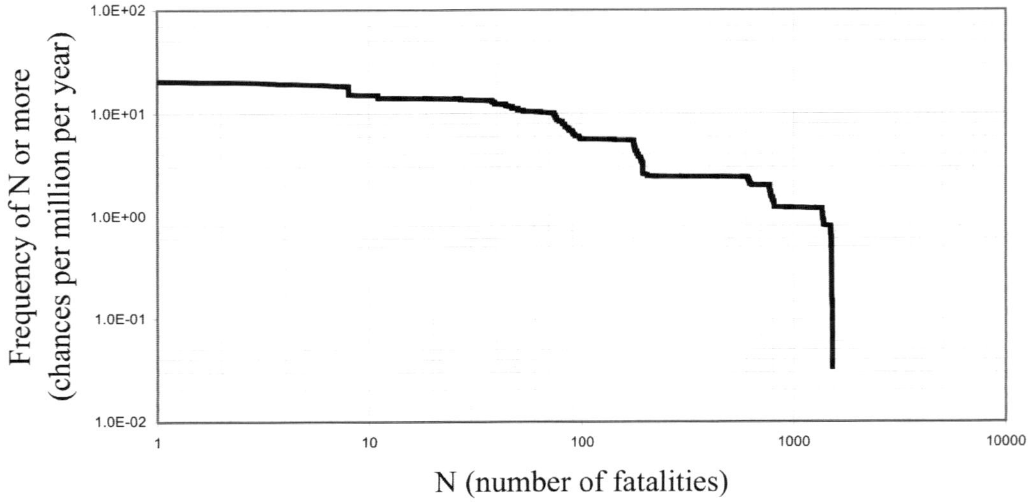

FN relationships can be used to analyse three types of major hazard risk. These have historically been recognised as; case, local and national risk to defined populations. (HSE 1989) This paper concentrates primarily on the risks to populations surrounding individual major on shore hazard installations, i.e. case.

2.6 Aversion and disutility:

Expectation value (EV) and predicted loss of life (PLL) provide measures of risk that take account of the number of people at risk from a hazard. (Hirst 1998) However, they do not distinguish between the relatively frequent accidents that could lead to one or two casualties and those that are rare but would result in considerable loss of life i.e. 10s, 100s or even 1000s of fatalities. Thus EV and PLL do not take account of society's aversion to catastrophic events. This is a subtle but important difference.

Evans and Verlander (1997) examined FN plots and reported drawbacks in using them with criterion lines for making decisions on the tolerability of societal risks. They demonstrated that inconsistent decisions could be made using FN criterion lines. They present arguments to support the use of a 'disutility' function that raises n (the number of people harmed at frequency of f) to some power depending on the level of aversion required. Aversion can be considered the weight given to large accidents. They suggested that decisions should be made by minimizing the expected disutility i.e. reducing the average harm from accidents.

Hirst (1998) developed this approach into weighted risk indicator based upon the mathematical relationships of f and n. The outcome of the disutility function was termed the "risk integral". This was an extension of the simple measure of expected number of fatalities per year:

$$(EV) = \sum f(N).N \quad \text{where } N=1 \text{ to } N_{max}$$

but weighted on the basis of N by an aversion multiplier 'a' of between 1 and 2 e.g. $N^{1.2}$, $N^{1.4}$, N^2, etc. The expression given by Hirst and Carter (2002) for risk integral COMAH (RI) was therefore:

$$RI_{COMAH} = \sum f(N).N^a \quad \text{where } N = 1 \text{ to } N_{max}$$

2.7 Criteria for judging risk – the use of risk integrals.

Developing criteria on tolerability of risks for hazards giving rise to consequences on a defined community is difficult. Nevertheless, HSE has developed criteria for when there is a risk of multiple fatalities occurring in one single event. These were developed through the use of FN-curves. The criteria are based on an examination of the levels of risk that society was prepared to tolerate from a major accident affecting the population surrounding the industrial installations at Canvey Island on the River Thames in the 1970s and 1980s. Reports on the risk from the installations at Canvey Island were discussed in Parliament, and (after improvements) the risk was deemed by Ministers to be just tolerable. (HSE 1978, HSE 1981) The HSC's Advisory Committee on Dangerous Substances subsequently endorsed the limit in the context of major hazards transport (HSC 1991). These criteria are, however, directly applicable only to risks from major industrial installations.

HSE in 2001 revisited its decision making process and published a figure of an accident causing the death of fifty people or more in a single event should be regarded as intolerable if the frequency is estimated to be more than one in five thousand per annum. (HSE 2001a) HSE's Hazardous Installations Directorate (HID) has proposed that this figure could be used as a basis for its operational criteria for case societal risk. This is consistent with the approach outlined in earlier HSE publications (HSE 1989, 1992) but develops the idea of risk integral criteria as mentioned above. The top criterion is the area bounded by a line of slope of –1 drawn through the point (50 deaths at 1 in 5000 years) to where N equals 1000. The bottom criterion is the area underneath the line two orders of magnitude below. Hirst and Carter (2002) describe this in more detail.

For the purpose of accident prevention, an aversion multiplier of 1.4 has been proposed for large industrial accidents. Using this aversion multiplier and the FN lines described above, criteria of 500,000 for the upper boundary and 2000 for the lower boundary have been proposed. If an installation's risk integral were to fall within the two levels, then risks would need to be reduced ALARP. If the value were above the upper boundary it would indicate that risks might have to be reduced and if it were below 2000, then risks might be considered broadly acceptable.

Time consuming and expensive QRAs are required to construct FN curves and are not always appropriate if a site presents minimal risk. Therefore, a rapid estimation of the risks installation posed to its surroundings could be used as a screen to enable resources to be targeted to further refine the risk picture where this was necessary. HSE has developed such a screening tool.

2.8 Approximate Risk Integral screening tool:

In 2002, the Health and Safety Executive published a methodology (Hirst & Carter 2002) to provide a rough but rapid indication of the magnitude of societal risks at and in the vicinity of a major accident hazard installation. This methodology has been named ARI_{COMAH} (the COMAH Approximate Risk Integral). Although coarse, the use of this methodology requires only the frequency of the 'worst-case' scenario and the maximum number of persons harmed (N_{max} and $F(N_{max})$. Using two algorithms for either a unidirectional or omni directional event, it provides a figure that can be compared for ranking purposes.

Experience with the application of ARI_{COMAH} has indicated, that for cases with large N_{max} from a uni-directional event, the method for predicting the frequency of N_{max} tends to consistently overestimate the result, increasing with increasing N_{max}. As ARI_{COMAH} is only used as a screening and ranking tool the absolute value of the result is not an important factor. However a more accurate method is needed for assessments where the absolute value is of importance, for example where risk reduction options are being considered. Such a method should not rely on the estimation of the frequency of N_{max}.

2.9 Quick F/N:

HSE commissioned work to develop an intermediate societal risk tool that could provide improved 'resolution' compared to ARI_{COMAH}, and would be less resource intensive than the calculation of 'full' FN curves. (Fowler *et al.* 2004 & HSE 2004)

The work involved taking realistic and representative installations for hazards including liquefied chlorine and anhydrous hydrogen fluoride along with a population distribution that reflected a real situation. The full F/N relationships were determined before sub sets of scenarios were selected and the F/N plots compared.

The authors concluded that the 'right-hand' (high N, low F) part of a FN societal risk curve could be closely replicated by inclusion of fewer scenarios and hazard sources than the full set typically used when calculating societal risks. If a close approximation is made, then the 'left-hand' (low N, high F) part of the societal risk curve can be extrapolated. (Figure 4) Inclusion of fewer scenarios and hazard sources reduces the resources needed to estimate societal risks. For the bulk chlorine installation the representative case only requires 5 scenarios and 1 hazard source rather than 51 scenarios and 7 hazard sources for the full FN base case. Similarly for the hydrogen fluoride installation only 12 scenarios and 5 hazard sources are required in the representative case, compared to 69 scenarios and 21 hazard sources in the base case.

This method negates the need for full QRA and can be used to quickly build up a more reliable picture of societal risk posed from a site.

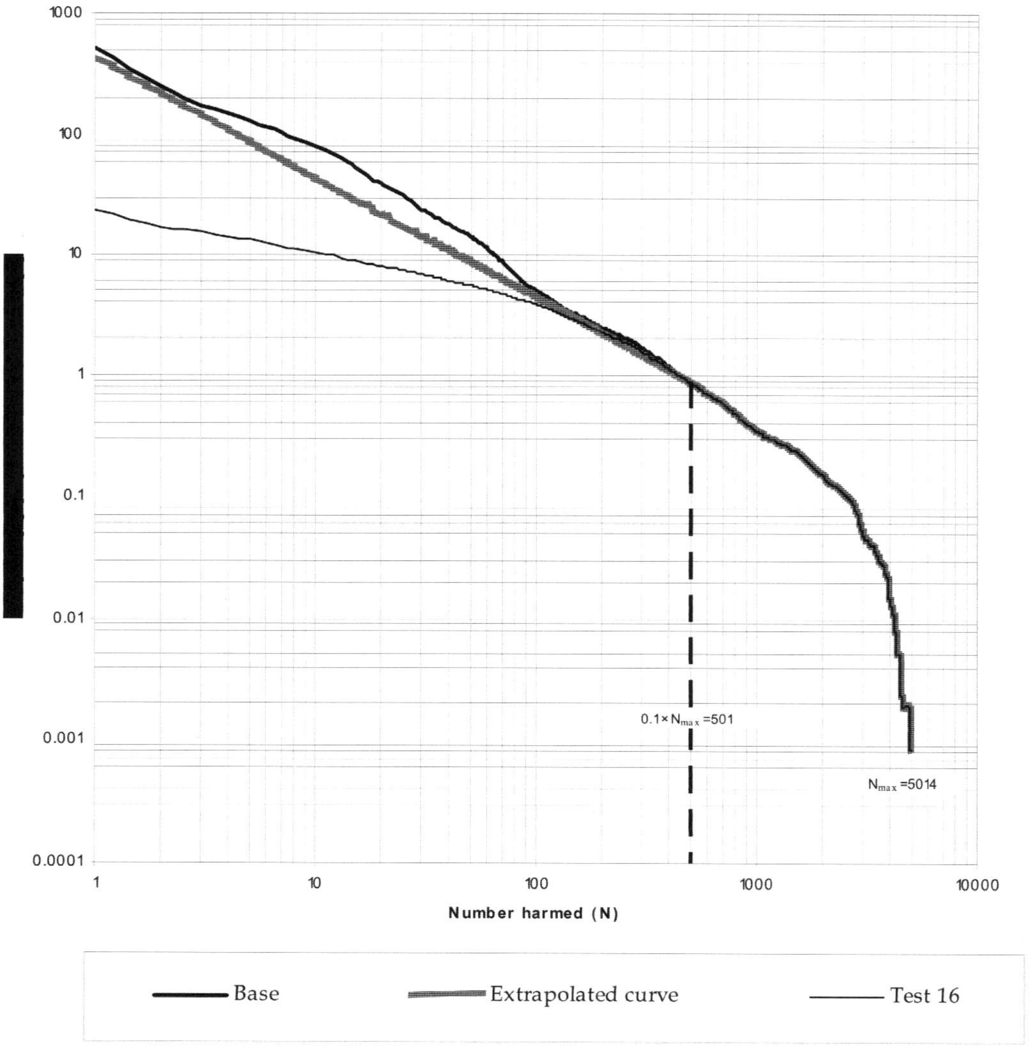

Figure 4. FN plot for Quick FN method showing results from a full FN analysis and sub set of high N scenarios with extrapolation to N = 1.

From the FN plot, the risk integral can then be determined for comparison with criteria.

2.10 Population tool

Details of populations are needed when assessing risks surrounding an installation. It is a fundamental requirement of the Seveso II directive and the COMAH regulations that the extent and severity of the identified major accidents are provided. This includes the number of people potentially harmed in the event of an accident. (HSE 2002) Nothing was easily available in the UK to derive information on the number of people living or working in the hazard ranges of hazardous installations. HSE therefore sponsored research by the UK's Staffordshire University to develop a tool that would give population data. (HSE 2005) This work was completed in 2004 and the UK's Health and Safety Laboratory on behalf of HSE now have a very powerful, accurate and rapid tool for deriving population data.

The tool uses GIS technology along with many databases available to the public and government. Figure 5 shows a screen shot from the tool. The software provides a realistic estimate of the number of people under the area marked on the map. The population can be broken down into different social groups. For instance, number of house residents, or occupants of a school or hospital can be established. This tool can provide populations for use with the Quick FN methodology, especially where the hazard range from toxic releases may extend to several kilometres. The tool can provide a population grid in a matter of minutes. Previous determinations of populations were carried out by hand and eye using often subjective judgments and many hours of effort.

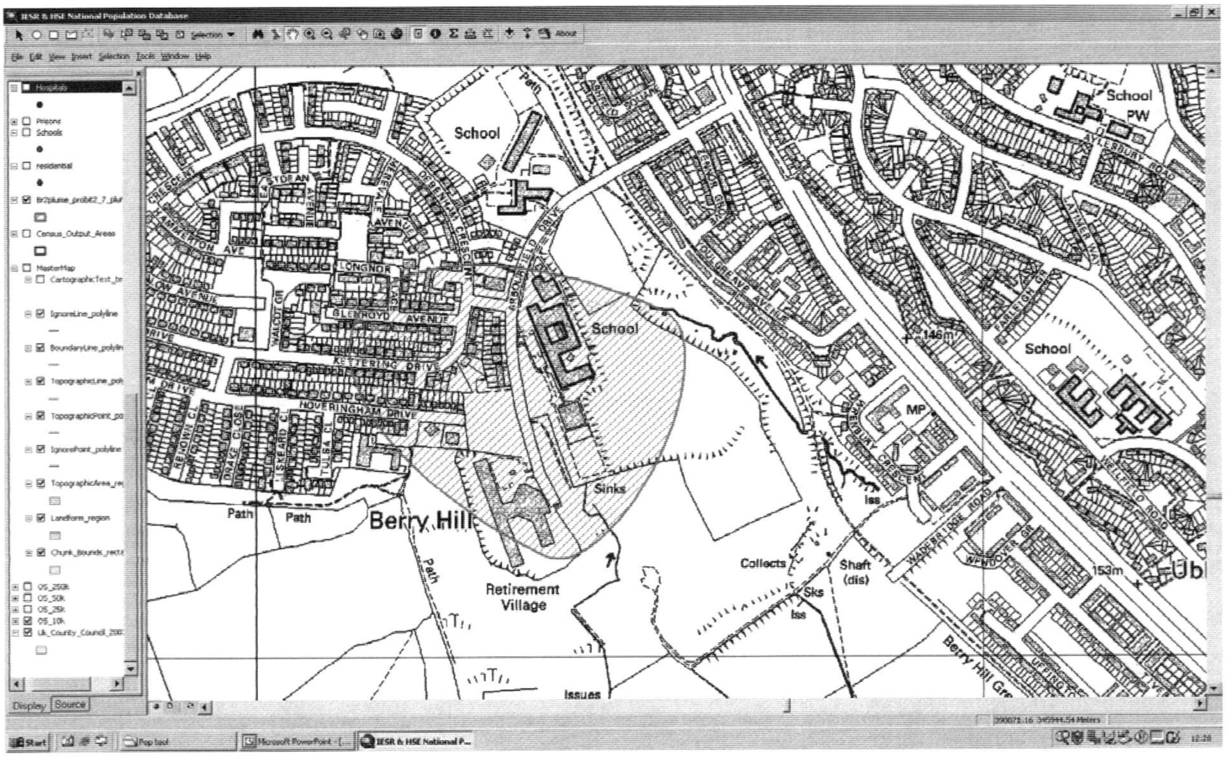

2.11 Risk Reduction Measures and Costs Databases.

HSE is developing two other areas of research to provide important guidance to our inspectors. Firstly, work has been carried out to produce a searchable list of risk reduction measures using a variety of sources from published literature to brainstorming. These, generally hardware measures, are known to reduce risks by for instance improving integrity, reducing inventories, providing mitigation. The database will not be a tick list, but a resource used by inspectors in dialogue with duty holders when discussing their demonstration that they have reduced risks as low as reasonably practicable.

As part of the ALARP determination, the cost of measures is a crucial ingredient. If a duty holder proposes that a certain risk reduction measure is too expensive in view of the amount of risk reduction achieved, the HSE inspector needs to be aware of the approximate realistic costs of that or similar measures. HSE has therefore commissioned research to look at real capital and on going costs that companies had invested. This will provide examples and ideas of realistic costs to HSE inspectors. It is HSE's intention to publish this work on our website.

2.12 Guidance on Cost Benefit Analysis

A duty holder is seen to have discharged their legal duty under COMAH if they have weighed up the costs (whether in money, time, trouble and effort) against the benefits of further measures that might reduce risks. (HSE 2001b) In order to do this a cost benefit analysis might be pursued. It might be appropriate for a simple qualitative analysis on the other hand; significant multi million pound expenditure might necessitate a full-quantified economic argument. Whatever the situation the analysis should be proportionate. HSE is about to publish brief guidance on cost benefit analysis in the form of principles and a checklist, aimed mainly at our own staff. Table 1 summarised a few of the points to consider when considering what are costs and benefits of a risk reduction measure.

Table 1. Examples of costs and benefits to be considered in a CBA for risks reduction measures.

COSTS	BENEFITS
• costs of installation, operation, training and any additional maintenance, and the business losses that would follow from any shutdown of the plant undertaken solely for the purpose of putting the measure into place. • All claimed costs must be those incurred by the duty • Sacrifice implies non-recoverable cost • If lost production is actually deferred production then it should only take account of "interest" on the lost production. • If the lost production costs are a strong	• The benefits should include all reduction in risk to members of the public, to workers and to the wider community. I.e. benefits can be broken down into prevented: o Fatalities o Injuries (major to minor) o Ill health o Environmental damage if relevant (e.g. COMAH) • Benefits can include avoidance of deployment of emergency services and avoidance of countermeasures such as evacuation and post accident decontamination if appropriate.

influence on a decision not to implement, the duty holder should show that phasing or scheduling the work to coincide with planned downtimes (e.g. for maintenance) would not change the balance. • The costs considered should only be those necessary and sufficient for the purpose of implementing the risk reduction measure (no gold plating or deluxe measures) • Ongoing production losses as a result of the measure can be counted. • Any savings as a result of the measure should be offset against the above costs. These are not considered safety benefits but are counted as ' cost savings' i.e. they reduce the overall cost of implementing a measure. • The costs claimed should be shown only to relate to the measure being implemented for safety. • Translation into monetary costs is often uncertain and all should be justified.	• The cash valuations of preventing health and safety effects on people can be estimated e.g. for a fatality £1,336,800 (2003 rates). • All benefits of a measure should be included. If a risk reduction measure is identified for one type of accident but reduces other risks as well e.g. health risks, all benefits should be counted. • It should be noted that duty holders might need to treat re-instatement costs as a benefit rather than offsetting them against costs. This would be the case if the plant being re-instated were a safety related plant (e.g. one that treats hazardous waste). This can represent a bias in favour of safety. This is because the gross disproportion factor is applied to all benefits prior to them being compared to the costs.

A simple theoretical example of a CBA method for coarse screening of measures is presented below. This puts the costs and benefits into a common format of '£s per year' for the lifetime of a plant.

Consider a chemical plant with a process that if it were to explode could lead to:

	Suggested monetary cost in £s (2003 data)
• 20 fatalities	£1,336,800 per fatality
• 40 permanently injured	£207,200 per injury
• 100 seriously injured	£20,500 per injury
• 200 slightly injured	£300 per injury

The rate of this explosion happening has been analysed to be about 1×10^{-5} per year, which is 1 in 100,000 per year. The plant has an estimated lifetime of 25 years.

How much could the company reasonably spend to eliminate (reduce to zero) the risk from the explosion?

If the risk of explosion were to be eliminated the benefits can be assessed to be:

Fatalities:	20	x 1,336,800	$\times 1 \times 10^{-5}$	x 25 yrs	= 6684
Permanent injuries:	40	x 207,200	$\times 1 \times 10^{-5}$	x 25 yrs	= 2072
Serious injuries:	100	x 20,500	$\times 1 \times 10^{-5}$	x 25 yrs	= 512
Slight Injuries:	200	x 300	$\times 1 \times 10^{-5}$	x 25 yrs	= 15
Total benefits:					= £9,283

The sum of £9,283 is the estimated benefit of eliminating the major accident explosion at the plant on the basis of avoidance of casualties. (This method does not include discounting or take account of inflation.)

For a measure to be deemed not reasonably practicable, the cost has to be grossly disproportionate to the benefits. This is taken into account by the disproportion factor (DF). In this case, the DF will reflect that the consequences of such explosions are high. A DF of more than 10 is unlikely.

Therefore it might be reasonably practicable to spend up to somewhere in the region of £93,000 (£9300 x 10) to eliminate the risk of an explosion. The duty holder would have to justify use of a smaller DF

This type of simple analysis can be used to eliminate or include some measures by costing various alternative methods of eliminating or reducing risks.

3. Conclusions.

This paper has outlined a number of tools and techniques that HSE has developed for assessment of risks surrounding hazardous installations. The work has shown that there is no one method that fits all and that risk assessment in the major hazard context requires a suite of tools and techniques to make decisions on the level of risks a site may pose, the tolerability of these risks and whether further measures to reduce the risk are worth doing.

4. Disclaimer

The views expressed in this paper are those of the authors and should not be used as a definitive statement of HSE policy.

References:

CCPS (2001) Layer of Protection Analysis – Simplified Process Risk Assessment, AIChE., New York.

Evans A. W. and Verlander N. Q. (1997). What is wrong with criterion FN-lines for judging the tolerability of risk? Risk Analysis Vol. 17 No. 2 1997, pp 157 – 168.

Fowler A.H.K., Reston S.D., Carter D. A., Quinn D.J., Davies P.A. (2004). QuickFN – a less resource intensive methodology for determining the magnitude of societal risks at major accident hazard installations. IChemE Symposium Series No. 150, pp 627 – 634. ISBN 0 85295 460 3

Health and Safety Executive (1978) Canvey: summary of a review of potential hazards from the operations in the Canvey Island/Thurrock area. HSE Books ISBN 0 11 883203 4

Health and Safety Executive (1981). Canvey: a second Report – a summary of a review of potential hazards from the operations in the Canvey Island/Thurrock area three years after publication of the Canvey Report. HSE Books ISBN 0 11 883618 8

Health and Safety Executive (1989) Risk criteria for land use planning in the vicinity of major industrial hazards. HSE Books 1989 ISBN 0 11 885491 7.

Health and Safety Executive (1992). The tolerability of risks from nuclear power stations. HSE Books 1992, ISBN 0 11 1411439 X

Health and Safety Executive (1999). Preparing Safety Reports: Control of Major Accident Hazards Regulations 1999. HSG 190, HSE Books, ISBN 0 7176 1687 8.

Health and Safety Executive (2000) Lines of Defence/Layers of Protection Analysis in the COMAH Context. Prepared by Amey VECTRA Limited for HSE, http://www.hse.gov.uk/research/subject/c/comah.htm

Health and Safety Executive (2001a) Reducing Risks Protecting People, HSE's decision making process. HSE Books. ISBN 0 7176 2151 0 http://www.hse.gov.uk/risk/theory/r2p2.htm

Health and Safety Executive (2001b) Principles and guidelines to assist HSE in its judgements that duty-holders have reduced risk as low as reasonably practicable. http://www.hse.gov.uk/risk/theory/alarp1.htm

Health and Safety Executive (2002) COMAH safety reports; Information about the extent and severity of identified major accidents. HID SPC/Permissioning/06 http://www.hse.gov.uk/comah/circular/perm06.htm.

Health and Safety Executve (2004) RR176 - Development of an intermediate societal risk methodology, An investigation of FN curve representation. HSE Books. http://www.hse.gov.uk/research/rrhtm/rr176.htm

Health and Safety Executive (2005) A National Population Data Base for Major Accident Hazard Modelling. Research Report 297 - HSE Books, http://www.hse.gov.uk/research/rrhtm/rr297.htm

Hirst I. L. (1998) Risk Assessment, A note on F-n curves, expected numbers of fatalities, and weighted indicators of risk. Journal of Hazardous Materials, 57 (1998) pp 169 – 175.

Hirst I.L. and Carter D.A.(2002) A "worst case" methodology for obtaining a rough but rapid indication of the societal risk from a major accident hazard installation, Journal of Hazardous Materials, Volume 92, Issue 3, 10th June 2002, pages 223-237

HMSO (1999) The Control of Major Accident Hazards Regulations 1999. Statutory Instrument No. 743 1999. The Stationery Office Limited. ISBN 0 11 082192 0. http://www.hmso.gov.uk/si/si1999/19990743.htm

HSC (1991), Advisory Committee on Dangerous Substances, Major hazard aspects of the transport of dangerous substances, 1991, ISBN 0 11 885676 6.

Middleton, M. and Franks, A. (2001) Using Risk Matrices. The Chemical Engineer, Sept 2001.

Naylor, P. J., Maddison, T. E., Stansfield, R. (2000) TRAM: Technical Risk Audit Methodology for COMAH. IChemE. Hazards XV, Symposium No. 147, pp 491 – 503. ISBN 0 985295 429 8.

Uncertainties in Evaluating Public Response to Toxic Exposure

Krishna S. Mudan
MSA Risk & Economic Consultants
Bethlehem, Pennsylvania
kmudan@msa-rec.com
and
Joseph R. Natale
Baker Engineering and Risk Consultants, Inc.
San Antonio, Texas
jnatale@bakerrisk.com

January 24, 2005

ABSTRACT

Accidents involving hazardous chemicals are rare, but they do happen, and sometimes with devastating outcome. In particular, releases of toxic chemicals have the potential to cause death and disability not only to onsite employees, but also to the offsite population. The worldwide regulatory authorities, have responded to chemical accidents by establishing risk criteria for public exposure and zones for emergency response. The chemical and petroleum industries have responded to the challenge by adopting a wide range of engineered solutions to mitigate accidental releases. A key parameter affecting the establishment of emergency zones and complying with risk management guidelines is the consequence and impact calculations associated with a potential release. The risk assessment models that are currently used to predict mortality in the exposed population appear to significantly over-predict the consequences of the toxic releases. This, in part, is due to the uncertainties in modeling the release. But a larger uncertainty stems from the use of toxicological end point criteria, which is based on animal studies and extrapolated to human exposure.

This paper reviews the toxic end point criteria established for selected chemicals that have been implicated in some of the major toxic releases. The scenarios are modeled and the estimated mortality is compared with actual outcome. An attempt has been made to use publicly available data on accident description, release conditions, prevailing meteorological conditions, and population exposure. The "Human Toxic Exposure Response Factor," derived from a comparison of predicted mortality and actual mortality, may account for some of the non-quantifiable, but nevertheless, significant issues associated with human behavior and response.

1. Introduction and Background

Accidents involving hazardous chemicals are rare, but they do happen, and sometimes with devastating outcome. In particular, releases of toxic chemicals have the potential to cause death and disability not only to onsite employees, but also to the offsite population. In the wake of the 1984 methylisocyanate incident in Bhopal, India, the regulatory authorities have responded to chemical accidents by establishing risk criteria for public exposure and zones for emergency response. The chemical and petroleum industries have responded to the challenge by adopting a wide range of engineered solutions to mitigate accidental releases.

While the approaches taken by many of the regulatory agencies and industry associations differ in detail, the underlying emphasis is on identifying the toxic hazards, modeling the dispersion of released chemicals, determining the toxic end points at which such chemicals can cause serious damage to population, and developing an effective emergency response program. All these regulations and industry association-sponsored research activities have led to effective means of identifying hazards presented by various chemicals and processes, collection of historical data on likelihood of failures leading to release of chemicals, and development of sophisticated dispersion programs to compute the extent of toxic dispersion including validation of model results with field experiments.

A key parameter affecting the establishment of emergency zones and complying with risk management guidelines is the consequence and impact calculations associated with a potential release. The risk assessment models that are currently used to predict mortality in the exposed population appear to significantly over-predict the consequences of the toxic releases. This, in part, is due to the uncertainties in modeling the release. But a larger uncertainty stems from the use of toxicological end point criteria, which are based on animal studies and are extrapolated to human exposure. Ferguson and Hendershot (2000) conducted a simple study of a chlorine release to observe that the predicted fatality results differed by orders of magnitude depending on the underlying toxic dose response relationship. Clearly, such a large variation in results can at best lead to an absence of risk-based decision making to improve process safety. At its worst, such variation can lead to misleading interpretation of results and wrongful decisions.

This paper reviews the animal toxicological data and toxic end point criteria established for selected chemicals that have been implicated in some of the major toxic releases. The scenarios are modeled and the estimated mortality is compared with actual outcome. An attempt has been made to use publicly available data on accident description, release conditions, prevailing meteorological conditions and population exposure. The "Toxic Exposure Response Factor", derived from a comparison of predicted mortality and actual mortality may account for some of the non-quantifiable, but nevertheless, significant issues associated with human behavior and response.

2.0 Sources of Uncertainties in Interpreting Animal Toxicology Data

Toxicological research and review of existing animal test data has been conducted to help determine the levels of exposure to chemicals at which serious injury or even death can be caused to the exposed population. Government agencies such as EPA's National Research Council for AEGLs and industry-sponsored groups through the American Industrial Hygiene Association for development of ERPGs have addressed identified exposure guidelines. Three toxic exposure levels were identified, with level 1 being the lowest concentration at which no significant damage is caused. A level 2 is meant to indicate absence of irreversible health effects or serious symptoms, whereas exceeding a level 3 may indicate a life-threatening condition.

The toxic exposure levels defined above are useful in determining an emergency planning zone (as under EPA's Risk Management Program rule), and preparing an emergency preparedness program. However, for the purposes of a Quantitative Risk Assessment (QRA), it is desirable to define toxic exposure levels at which there can be potential fatalities. In this context, the toxicological data are further reviewed to reduce the uncertainty and determine levels of exposure at which there is a "probability" of fatality among the exposed population. This probability, often termed as "probit," is widely used in QRA's in many parts of the world, and in some countries such as UK, The Netherlands or Australia, the use of such assessments are mandated by the regulatory agencies of those countries.

While it is tempting to convince oneself that the results of all this research and analysis can lead to absolute and accurate estimation of toxic impacts, the reality is far from it. The results, as in any analysis, are only as good as the data used to produce the results. However, even with the best available data, there are several sources of "systemic uncertainty," and caution must be exercised while interpreting the results of the analysis. One of the sources of uncertainty comes from the application of toxicological end point criteria, which are usually based on animal studies and extrapolated to human exposure, usually with a safety factor. Under ideal conditions of exposure (to which animals are subjected to during the testing procedure), these extrapolations are perhaps valid. Such an extrapolation may represent exposure to people working outside, not taking any evasive action, and continuing to breathe normally. In reality, however, the human response could be significantly different. Some of these "un-quantifiable" differences are discussed below:

- Unlike laboratory test animals exposed to a toxic chemical in a closed environment, human beings are free to move about, and exercise caution.
- Human beings can "smell" danger associated with a chemical release (this is particularly true of chemicals such as ammonia and chlorine that have a very low odor threshold). They can respond by holding their breath, covering their eyes, nose and mouth, and attempt to move away from or outside a dangerous cloud. If the opportunity exists, it may be possible for people to breathe through moist clothing.
- Sometimes, it is possible for people outdoors to move inside buildings, close doors and windows, or stay inside a car with windows closed.
- The shelter-in-place concept, where the response of indoor population has been studied, indicates that the concentration indoors can be significantly lower than concentrations outdoor, sometimes by an order of magnitude, and thus can provide a safe haven for the public.
- The emergency response can be very effective in preventing exposure of large groups of people by warning them of the impending danger, and recommending appropriate mitigating measures.
- The medical technology has come a long way in helping victims survive incidents by rapidly deploying antidotes with an improved ability to provide large groups of public with emergency treatment.
- The hospitals and physicians, in particular in industrialized regions, are specially trained to deal with complex situations involving damage to the respiratory tract and other key organs, and with the advent of advances in medicine and surgical techniques; the recovery probability is increased significantly.

While all these factors do influence the overall mortality rate resulting from a toxic chemical release, many factors cannot be quantified. With the exception of models to predict indoor concentration, where the building volume, ventilation rate and outside concentration are the input, none of the other parameters have been studied. The approach taken in this study is to model specific accident scenarios (for which we have reasonable accident descriptive data) with the predicted toxic endpoints, and compare the predicted mortality with actual mortality. Based on these comparisons, we derive a "Human Toxic Exposure Response Factor," to account for the differences between predicted and actual values. This factor is intended to take into account many, if not all the non-quantifiable issues described above, and reduce the uncertainty associated with the prediction of mortality.

3.0 Acute Inhalation Toxicity Data for Selected Chemicals

To make this comparison, we have selected three chemicals: ammonia, chlorine and methylisocyanate (MIC). Ammonia and chlorine are basic chemicals that are used in large quantities, transported extensively, and have been involved in several accidents. Both chemicals have been studied for their toxicological effects for several decades, and chlorine was used as a warfare agent by the German army during World War I. The toxicity of MIC was studied extensively following the Bhopal tragedy. We also chose a total of five accidents to analyze; two involving an ammonia release, two involving a chlorine release, and one involving a release of MIC. These were the accidents for which we had reasonably good description of precursors to the accident, prevailing conditions, quantity of chemical released, and population exposure. The number of fatalities and injuries were also available in the database. The endpoint criteria, for modeling purposes, were determined from Fatality Probits taken from literature as described below.

3.1 Ammonia Toxicity

There are several sources of animal toxicology data, but the most frequently cited sources appear to be the experiments conducted by Appleman et al (1982) and Schorsch et al (1995) involving rats exposed to ammonia concentration over differing duration. Tissot and Pichard (2003) of the French Ministry of Ecology and Sustainability reviewed the inhalation toxicity data for ammonia with a view of establishing acute toxic levels for human exposure. In order to do this, they also conducted extensive statistical analysis of all available data. Using the results of data for rats, we obtain the following probit equation:

$$\text{Fatality Probit for Rats} = -42.4 + 2.17 \,\text{Ln}\,[(Cv)^{1.834}\, T] \qquad (1)$$

Where Cv is concentration of ammonia in PPM and T is exposure duration in minutes.

The concentration-duration response given by Equation (1) is very close to a square of concentration relationship, and has been used by several regulatory agencies to establish emergency exposure limits.

3.2 Chlorine Toxicity

Like ammonia, various researchers have studied the acute toxicology of chlorine for over eight decades. A detailed review of chlorine toxicity was presented by Withers and Lees (1985). A more recent source of animal toxicology data is based on experiments conducted by Zwart and Woutersen (1988) involving rats exposed to chlorine concentration over differing duration. A second study by Demanti et al (1995) addressed very short duration exposure of rats to chlorine. Based on a review of these, Tissot and Pichard (2001) gave the following expression to evaluate rat mortality:

$$\text{Fatality Probit for Rats} = -8.58 + 1.13 \,\text{Ln}\,[(Cv)^{1.29} \cdot T] \qquad (2)$$

3.3 Toxicology of MIC

MIC was not a well-researched chemical from a toxicological perspective. After the Bhopal incident, ten Berge (1985) of Dutch State Mines compiled the limited data of Kimmerle and Eben (1964) to provide guidance on acute inhalation toxicology of MIC. Using only rat data, exposure duration from 60 minutes to 240 minutes, and probit analysis, ten Berge gave the following expression:

$$\text{Probit} = -6.64+1.637 \text{ Ln } [C^{0.653} * T] \quad (3)$$

Where the concentration was measured in mg/m^3, and exposure time was in minutes.

More recently, several experiments on the acute inhalation toxicology of MIC using different animal species have been conducted. The extensive data has been reviewed by the National Research Council for the development of AEGLs (2003). This paper is limited to a brief review of only short duration inhalation data for rats. Dinsdale et al (1987) conducted a whole-body exposure of rats to MIC concentrations ranging from 8.4 PPM to 420 PPM for a one-hour duration. Dutta et al (1988) conducted a very short duration (8 minutes) exposure to intense concentration of MIC ranging from 1344 PPM to 9219 PPM on groups of six rats. Jeevaratnam and Sriramachari (1994) exposed 8 to 12 rats to concentrations of 233 PPM to 930 PPM in a static exposure chamber, and estimated a 30-minute LC50 of 465 PPM. The data from these sources were combined to obtain the following regression equation:

$$\text{Fatality Probit for rats} = -20.008+2.5837 \text{ Ln } \{Cv*T\} \quad (4)$$

The regression coefficient was 0.867.

The resulting concentrations (in ppm) for 30-minute, 60-minute and 120-minute exposures are shown in Table 1.

Table 1: Acute Exposure Concentration (in ppm) for MIC

Fatality Probability	30-minute Exposure	60-minute Exposure	120-minute Exposure
99%	1307	1120	560
90%	872	436	218
50%	531	266	133
10%	323	162	81
1%	216	108	54

4.0 Influence of Shelter-in-Place on Toxic Impact

Accidental release of toxic chemicals can have a devastating impact on the exposed population. However, the severity of impact appears to depend upon not only the concentration of the chemical and the exposure duration, the two aspects already discussed in the development of the toxic probit, but also depends on whether people are inside buildings or outdoors. People staying outdoors are subjected to the toxic gas exposure immediately. However, people remaining inside buildings would experience an increasing concentration over a period of time that could reach concentrations found outdoors.

4.1 A Simple Model for Indoor Concentration

The time period over which the concentration builds up inside a house or building depends upon the following two factors:

- Infiltration rate of outside air into the buildings, expressed as number of Air Changes per Hour (ACPH), and
- Elapsed time.

The former parameter is dependent on ventilation rate, presence of doors and windows, and the location of heating and air conditioning system inlets, to mention a few. A simple infiltration model suggests that the interior concentration buildup is given by the following equation:

$$Ci\ (t) = Co\ [1\text{-}Exp\ (-L^*t)] \qquad (5)$$

Where: Ci(t) = Time dependent Indoor concentration, ppm
Co= Constant Outdoor concentration, also in ppm
L= Ventilation rate, expressed in number of air changes per minute
t= time, in minutes

The indoor concentration increases over a period of time, and if the duration of release is long, the indoor concentration will be same as outdoor concentration. However, once a source of toxic release is controlled or stopped, the outdoor concentration essentially becomes zero. The indoor concentration, on the other hand, starts to decrease slowly, depending upon the ventilation rate. The indoor concentration in given by:

$$Ci = Cir^*\ Exp\ (-L(t\text{-}Tr)) \qquad (6)$$

Here, Cir = Indoor concentration at the end of release duration, and
Tr = Release duration

A typical profile of indoor concentration is shown in Figure 1.

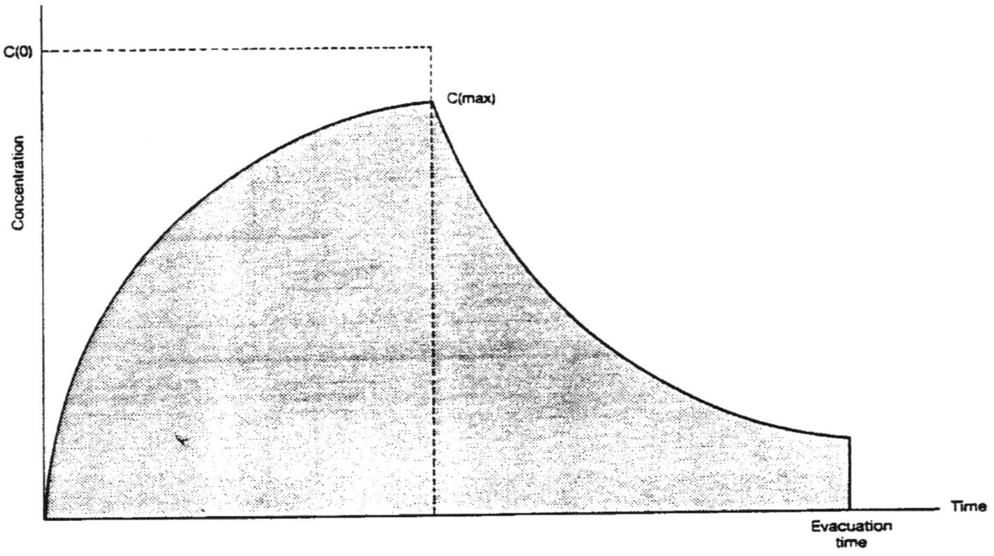

Figure 1. Typical Indoors Concentration Profile

4.2 Indoor Toxic Load and Probits

It was discussed in Section 3 that concentration with exposure duration defines the toxic impact. It was also established that different chemicals exhibit differing toxic impact relationships with MIC showing a linear dose response, chlorine with a near-linear dose response relationship, and ammonia showing a near square of concentration dependency. Therefore, the indoor dose will also be a function of this relationship. In general, the indoor toxic load can be derived by integrating the time-dependent concentration over the exposure duration, Td. For the linear toxic load relationship, the indoor toxic dose is given by the following equation:

$$\text{Toxic Load} = Co^* [Td - 1/L\{1 - Exp(-L^*Td)\}] \quad \text{for } Td < Tr \tag{7}$$

$$\text{Toxic Load} = Co^* [Tr - 1/L\{Exp(-L^*Td)[Exp(L^*Tr)-1]\}] \quad \text{for } Td \geq Tr \tag{8}$$

The outdoor toxic load is given by the product of outdoor concentration and the exposure duration. Therefore, the ratio of indoor toxic load to the outdoor toxic load is given by the following expression:

$$Rn1 = 1 - 1/(L^*Td)\{1 - Exp(-L^*Td)\} \quad \text{for } Td < Tr \tag{9}$$

$$Rn1 = 1 - 1/(L^*Tr)\{Exp(-L^*Td)[Exp(L^*Tr)-1]\} \quad \text{for } Td \geq Tr \tag{10}$$

The indoors probit is given by the following equation:

$$\text{Indoor Probit} = \text{Outdoor Probit} - B\ Ln(Rn1) \tag{11}$$

A similar expression can be derived for chemicals with a square concentration relationship, like ammonia and chlorine. The toxic load can be used to determine the appropriate indoor probit. For other (fractional) dose response relationships, a numerical assessment will be required to determine the indoor equivalent probit.

4.3 Ventilation Rates

As mentioned before, the ventilation rates depend upon a host of parameters, including the degree of forced ventilation with air conditioners and heaters, the degree of natural ventilation with doors and windows, leakiness of windows, etc. A convenient parameter for measuring the ventilation rates is the number of Air Changes Per Hour, or ACPH. For many commercial buildings, the ACPH is primarily the make up air from the heating or cooling system. The ACPH can be as low as 0.1 or one air change (of the complete building volume) in about 1 hour. For older buildings, or homes in tropical areas where the windows and doors are likely to be open most of the time, the ACPH could be 2 or 3. Lees suggests the ACPH range to be between 0.06 and 3. Once a ventilation rate is chosen, the toxic load ratios can be computed. In Tables 2 and 3, the toxic load ratios for the two relationships are shown when the release duration and exposure duration are same.

The effect of these factors on potential mortality rate is significant. For instance, a five-minute release of MIC with an outside toxic load corresponding to a 99% fatality would lead to less than 1% fatality indoors, even when the ventilation rate is about 2 ACPH. In the following section, we have used these relationships to estimate the actual fatality distributions in chemical accidents.

Table 2: Toxic Load Ratio for Linear Concentration Relationship

Duration, Minutes	ACPH=2	ACPH=1	ACPH=0.5	ACPH=0.25
2	0.03	0.02	0.01	0
5	0.08	0.04	0.02	0.01
10	0.15	0.08	0.04	0.02
30	0.37	0.21	0.12	0.06
60	0.57	0.37	0.21	0.12

Table 3: Toxic Load Ratio for a Concentration Square Relationship

Duration, Minutes	ACPH=2	ACPH=1.0	ACPH=0.50	ACPH=0.25
2	0.001	0.0004	0	0
5	0.008	0.002	0.0006	0
10	0.03	0.008	0.002	0.0006
30	0.17	0.06	0.017	0.005
60	0.38	0.17	0.058	0.017

5.0 Analysis of Accidents

5.1 Houston Ammonia Incident

5.1.1 Accident Description

On May 11, 1976, a tank truck carrying approximately 6500 gallons of anhydrous ammonia overturned at the intersection of I-610 and Route 59 and fell about 20 to 30 feet. About 17 tonnes of ammonia was released immediately. The photograph taken about one minute after the crash indicated a large, visible cloud approximately 300 ft (100 m) in diameter and 60 ft (20 m) in height. The visible cloud extended to about 3000 ft (1000 m) in about 5 minutes. This is consistent with an observed wind speed of about 3 m/s. Eyewitness accounts indicated that the cloud was visible for about 5 to 10 minutes.

5.1.2 Impact on Public

In all there were six fatalities and 178 injuries. Inhalation of ammonia was the cause of five fatalities. The sixth fatality was the driver of the tank truck, who was killed upon impact. All fatalities were within 200 feet or about 65 m of the release location. Figure 2 shows the location of all fatalities, including the truck driver. There were a total of 178 injuries within about 1000 ft (300 m) of the release. Of these, 78 persons required hospitalization. The symptoms indicated that the majority of the patients had been exposed to very high concentrations of ammonia for about 5 minutes to 10 minutes. There were exceptions and these are described later.

The exact number of persons exposed to the ammonia cloud is not known. The NTSB Report on the accident investigation merely states that there were several people in the vicinity of the accident. It is conceivable that many were inside buildings like the Houston Post Office Building. As stated earlier, buildings can provide an effective shelter from toxic releases.

The actual experience of patients surviving the accident, and incurring severe respiratory tract injuries is as follows:

- Three patients had crashed their car directly into the ammonia tank truck. They were exposed to very high concentration of ammonia for about 20 minutes. They all had second to third-degree burn injuries on their faces and extremities, consistent with an exposure level of 20,000 to 30,000 ppm. Of these three patients, a 17-year old girl and a 6-month old baby survived after extensive medical treatment including tracheotomy and application of corticosteroids. The 26-year old man, however, died later from injuries sustained in the accident (included in the fatality data).
- Two women, a 23-year old and a 30-year old, had driven their car into the ammonia cloud and the car engine stopped (presumably due to lack of oxygen) and stalled. They were rescued by the fire department, but it took nearly 30 minutes. They had mild burns on the face consistent with an exposure level of about 10,000 ppm. The medical treatment consisted of intubations and administering corticosteroids. Both women survived.

Of the 78 persons admitted to the hospital, it appears that they were within 1000 ft from the tank truck. It also appears that with the exception of twelve patients, the remaining did not require further hospitalization or extensive medical treatment.

The remaining 100 persons were treated locally. It would appear that these injuries did not require hospitalization.

A schematic of the accident, indicating location of fatalities, initial size of the vapor cloud, and location of various buildings is shown in Figure 2.

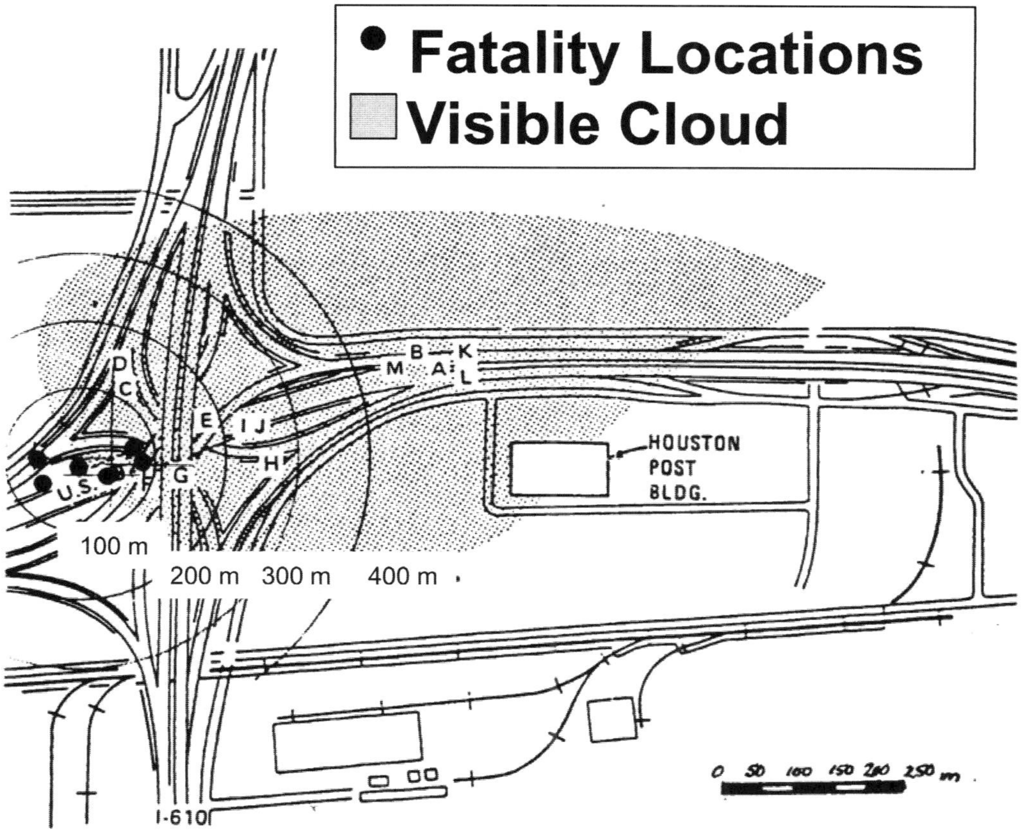

Figure 2: Ammonia Truck Accident in Houston

<u>*5.1.3 Mortality Statistics*</u>

The traditional means of deriving mortality statistics is based on the total number of fatalities in a given region, divided by the total number of persons exposed to the toxic incident. A clear distribution of all persons within the given region is not known. However, based on the location of victims, some conclusions can be drawn. Since the total number of persons is not known, and only the number of persons injured is available, the mortality ratio should be construed as an upper bound for mortality statistics. Based on the above information, one may infer the following:

- The total number of persons within the 200-foot cloud is, at a minimum, eleven. Of these, there were 6 fatalities, including the driver and the 26-year old man who died later in the hospital. Although the cause of the driver's death was the impact of the accident, it is also reasonable to assume that the driver would have died due to ammonia inhalation. This leads to a fatality probability within the 200-foot (65m) distance of about 55%.
- The duration of exposure was more than about 5 minutes, but less than about 20 minutes, largely due to immediate response from rescue personnel. This exposure duration was also consistent with the burn injuries suffered by some of the patients admitted to the hospital. Therefore, given the square of concentration relationship for acute toxicity of ammonia, assigning exposure duration of 5 minutes should provide a slightly conservative estimate of the mortality. There were no fatalities beyond the 200-foot (65m) radius, but injuries persisted up to a distance of about 1000 feet (300m) from the accident location. Some of these injuries were serious, requiring hospitalization.

5.1.3 Dispersion Analysis

A toxic dispersion analysis using HGSYSTEM was conducted to estimate the time-averaged concentration at these locations. Based on the description of the accident, the following conditions were used for exercising the HGSYSTEM dispersion model:

- Steady state release of 17,200 kg ammonia over a 5-minute period
- Weather state B, wind speed 3 m/s (based on NTSB Reports)
- Humidity: 70 %
- Temperature: 26 °C
- Surface Roughness: 0.1 m (assuming release in semi-suburban area)
- 5-minute exposure duration.

The 5-minute release duration is consistent with eyewitness accounts indicating the visible cloud vanished after about five minutes. The five-minute exposure duration is also consistent with the hospital records that indicate that the burn injuries were consistent with roughly a 5-10 minute exposure. Given the toxic dose relationship for ammonia, the assumption of shorter release duration should result in a conservative estimate of the hazard. In order to estimate the impact of toxic exposure for persons indoors, we need to make an additional assumption on the ventilation rates. Since this incident occurred on a highway, and many of the victims were passengers in nearby vehicles, it is reasonable to assume that the number of ACPH is high. However, even with an ACPH of 2, the indoor concentration is well below the 1% fatality level.

In Figure 3, the predicted fatality probability distribution is shown as a function of downwind distance from the location of release. The predicted 100% mortality region extends to about 25 m and the predicted fatality probability is less than 1% beyond a distance of 200 m. Also shown in Figure 3 is the estimated actual fatality in the region based on 6 fatalities out of 11 exposed persons. It is important to note that the predicted mortality probability is greater than the actual upper limit.

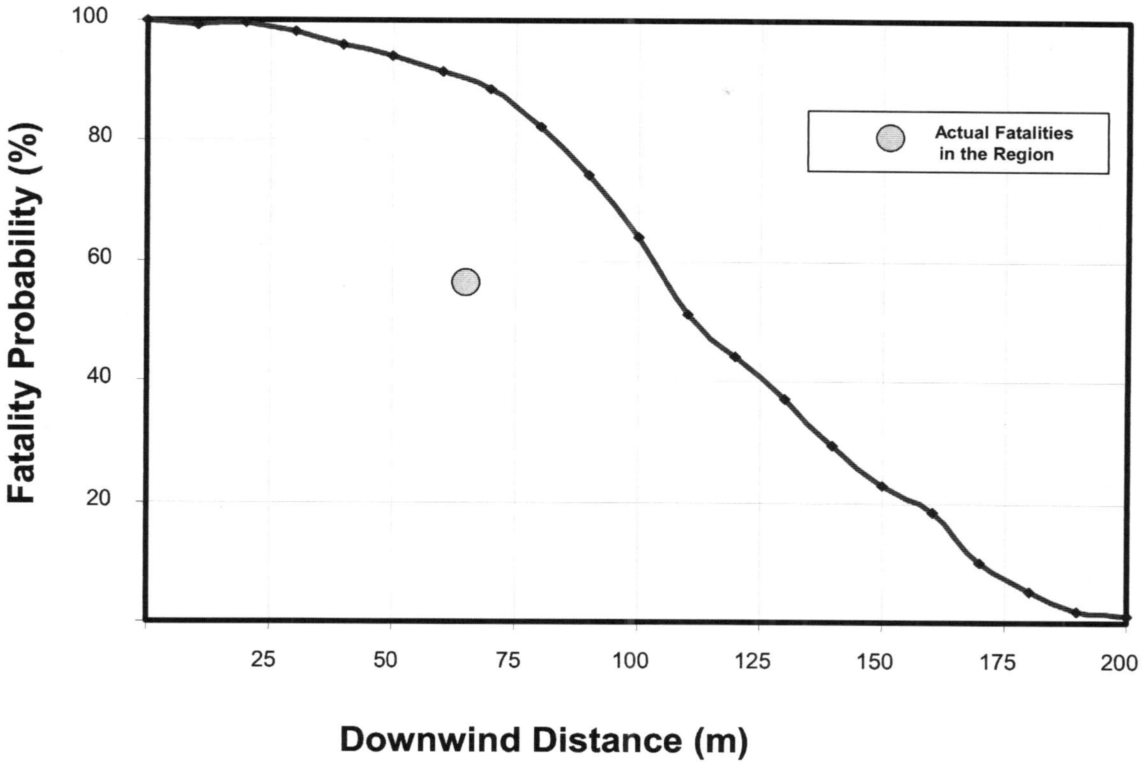

Figure 3: Estimated Fatality Probability Distribution for the Houston Ammonia Accident

In addition to the downwind dispersion, there was also an upwind dispersion to a distance of about 100m.

5.2 Potchefstroom Ammonia Incident

5.2.1 Accident Description

On the afternoon of July 13, 1973, a major release of ammonia occurred at a fertilizer plant in Potchefstroom, South Africa. A pressurized storage bullet containing about 38 tons of anhydrous ammonia failed suddenly, releasing all of its contents. The incident resulted in 18 fatalities, including six members of the public, and 65 persons were treated at the local hospital. Other relevant conditions at the time of the incident are as follows:

- Tank pressure: 90 psig
- Tank Temperature: 15 °C (59 °F)
- Ambient temperature: 19° C (66 °F)
- Relative Humidity: about 35%
- Wind speed: No wind initially, slight breeze towards town soon after release.

The time of the incident was 4:15 in the afternoon, and at the time of the accident the tank was being filled from a nearby road tank car. The investigation indicated that the opening in the tank measured nearly 6 square feet. Such a large opening would cause most if not all of the contents of the tank to be discharged into the atmosphere immediately. Eyewitness accounts of the incident indicate that the initial size of the ammonia vapor cloud was roughly 150 m (about 500 feet) in diameter, and engulfed most of the process area. There was no prevailing wind at the time of the incident, but soon there was a slight breeze moving the cloud towards the nearby township. The cloud reached about 300 m (about 1000 feet) in width and was visually tracked to about 450 m (1500 feet) downwind from the plant.

5.2.2 Impact on Plant Personnel and Public

There were about 350 employees in the plant at the time of the incident. In Figure 4, the plot plan is shown indicating the location of the storage tank, position, and location of several workers at the time of the incident. Given the initial size of the cloud, it is likely that a good part of the facility was engulfed within the ammonia vapor cloud.

- Within 50 meters of the tank, there were several workers exposed to very high concentrations of ammonia. Of these, two were found dead and five were rescued, but died later. Only four persons in this area managed to escape.
- Between 50 meters and 100 meters, one died immediately, but five died after rescue. Two of the deceased were outside the fence line in a public area. There were about 20 survivors in this region and many sought shelter inside the control room for nearly 30 minutes. There were survivors inside the maintenance office, washrooms, granulation plant, and the mill house.
- Between 100 m to 150 meters, one person died after being rescued, an employee who came out of the washroom and probably collapsed immediately. There were two others who also came out of the washroom with him, but both managed to escape. There was one employee at the sulfuric acid plant, and four more at the granulation plant. This region also covered the roads outside the facility, but the population exposure is unknown.
- The remaining four fatalities were all public, and were between 150 meters and 200 meters from the release scenario. One of the deceased appeared to be inside a shop, but went outside and collapsed. The other residents also appeared to have come outside their houses, and died.

An unknown number of persons escaped from the cloud beyond the 200-meter location. The local hospital reported 65 cases on "non-lethal" gassing. Since the plant personnel location was fairly well documented, the remaining cases of "non-lethal" gassing were more likely members of the public. On this basis, one can estimate the number of public exposed to be 38 (4 fatalities and 34 non-lethal injuries).

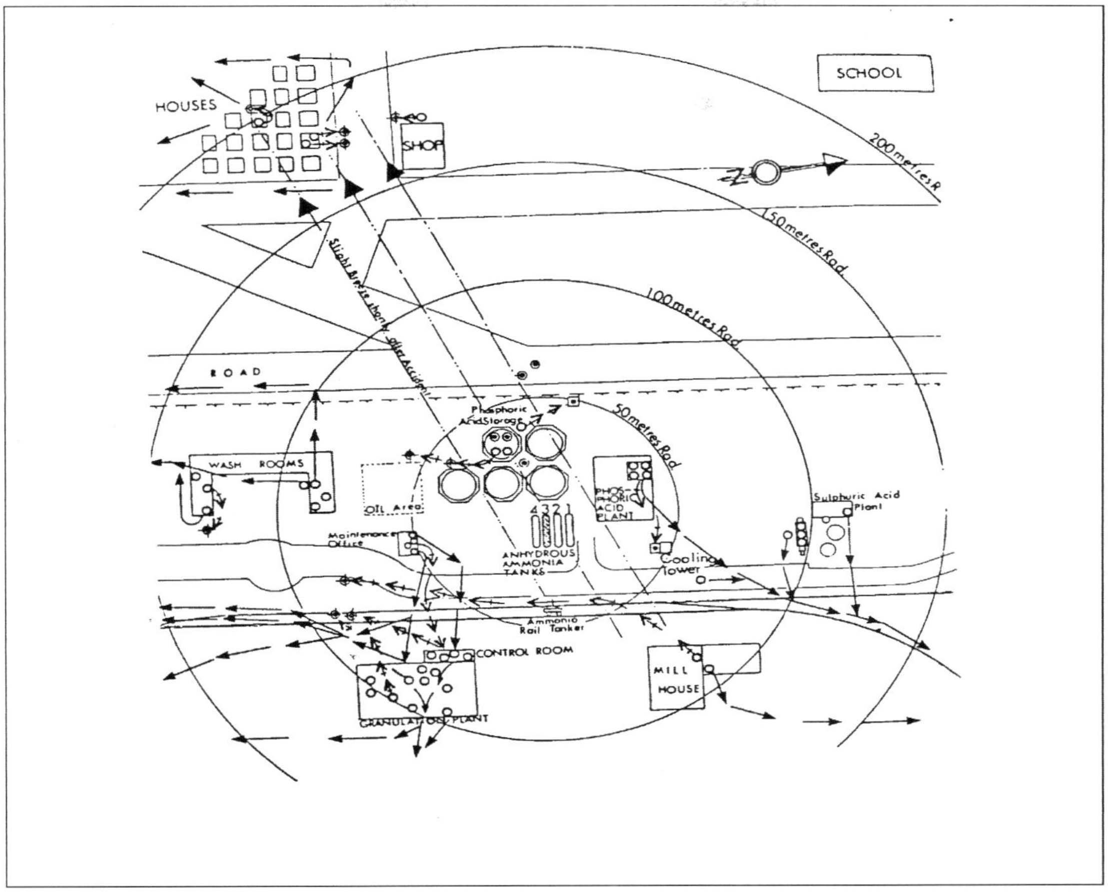

Figure 4. General Layout of the Potchefstroom Plant and Location of Victims

5.2.3 Mortality Statistics

An important denominator in computing mortality statistics is the total number of persons exposed to a hazard. We know that there were a total of 350 persons within the plant. But the number of off-site personnel was not known. If one were to go strictly by the body count, the mortality rates at various zones are indicated in the following Table 4.

Table 4. Mortality Statistics for the Potchefstroom Incident

Hazard Zone	Initial Number Of People Admitted To The Hospital	Fatalities, Immediate And Delayed	Mortality Rate, %
Less than 50 meters	11	7	64%
50 m to 100 m	26	6	23%
100 m to 150 m	8	1	13%
150 m to 200 m	38	4	11%
Total people in plant exposed to ammonia cloud (excluding above)	305	0	0%

5.2.4 Dispersion Analysis

Once again, the dispersion model HGSYSTEM was used to estimate the short-term averaged concentration of ammonia and the resulting toxic dose. The following initial conditions were used for modeling this incident:

- Steady release of 38,000 kg ammonia over 5 minutes
- Weather state D stability
- Wind speed 2 m/s
- Humidity 35 %
- Temperature 19 °C
- Surface roughness 0.1m (assuming release in semi-suburban area)
- 5-minute exposure duration

The dispersion results indicated that there was upwind dispersion of ammonia and the entire plant was covered by the ammonia cloud. This result is consistent with the eyewitness account that suggested that within minutes of release, the white cloud covered nearly all of the facility and extended about 450 m (1500 feet) downwind. The fatality probability assessment, shown in Figure 4, predicts that the 99% fatality zone extends up to 150 m, covering a good part of the facility and the adjacent road, but shy of the residential areas, shops, and the local school.

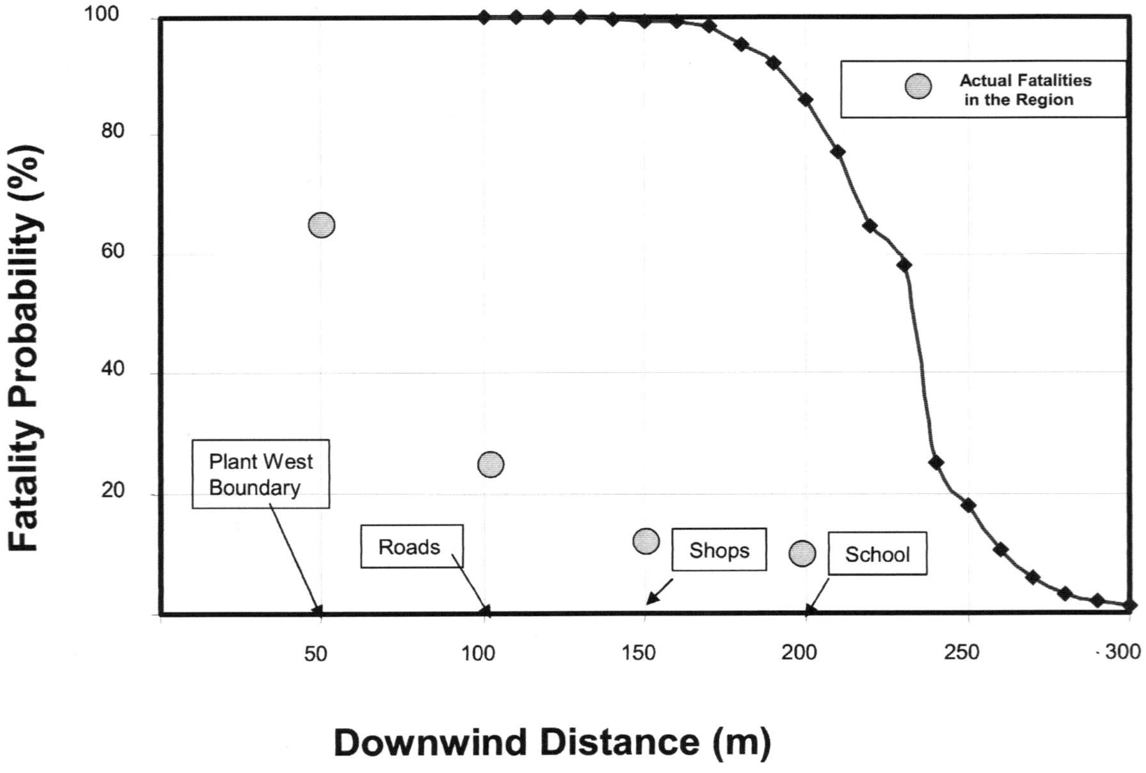

Figure 5: Estimated Fatality Probability Distribution for the Potchefstroom Ammonia Accident

There were, in all, 350 persons within the plant. Given the small size of the facility and large size of the cloud dispersing both upwind and downwind, it is reasonable to assume that the entire site was engulfed within the ammonia cloud. Hence, all plant personnel were exposed to the ammonia cloud. However, it is also reasonable to assume that many were within buildings, control rooms, or other forms of shelter. An indoor concentration and toxic load assessment for this release gave results similar to the Houston release. The indoor dosage was well below the 1% fatality level, even when the predicted fatalities for outdoors were at 99%. This is also consistent with Figure 3, where all the recorded fatalities are outdoors.

In Figure 5, the actual mortality ratios at various distances is shown, as indicated in Table 4. The analysis-based fatality probabilities are greater than the actual mortality ratio.

5.3 Montana, Mexico Chlorine Incident

5.3.1 The Incident

The accident occurred on August 1, 1981, and involved a runaway train. The train was in transit between Pajaritos and Tampico, Mexico carrying a large quantity of chlorine. In all there were 32 cars, with each one carrying about 50 tons of chlorine under pressure. The brake system failed rather suddenly while the train was descending a 3% gradient, with a speed restriction of 28 km/h (17 mph). The train radio was used to warn other trains on the single track. A passenger train and a freight train were pulled into the loop lines at Montana Station. The chlorine train passed these at a speed of about 80 km/h (50 mph) and derailed in a curve. Two locomotives and the first five cars continued to travel at great speed downhill for several miles.

There was a large pile up of 28 chlorine tank cars, a boxcar, a flat car loaded with steel pipes, an empty sulfuric acid tank car, and the caboose. Many cars were badly damaged. One of the tank cars lost its head, and the shell was propelled about 200 m (650 feet). A second tank car suffered a large 20" (50 cm) diameter hole. A third car was split along the sides, and came to rest with the split facing upwards. Four other tank cars suffered severe damage to their relief valves. In some cases the valves were completely ripped off, and in others they were leaking.

5.3.2 Public Impact

Figure 6 shows the sketch of the village of Montana railroad station, location of derailment, and location of population. The derailment was about 350 m (1150 feet) from the railroad station. The population consisted of the village of Montana (roughly 400 persons) and approximately 300 people from another small village, but their exact location was uncertain. The number of people in the passenger train was estimated at about 300.

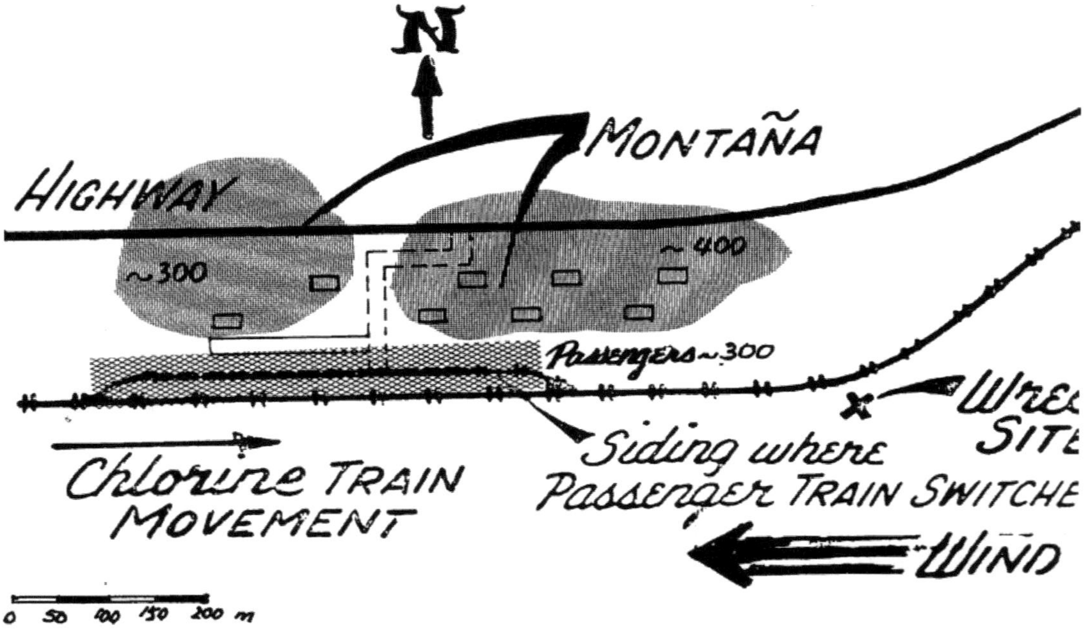

Figure 6: Chlorine Tank Car Location for the Montana Accident

There was very little, if any, warning. A large crowd had gathered at the station to watch the "runaway train." All of the villagers were exposed to a very high concentration of chlorine. About 1000 persons were treated at the local hospital, and 256 were detained. There were 17 fatalities, 4 in the caboose who died as a result of the crash, and 13 who died as a result of exposure to chlorine. The town was evacuated for several days, over which time chlorine continued to leak from other damaged cars, but at a much lower rate.

5.3.3 Mortality Statistics

While it is difficult to know the exact location of all the people, Figure 6 does indicate approximate population densities. The derailment took place about 350 m from the Montana station. The wind was blowing towards the village, and all the population appeared to be to the right of the vapor cloud.

It is reasonable to assume that all 17 fatalities were the result of chlorine exposure. If the crash had not killed the crew, the extreme concentration of chlorine would clearly have killed them all. If we limit the population exposure to the village of Montana and the passenger train, the total exposure was 700 persons. However, it is important to note that nearly 1000 persons went to the hospital (this would account for nearly all the population of Montana, train and other villages), but only about 256 were detained. This leads to a 25% injury rate requiring hospitalization.

Of the nearly 1000 persons admitted to the hospital, about 300 were at the railway platform, clearly considered an outdoor area. The others were most likely inside the train, or inside houses or buildings in the village. Given the season and time of day, it is more likely that the ventilation rates would be relatively high. But even with an ACPH of 2, the indoor toxic load shows a probability of fatality of less than 1%. With a possible

300 outdoor persons exposed to chlorine gas, the mortality rate for this incident becomes 5.7%.

5.3.4 Dispersion Analysis

The prevailing weather conditions were as follows:

- Local time 5.52 PM
- Wind speed warm wind blowing up the valley
- Ambient temperature 30 °C (86 °F)

Given the time of the day, it is more likely that the stability was close to neutral. The wind speed can be assumed to be 5 m/s, typical of neutral stability. In all, about 300 tons of chlorine was released. However, the release continued for one or two days. The larger of the release took place immediately following the derailment, most likely from the cars that suffered direct damage to their shell. Since there were three cars, it is reasonable to estimate the total quantity to be about 100 to150 tons. For the purposes of the analysis, we will assume the lower quantity of 100 tons. It is also important to note that not all chlorine is airborne. Part of chlorine under pressure will flash, and entrain some liquid droplets along with. It also has been noticed in actual accidents that evaporative cooling takes place with the remaining chlorine, and the sub-cooled chlorine evaporates over a period of time. Once again, for the purposes of the analysis, we assumed that 50%, or 50 tons of chlorine is airborne, and released with in a very short period of time of two minutes.

Since a downwind observer will be exposed to both releases, an additive dosage will be calculated. The initial 50,000 kg release (2 minute duration) was modeled as a transient release, and the following 50,000 kg release (2 hour duration) was modeled as steady state. The resulting dosage for each release was then added to generate the final combined impact. The initial 50,000 kg release was the driving force of this scenario. The computed chlorine concentration as a function of downwind distance is shown in Figure 7.

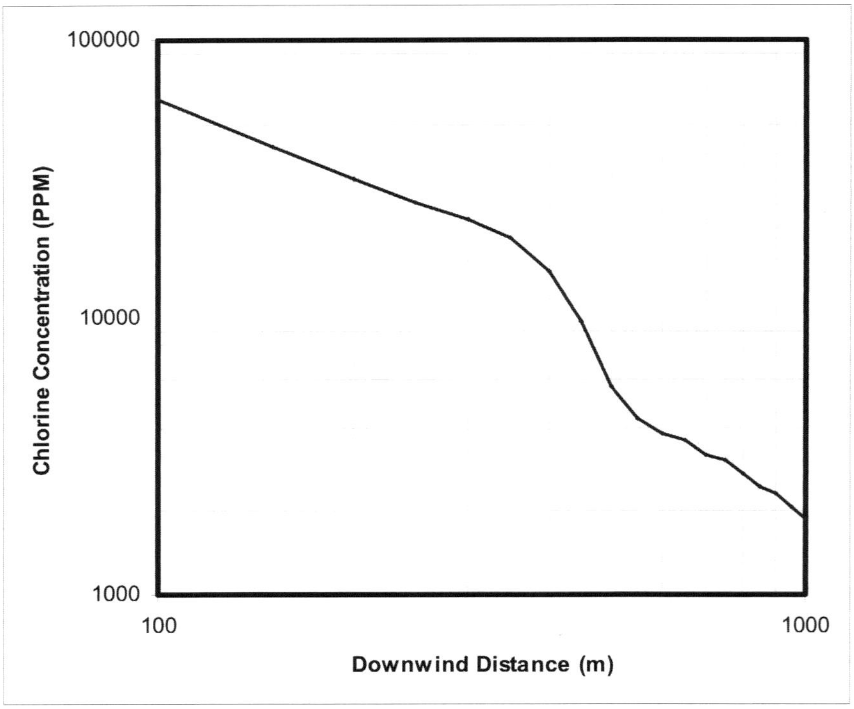

Figure 7: Predicted Average Chlorine Concentration for the Montana Tank Car Accident

The estimated Fatality probabilities are shown in Figure 8.

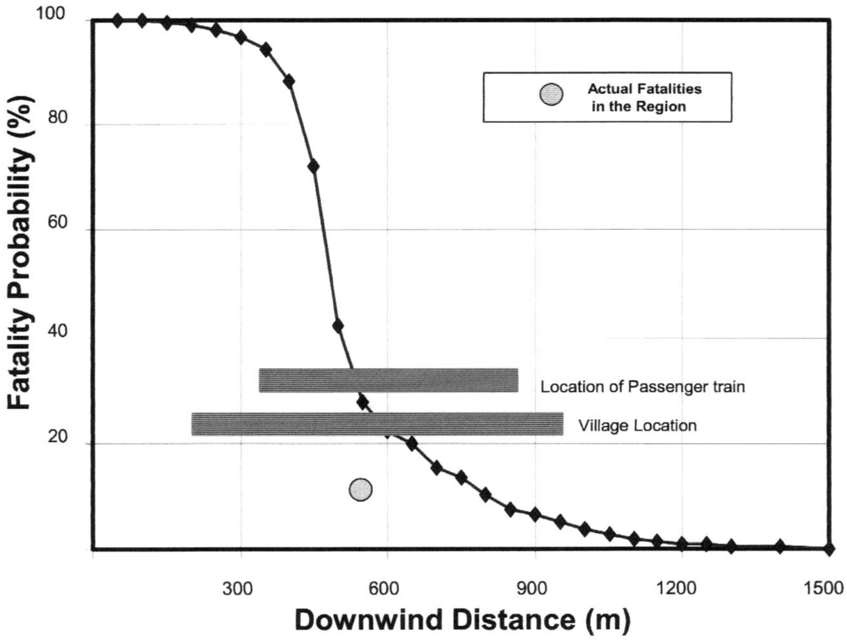

Figure 8: Predicted Fatality Probability Distribution for the Montana Chlorine Release

The results indicate that the dispersion is dominated by the initial release. The 99% fatality zone extends up to a distance of 300 m or about 1000 feet. Relative positions of the passenger train and the village, shown in Figure 8, indicates that both population centers were just outside this zone. The predicted fatality probabilities within the village ranges from 100% to about 5%, and for the passenger train, the probabilities range from 95% to about 8%. The weighted average fatality was predicted to be about 24%. Beyond a distance of about 1200 m, the fatality probability is below 1%. This distance clearly covered the entire passenger train, the train station, the village of Montana, and the adjacent highway. Therefore, it is reasonable to assume that many of victims admitted later to the hospital came from this area. The actual mortality rate, shown in Figure 8, is lower than the predicted fatality probability.

5.4 La Barre Chlorine Incident

5.4.1 Accident description

The incident took place on January 31, 1961 at approximately 8:30 AM. A freight train derailed, and a chlorine tank car was damaged. The extent of damage to the tank car is not known. The capacity of the tank car was about 6000 gallons, which translates to about 35 tons of chlorine. The cloud appeared to have been blocked on the northern side by one of the levees of the Mississippi river. An air-monitoring program was initiated, and a chlorine monitoring system was installed in about 3 hours following the incident. The detector recorded a level of 400 ppm at a distance of 75 m from the spill site up to 6 hours following the accident. The concentration started decreasing afterwards, and only about 8 ppm was measured by 5:00 PM. This indicates that the wreck probably caused severe damage to one of the relief valves, and the release continued for several hours. Given the typical size of 1" angle valves or 2" relief valves installed on chlorine tankcars, such extended release duration is not uncommon. In Figure 9, the accident site, location of some of the houses, and the approximate dimensions of the cloud are shown, based on information gathered after the release was over.

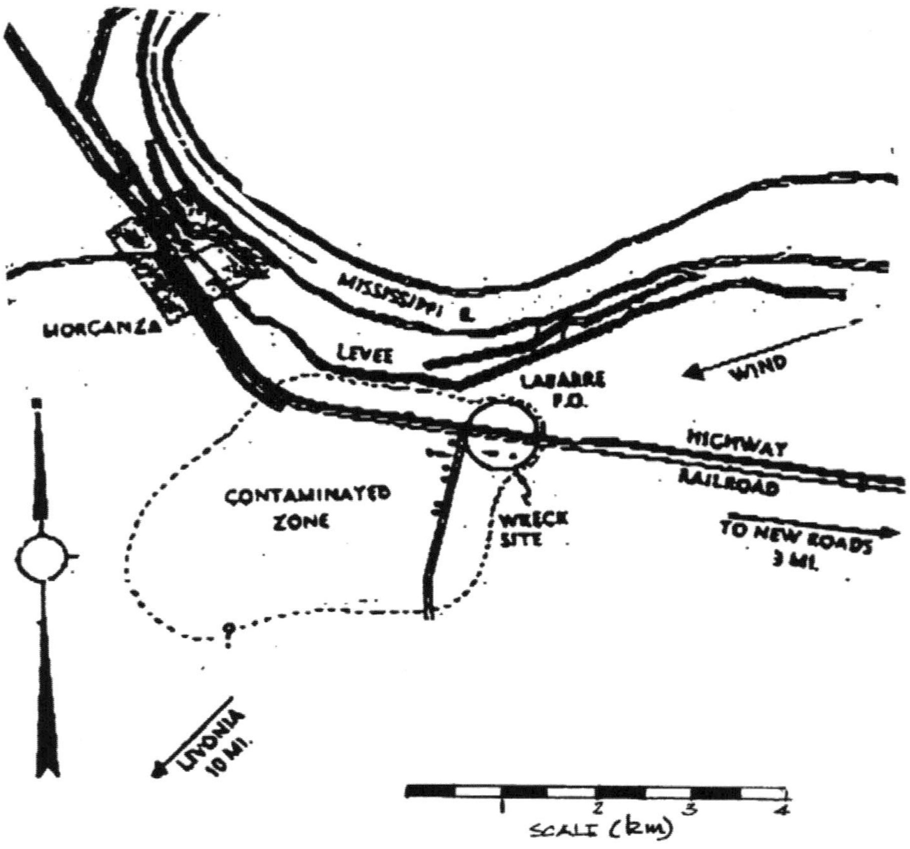

Figure 9: Location of the Chlorine tank-car and Residences for the La Barre Accident

5.4.2 Accident Impact

The cloud was evidently rapidly lethal to animals up to 2 km from the site of derailment. The area of animal fatality was directly in the downwind of the release (since the upwind region was blocked). In all, 49 pigs, four mules, two cows, one horse, and 320 chickens died as a result of chlorine exposure.

Roughly 100 persons were affected and 65 were taken to the New Roads Hospital. The remaining persons were treated by local physicians. Fifteen of the 65 patients were detained at the New Roads Hospital for further observation and treatment. One child of 11 months died five hours after the exposure. Ten of those admitted to the hospital developed pulmonary edema. Three other children and one adult were unconscious on admission. All patients regained consciousness within one hour. The child who died was in a house situated 50 m from the accident. The child's father had tried to relieve the infant's choking and gasping by taking him out into the open air, but unfortunately the conditions outside were worse than inside the house. A second child, a 2-month old infant, who was left inside the house as well as the father who returned into the house, both survived the exposure.

It is also significant to note that an air monitoring system was installed, and first samples were taken at about 11:15 AM, roughly 3 hours after the release. After 7 hours into the spillage, the chlorine monitoring system recorded 400-ppm concentration, and by

5:30 PM the concentration had decreased to about 8 ppm. This is indicative of the continuous nature of the release. This may also provide a benchmark for the dispersion calculation. However, the report states that this concentration was measured in a "moderately" affected area and therefore does not represent concentrations in the denser parts of the cloud.

5.4.3 Mortality Statistics

It is difficult to derive mortality statistics for this incident. What is undoubtedly clear is the degree of protection offered by indoor sheltering. Given the date and time of the incident (January, morning time), it is reasonable to assume that the doors were closed and windows were probably shut. Since this was a farm community (clearly indicated by the animal fatalities), it is possible that some of the heavily exposed persons were working outdoors at the time of the incident. If we assume that ALL casualties admitted to the New Roads Hospital were exposed to higher concentrations of chlorine, including the child that later died, the number of outdoor personnel can be estimated to be 15.

It is reasonable to assume that the air monitoring began after evacuation of the public (by the emergency personnel with adequate breathing equipment). The air monitoring did not start till about 3 hours after the incident. Since it was a relatively steady release, it is also reasonable to assume that people were exposed to chlorine concentration for a relatively long period of time, say, up to an hour or more. The exact location of all persons is not known. However, it is stated that most people lived within 250 m from the railroad track. This is also consistent with the map of the area shown in Figure 9. The family of the deceased child lived at a distance of about 50 m from the accident scene.

5.4.4 Dispersion Analysis

The prevailing meteorological conditions are not known. However, considering the duration of release, the prevailing metrological conditions become less of a factor. For the purposes of the analysis we assumed the following conditions, based on historical statistics for the area:

- Ambient temperature: 50 °F
- Prevailing Wind speed: 3 m/s
- Atmospheric Stability: D
- Relative Humidity: 70%

HGSYSTEM flow rate model was used to determine the release rate and average release rate. The maximum release rate lasted for about 2 hours, whereas the average release rate lasted about 7 hours. In reality, the release would start at maximum, and slowly reduce over a period of time. Since the exposure was limited to one hour, the maximum flow rate was used. The predicted mortality is indicated in figure 10.

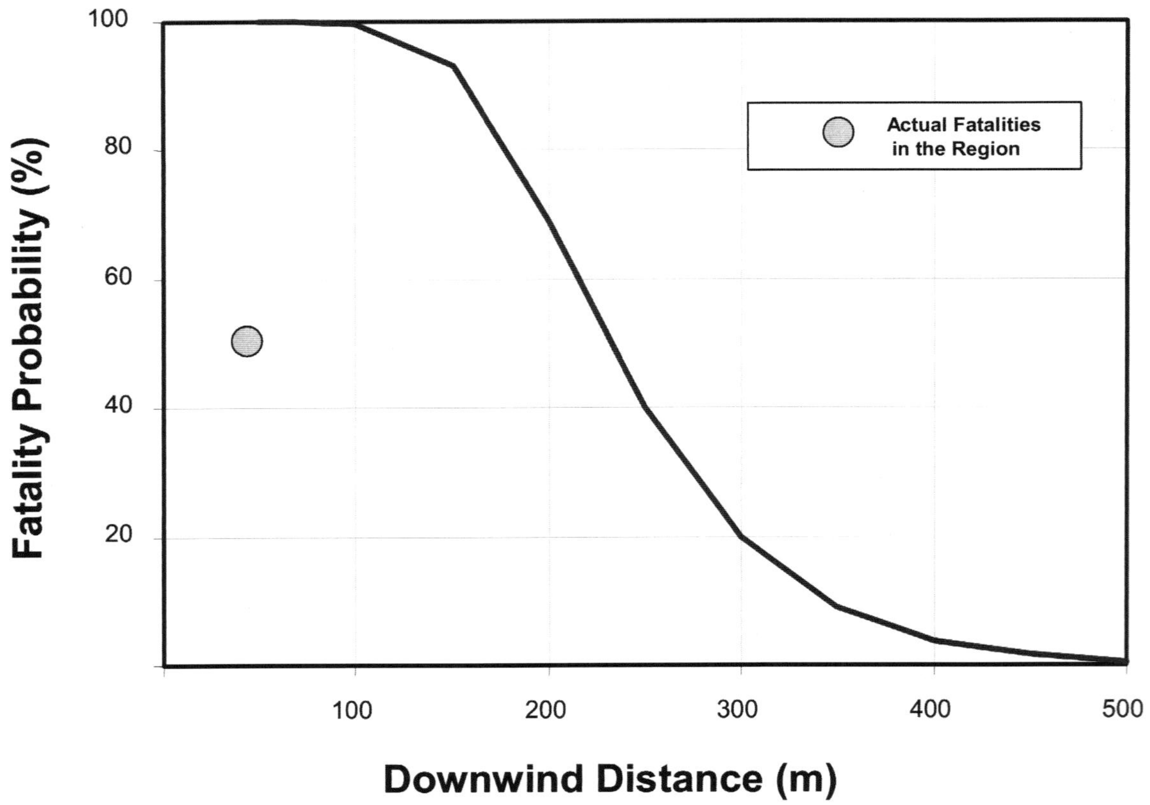

Figure 10: Predicted Fatality Probability for the La Barre Incident

These results also indicate that the toxic vapor cloud traveled considerable distance and affected a relatively large area. The fatality threshold zone (represented by 1% fatality probability) extends to nearly 500 m (about 1700 feet), and appears to cover all of the populated area surrounding the accident scene. It is reasonable to assume that all of the affected public came from this area, and were admitted to the hospital. This incident resulted in 100 casualties, 15 hospital admissions, and one fatality. However, the single fatality came from a household within a distance of 50 m from the release and involved an infant that died and a father who survived. With the assumption that both were exposed to high concentration of chlorine, the mortality index at 50 m would be 50%. The overall mortality index based on all 15 persons exposed to chlorine during the incident is 6.7%.

The La Barre and the Montana incidents, in addition to several other chlorine incidents, led the UK's Institution of Chemical Engineers (I.Chem.E.) to infer that fatalities due to chlorine releases are not likely to occur at distances exceeding about 500 m from the scene of the accident.

5.5 Bhopal Tragedy

Early in the morning of 3 December 1984, a relief valve lifted on a storage tank containing highly toxic methyl isocyanate (MIC) at the Union Carbide India Ltd. plant in Bhopal, India. A cloud of toxic MIC gas was released onto the residential areas surrounding the facility, including the shantytowns adjoining the site. Thousands of persons died within a short period of time and tens of thousands were injured. The casualty statistics are discussed later in this Section. The accident at Bhopal is considered by far the worst disaster that has ever occurred in the chemical industry.

We reviewed many of the available documents in developing the Bhopal accident scenario. Following the accident, the Government of India (GoI) set up an inquiry and a Report was issued in 1985. An independent investigation was conducted by the Union Carbide Corporation, the parent company of Union Carbide India Ltd, a report was issued in 1985, and later a modified version was presented in 1988. Another investigation commissioned by ICTFU-ICEF was also published in 1985. A recent review by Dhara and Dhara (2003) gives a description of prevailing health problems directly linked to the release.

5.5.1 Accident Description

MIC was an intermediate product in the reaction of monomethylamine (MMA) in excess phosgene in the production of proprietary agricultural products developed by Union Carbide Corporation of the Unites States. MIC was stored in one of the two storage tanks (tanks 610 and 611) and a third storage tank (tank 619) was to be used in the event of an emergency. The nominal capacity of the tanks was about 15000 US gallons, and the stainless steel tanks were designed for a pressure of 40 psig. The safety relief valve was set at the design pressure. On the night of the accident, the operator noticed that the pressure in storage tank 610 was 10 psig. While it was above the normal pressure, it was well within the range of 2 psig to 25 psig operating pressure in the tank. At 12:15 AM one of the field operators noticed and reported an MIC leak in the process area, and by this time the tank pressure had increased to about 30 psig. The supervisor went out to inspect the tank, and heard "rumbling" sounds from the tank and a "screeching" noise from the relief valve. Within about 20 minutes, the plant operations were suspended, and at 1:00 AM one of the operators turned on the toxic gas siren. After about 5 minutes the loud siren was turned off and a muted siren was kept on. The plant personnel also attempted to direct water towards the vent stack, but were unsuccessful. Sometime between 1:30 and 2:30 AM, the relief valve reseated and the release of MIC ceased. It was a still night, with very low wind blowing southeast towards the shantytowns in the vicinity of the plant. After roughly 45 minutes, the wind direction changed towards south for about 30 minutes. The wind shifted again to the southwest for an additional 15 to 20 minutes. A simplified plot plan (L. Everest 1985) of the area near the Union Carbide Plant is shown in Figure 1.

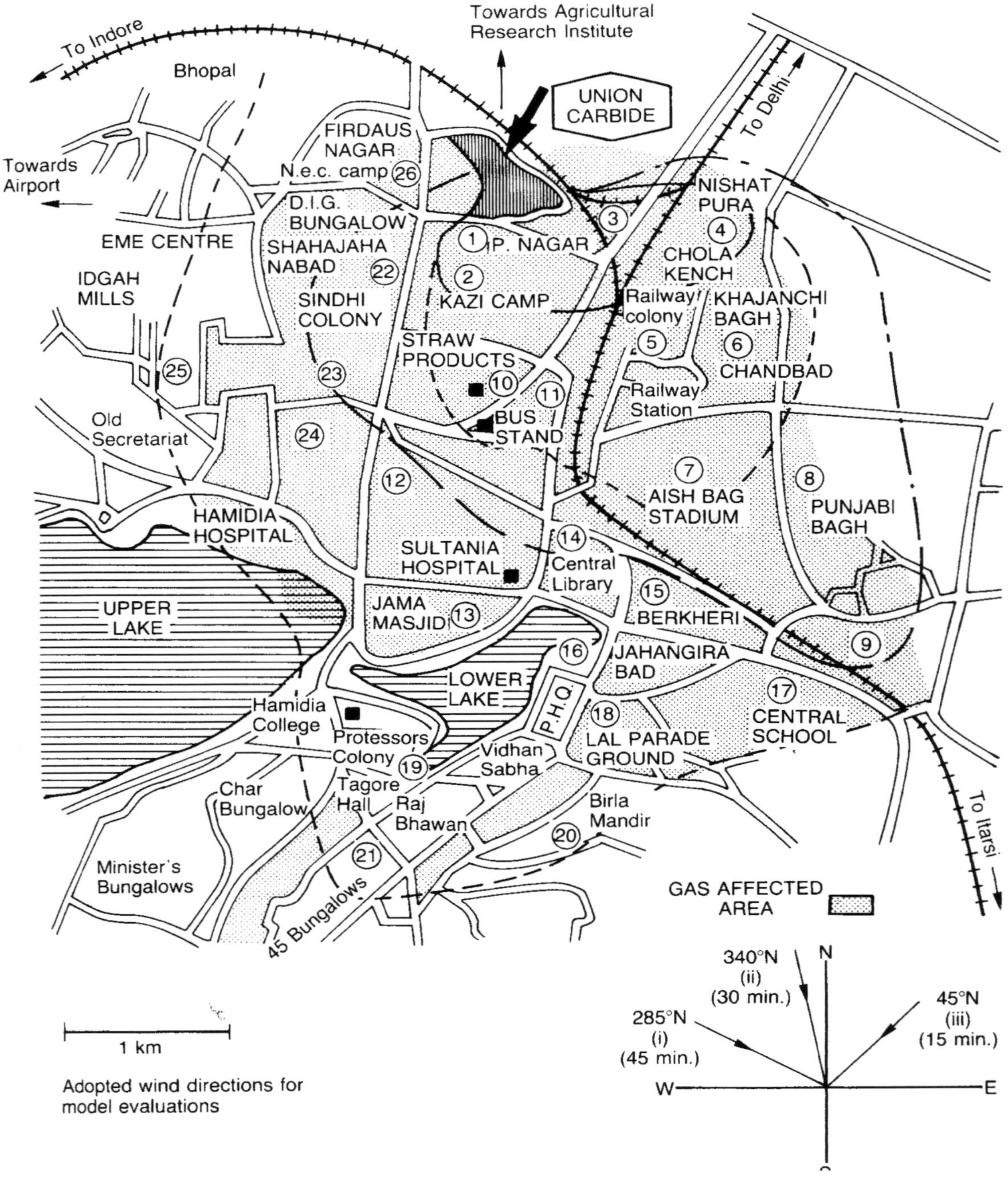

Figure 11: Bhopal MIC Release Plot Plan

The quantity of material released during the accident is fairly well established. However, not all the chemical released was MIC. It is estimated that about 90,000 lbs of MIC was in the tank prior to the accident. During the accident, nearly 80,000 lbs of the material was released. The reaction caused MIC to decompose, and the release contained about 68% of un-reacted MIC in vapor form, with the remaining 32% of reaction products. It is estimated that the total duration of release was about 90 minutes.

The shantytowns near the plant site were cramped and crowded. For instance, the town of J.P.Nagar in the immediate vicinity of the plant was about a quarter of a mile long, about 50 yards wide, and had about 500 households with 3000 people. The Kazi Camp, the next town had 580 households, and 3160 persons living in less than 6.3 hectares. The Railway Colony was about a mile away from the plant site, and was home to well over 10,000 persons. Some of the other shantytowns had a population density of about 925 people/hectare.

In all, about 200,000 people were exposed to the toxic effects of MIC over a 25 square kilometer area (approximately 10 square miles). The affected area, shown in Figure 11, is illustrative of widespread toxic impact due to shifting in wind direction. The average downwind distance appears to be about 6 to 7 km, and the mean width is about 4 km. The overall effects of gas poisoning were staggering:

- 170,000 persons were treated in local areas and 22 district hospitals outside Bhopal. Many were treated as outpatients.
- Nearly 65,000 persons inhaled large quantities of MIC, and had to undergo repeated treatments.
- The official death toll confirmed by Government of India was 1,754 reported two years after the incident. In 1985, the Director of Health Services for the State of Madhya Pradesh, Dr. Nagu, estimated the death toll to be about 2000. In 1989, the Government of India increased the death toll to 3150. The unofficial death toll ranges between 2000 and 5000. A generally quoted number of fatalities is 2500. The two local hospitals, Hamidia and Jayaprakash, were overwhelmed with casualties. The difficulties in treating the patients were further complicated by the fact that it was not known what the gas was and what its effects were. There was also a speculation about the gas, including suggestions that it was phosgene. There were also conflicting views on treatment of casualties.

5.5.2 Mortality Data

It is clear from various eyewitness accounts and reports that the effect of MIC poisoning was immediate and deadly. One report states that at Railway Colony, where about 10,000 persons were exposed, within a few minutes 150 had died, 200 were paralyzed, 600 rendered unconscious, and 5000 were severely affected. JP Nagar and Kazi Camp, two of the hardest hit areas, accounted for nearly 7000 casualties, including about 25% fatalities. Another area about one mile from the plant where nearly 1000 persons were exposed, 21 died, 200 were unconscious, and 600 were suffering from the inability to breathe, vomiting and choking.

The first significant factor influencing the large number of fatalities is the population density. It is estimated that within the first mile from the plant site there were close to 50,000 people, and possibly as many as 100,000 within the first two miles. The population density was significantly lower near the central library and the lower lakes region.

The second factor affecting mortality was the exposure concentration and duration. Based on wind shift data, it appears that the wind was blowing towards the southeast during the first 45 minutes of the release, and this clearly would have affected people living in J.P.Nagar, Kazi Camp, Railway Colony, and Railway Station. There was a shift in wind direction towards the South and this lasted another 30 minutes. For people in the immediate vicinity, it is likely that the exposure continued for another 30 minutes. The impact further downwind would have been severe, but not necessarily fatal. The change in wind direction could have spared the lives of people living further downwind.

5.5.3 Dispersion Model Results

There is some uncertainty about the actual quantity of MIC released during the incident. However, the range of uncertainty is relatively small. The storage tank contained little more than 90,000 lbs of MIC prior to the accident. About 10,000 lbs of residue was collected from the tank after the accident. This leads to a total release quantity of about 80,000 lbs.

The release started sometime after midnight, perhaps closer to 1:00 AM. The records indicate that the relief valve receded between 2 AM and 3 AM. This gives an average release duration of about 90 minutes, and an average release rate of about 14.8 lbs/sec or about 6.72 kg/s. There is some uncertainty about whether the release was pure MIC, or products of reaction or combination. The analysis of residue and independent reactivity tests conducted by Union Carbide Corporation in the U.S. suggests that of the 80,000 lbs released during the accident, about 54,000 lbs was un-reacted MIC. The remaining 26,000 lbs was the reaction products. Although the specific toxicology of reaction products is unknown, it is believed that they are also toxic, including chemicals such as phosgene and chloroform. For the purposes of this analysis, it was assumed the entire release consisted of unreacted MIC. Lees (1996) estimated that the pressure to sustain such a release would have to be about 180 psig, and the reaction temperature to increase the pressure would have to be about 200 °C. These estimates were also consistent with calorimetric testing done independently by UCC.

The wind speed was relatively low, with very little turbulence. These are characteristics of stable atmospheric conditions. The wind speed was assumed to be 2 m/s (or about 4.5 mph). As mentioned earlier, the wind direction was steady, blowing towards the southeast for the first 45 minutes, and south for the next 30 minutes. It is reasonable to assume that the residents of the shantytown areas of J.P.Nagar, Kazi Camp and the Railway Colony were exposed to toxic fumes for at least one hour, and a corresponding concentration was used in determining the dispersion distances. An increase in exposure duration would decrease the concentration, and hence lead to larger distances. In Figure 12, the predicted fatality probability distribution is shown for the MIC release at Bhopal.

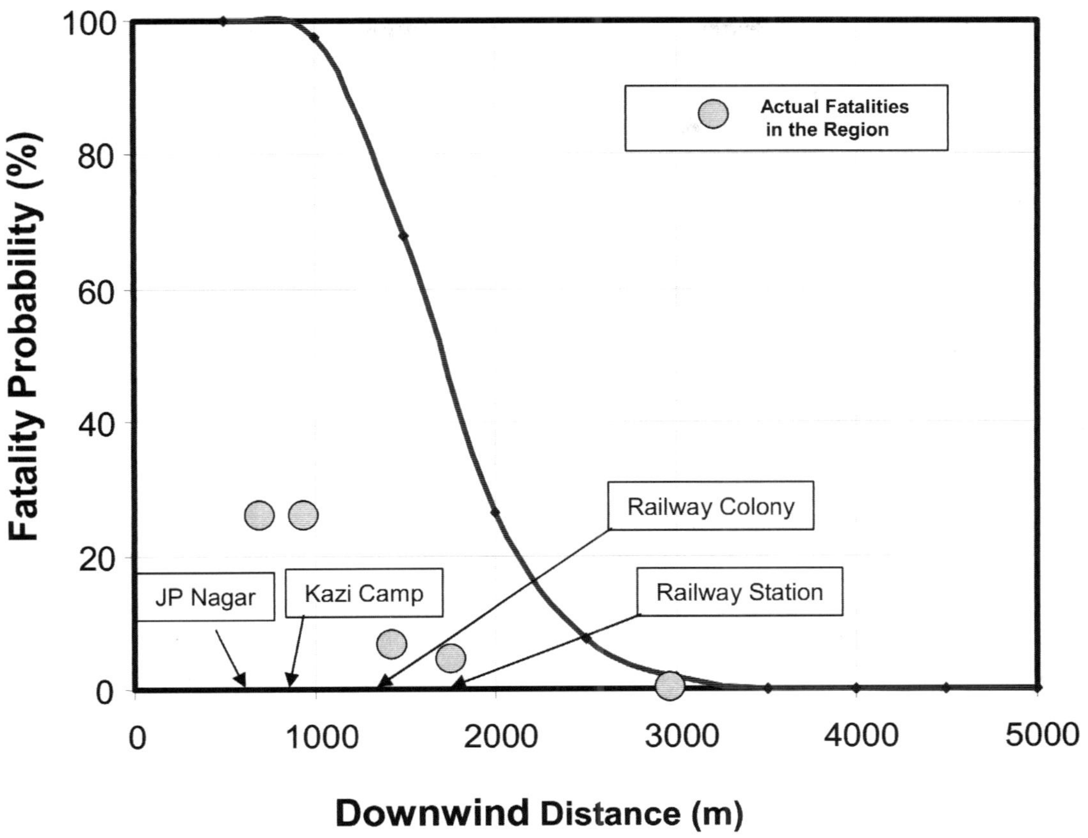

Figure 12: Computed Fatality Probability for the Bhopal MIC Release

As Figure 12 indicates, the dispersion model coupled with toxicity data computes a 100% mortality distance of about 1 km; the fatality probability is about 30% at 2 km and negligible at distances beyond 3.5 km. Also shown in Figure 12 are the actual fatalities in the region, and these, while lower than the predicted values, are reasonably consistent with the trend. The actual mortality information indicates the following:

- Within the first 1 km, covering both J.P.Nagar and Kazi Camp, the mortality was highest, and estimated at about 25%. The 25% mortality rate within the first zone accounts for nearly 1540 fatalities, a large fraction of the 1754 fatalities officially reported by Government of India. It is important to note that very few, if any, of the people in the immediate vicinity of the plant had the protection of indoor occupancy or shelter–in-place. The shantytown dwellings were essentially equivalent to open areas, and did not offer protection of any kind.
- At a distance of about 1.3 km covering the Railway Colony, of the total 10,000 persons exposed to MIC, an estimated 150 died, and 200 were paralyzed and unconscious. This would lead to a mortality ration of 3.5%.
- Near the Railway Station area, roughly 1.6 km away, there were about 1000 persons affected, and 21 fatalities were noted.

The gas-affected area covered about 6 to 7 km downwind and about 3 to 4 km crosswind, as shown in Figure 11. This encompassed a total of about 170,000 to 200,000 people. Nearly 170,000 persons were treated at 22 different hospitals, with about 65,000 treated for complications related to inhaling large quantities of MIC. It is not known how many fatalities were in this region. Since J.P.Nagar, Kazi Camp, Railway Colony and Railway Station seem to account for many of the immediate fatalities (total of 1911, consistent with GoI estimates), it is likely that there were fewer fatalities further downwind. It is also possible that at further downwind distances, the buildings were better designed and offered some degree of protection.

6.0 Regulatory Toxic Endpoint Criteria

In Section 3, probit equations were developed for ammonia, chlorine and MIC using the animal test data, and these were used in the dispersion modeling and fatality estimation described in Section 5. Fatality estimates were made using the risk assessment models for comparison with actual event fatalities. The results in the case for each chemical analyzed showed the model results to be slightly conservative, but very close to the actual event fatalities. It is important to note that the probits used were taken from the literature and are the same probits used by many regulatory agencies for development of emergency response guidelines and facility siting criteria. However, the probits used in the analysis were not modified with an interspecies factor to extrapolate from animals to humans, or an uncertainty factor to account for differences in human sensitivity.

Endpoint criteria established by regulatory agencies was reviewed and compared with results from the probits developed in Section 3. The agencies chosen are as follows:

- U.S. EPA's National Research Council program for the establishment of AEGLs.
- UK Health & Safety Executive's Specified Level of Toxicity (SLOT) and Significant Likelihood of Death (SLOD) used in the UK for land use planning purposes.
- The French Ministry of Ecology and Sustainable Development guidelines for potential lethality.
- The Dutch Ministry of Environment procedures as outlined in the "Green Book".

The AEGL3 represents a time-dependent airborne concentration of a substance above which it is predicted that the general population, including susceptible individuals, could experience life-threatening health effects or death. In the absence of chemical specific human exposure data, AEGL-3 is based on a concentration lower than the highest value that resulted in no deaths in rats, incorporating an interspecies factor of 3 to extrapolate from animals to humans, and an uncertainty factor of 3 to account for differences in human sensitivity, giving a total adjustment factor of 10.

The SLOT represents a level at which about 1% of the exposed population would experience fatalities and SLOD represents a 50% fatality level. Both levels are based on toxicological data from the "most sensitive" of the animal species.

The French Ministry's criteria are also based on animal data incorporating an uncertainty factor. However, a specific value is not given in the Environmental Ministry reports.

The Dutch Ministry's approach considers an interspecies factor (typically between 0.25 to 0.5) but an uncertainty factor of 10, to give a total adjustment factor between 2.5 and 5.

Although the approaches are slightly different, the majority of these criteria lead to roughly an order-of-magnitude reduction in the toxic exposure levels while extrapolating animal data to human beings. This suggests that if one uses regulatory criteria similar to that reviewed in this paper, the "Human Toxic Exposure Response Factor" may be applied to the resulting fatality estimates to account for the differences between predicted and actual values. This factor could account for many, if not all, the non-quantifiable issues associated with human behavior and described in Section 2, and reduce the uncertainty associated with the prediction of mortality.

7.0 Summary and Conclusions

In this paper, we have attempted to address the uncertainties associated with estimating the fatal impact of toxic chemical releases, using probits. The probits were taken from existing sources or were developed using animal exposure data. To the extent possible, only rat data were used in developing the fatality probits for ammonia, chlorine and MIC. These probits were used in conjunction with the publicly available and validated dispersion model, HGSYSTEMS, to determine the toxic concentration and dosage for selected accidents. The accident investigation information was used to estimate the actual fatality levels as a function of distance. Based on the probits developed in Section 3, dispersion modeling of actual incidents, and a comparison of predicted and actual fatalities described in Section 5, we draw the following conclusions:

1. Animal toxicological (rat) data provides a good basis for making fatality predictions in all of the five accidents studied in this article.
2. The probits derived from rat data, without any adjustments for interspecies effect, appear to predict slightly higher fatality probabilities.

The toxic endpoints specified by regulatory agencies do incorporate "uncertainty factors" up to an order of magnitude. They are particularly well suited for emergency response planning and facility siting. However, their use, without any adjustments to the probits, may lead to a significant over-prediction of fatality probabilities.

8.0 References

1. Acute Exposure Guideline Levels for Selected Airborne Chemicals: Volume 3 (2003) Board on Environmental Studies and Toxicology (BEST), published by the National Research Council.

2. Appleman, L.M., ten Berge, W.F., and Reuzel, P.G.J. (1982) "Acute inhalation toxicity study of ammonia in rats with variable exposure periods", American Ind. Hygiene Association, 43: 662-665.

3. Bunce, N.J and Remillard, R.B.J (2003) "Haber's Rule: The Search for Quantitative Relationships in Toxicology" Human and Ecological Risk Assessment, 9: 973-985.

4. Demnati R., Fraser R., Plaa G. and J.L. M. (1995) - Histopathological Effects of Acute Exposure to Chlorine Gas on Spargue-Dawley Rat Lungs. *Journal of Environnemental Pathology, Toxicology and Oncology,* **14,** 15-19.

5. Dhara, V Ramana and Dhara, R (2002) "The Union Carbide Disaster in Bhopal: A review of Health Effects" Arc. Environmental health, 57, 5:391-404.

6. Dinsdale, D., B.Nemery, and S.Sparrow. 1987. Ultrastructural changes in the respiratory tract of rats following methyl isocyanate inhalation. Arch. Toxicol. 59:385–390.

7. Dutta, K.K., G.S.D.Gupta, A.Mishra, A.Joshi, G.S.Tandon, and P.K.Ray. (1988). Inhalation toxicity studies of methyl isocyanate (MIC) in rats: Part I—Pulmonary pathology and genotoxicity evaluation. Ind. J. Exp. Biol. 26:177–182.

8. Everest, L. (1985) "Behind the Poison Cloud", Banner Press, Chicago.

9. Fairhurst S and Turner RM (1993) "Toxicological assessments in relation to major hazards." Journal of Hazardous Materials **33**, 215-227

10. Ferguson, J.S and Hendershot, D.C (2000) "The impact of Toxicity Dose-Response Relationships on Quantitative Risk Analysis Results", Process Safety Progress, 19, 2: 91-97.

11. Ghosh, B.B., S.Sengupta, A.Roy, S.Maity, S.Ghosh, G.Talukder, and A. Sharma. 1990. Cytogenetic studies in human populations exposed to gas leak at Bhopal, India. Environ. Health Perspect. 86:323–326.

12. Jeevaratnam, K., and S.Sriramachari. 1994. Comparative toxicity of methyl isocyanate and its hydrolytic derivatives in rats. I. Pulmonary histopathology in the acute phase. Arch. Toxicol. 69:39–44.

13. Kimmerle, G., and A.Eben. 1964. On the toxicity of methyl isocyanate and its quantitative determination in air [in German]. Arch. Toxikol. 20:235–241.

14. Schorsch, F (1995) "Etude de la toxicite aigue de l'ammoniac administer par inhalation chez le rat", INERIS, Verneuil-en-halate:1-19.

15. ten Berge, W.F (1985) "The Toxicity of Methylisocyanate for rats" J. Hazardous Materials, 12:309-311.

16. Tissot, S. and Pichard, A. (2001) "Seuils de Toxicité Aiguë CHLORE (Cl2)", Ministère de l'Écologie et du Développement Durable, France.

17. Tissot, S. and Pichard, A. (2003) "Seuils de Toxicité Aiguë AMMONIA (NH3)", Ministère de l'Écologie et du Développement Durable, France.

18. TNO (1990) " Methods for Determination of the possible damage to Humans and Goods by the release of Hazardous Materials" (Green Book), Dutch Ministry of Housing, Physical Planning and Environment, The Hague.

19. The U.K. Health & Safety Executive (2001) "Assessment of the Dangerous Toxic Load (DTL) for Specified Level of Toxicity (SLOT) and Significant Likelihood of Death (SLOD)" Assessment Methodology for COMAH Compliance.

20. De Weger, D, Pietersen, C.M, and Reuzel, P.G. (1991) "Consequences of Exposure to Toxic Gases following Industrial Disasters" J. Loss Prev. Process Ind., 4: 272-276.

21. Withers R.M.J. and Lees F.P. (1985a) - The assessment of major hazards: the lethal toxicity of chlorine Part 1, review of information on toxicity. *Journal of Hazardous Materials,* 12, 231-282.

22. Withers R.M.J. and Lees F.P. (1985b) - The assessment of major hazards: the lethal toxicity of chlorine Part 2, Model of toxicity to man. *Journal of Hazardous Materials,* 12, 283-302.

23. Withers R.M.J. and Lees F.P. (1985c) - The assessment of major hazards: the lethal toxicity of chlorine Part 3, Crosschecks from gas warfare. *Journal of Hazardous Materials,* 12, 301-342.

24. Zwart A. and Woutersen R.A. (1988) - Acute inhalation toxicity of chlorine in rats and mice : concentration-mortality relationships and effects on respiration. Journal of Hazardous Materials, 19, 195-208.

Modeling of a Warehouse Fire

A Case Study

Ertugrul Alp, Ph.D., P.Eng.
Robert Michalowicz, SM, P.Eng.

Alp & Associates Incorporated
Toronto, Ontario, Canada
Ertugrul.Alp@rogers.com

ABSTRACT

This paper describes the modeling study undertaken for an industrial site, regarding worst case scenarios for a calcium hypochlorite storage warehouse.

Calcium hypochlorite produces its own oxygen and results in a very hot burning fire as it decomposes. Products of combustion include HCl, Cl_2, CO, CO_2, C, and H_2O. The material is stored in polypropylene pails on wooden pallets, shrink-wrapped with polystyrene.

The modeling started with the estimation of rate of development of the fire and the source terms using basic principles of heat transfer, thermodynamics and chemistry. This source term was then used in a dispersion model to estimate the toxic plume impact zone, taking into account plume rise. Effect of different weather conditions was investigated by repeating the dispersion simulation for every hour of a five-year meteorological data set. Thermal radiation impacts were also modeled. Effect of the fire suppression system was also investigated, as this formed an important part of a fire response strategy.

1. INTRODUCTION

Modeling potential impacts of worst case scenarios has been a requirement under Responsible Care, and under the 1996 EPA Risk Management Program Rule in the US. An appreciation of the size of potential impact zones of worst case scenarios help emergency responders develop appropriate plans for responding to such emergencies.

Many site risk assessments have identified warehouse fires as among the more significant scenarios that could occur at those sites. However, due to the difficulties in modeling these "difficult-to-model scenarios", these sites had not included these scenarios in the quantitative risk analysis programs in preparation for the initial rounds of verification audits. Some of these sites later decided to undertake these modeling studies in response to findings of the verification audits.

This paper describes the modeling study undertaken for one of these sites, regarding worst case scenarios for a calcium hypochlorite storage warehouse.

2. FACILITY DESCRIPTION

The facility is in an industrial area, surrounded by a major highway to the north, railway tracks to the west, and other industrial buildings to the south and east. The nearest residences are to the north, across from the highway, at approximately 350 m (Figure 1).

Figure 1 Residences to the North, from warehouse rooftop

The building in which the calcium hypochlorite ($Ca(OCl)_2$) product is stored has concrete floors, with steel walls and roof. It has a floor area of approximately 265 ft (E-W) x 180 ft (N-S) = 47,700 ft^2. The building is further subdivided into three smaller areas with two 2-hour rated firewalls. Passage is provided from one area to the other with a fire door (also 2-hour rated). Only one of these areas (approximately 96 ft x 180 ft) houses the calcium hypochlorite product.

The calcium hypochlorite inventory at any given time can be up to 2000 tonnes, with an average inventory of approximately 1000 tonnes. A value of 1400 tonnes was used as the basis of the modeling in this study.

The calcium hypochlorite is stored in various size containers. A typical container is a 100 lb pail made of polypropylene. This container was taken as the basis of fire modeling. Its dimensions are approximately 14" average in diameter, and a height of 16".

The pails are palletized, 16 pails to a pallet, stored two pails high. Each pallet is shrink-wrapped with polystyrene. The pallet dimensions are 48" x 40". The typical weight distribution of product and packaging on each pallet is shown in Table 1.

Table 1 Weight distribution of product and packaging on each pallet

Component	Material	Weight (kg/pallet)
Container	Polypropylene	48
Stretch wrap	Polystyrene	0.5
Pallet	Wood	15 to 18
Product	$Ca(OCl)_2$	725

The pallets are stored in rows along their longer dimension. The rows are 8 to 11 pallets deep in an east-west direction, and three pallets high. Typical distance between the rows is 8". The typical storage pattern in the warehouse is three blocks of rows against the west wall, and one block of rows against the east wall, with a north-south isle of sufficient width in the middle for forklift movement.

The building is equipped with roof sprinklers with automatic control valves (0.23 gpm/ft^2 of water over 2,500 ft^2). The area housing the calcium hypochlorite and the adjacent area is surrounded by an 8" x 8" concrete perimeter dyke to contain the sprinkler water in case of activation

The typical composition of the product is shown in Table 2.

Table 2 Composition of product

Component	Weight %
Calcium hypochlorite	65
Calcium Hydroxide	3 - 7
Calcium Carbonate	1 - 2.5
Sodium Chloride	16.5 - 25
Water (hydrolized)	6 - 9

3. MODELING METHODOLOGY

If involved in a fire, calcium hypochlorite produces its own oxygen and results in a very hot burning fire as it decomposes. Products of combustion include HCl, Cl_2, and CO.

3.1 Processes

During a warehouse fire involving calcium hypochlorite, the following processes are all occurring simultaneously.

1. Calcium hypochlorite is heated to ignition
2. Calcium hypochlorite decomposes as follows:
   $Ca(OCl)_{2\,(s)} \rightarrow CaCl_{2\,(s)} + O_{2\,(g)}$ $\quad\quad \Delta H_R^o = -20.8$ kcal/gmol
3. Combustibles (polypropylene, polystyrene, wood) are heated
4. Combustibles pyrolyze and release flammable gas

5. Oxygen diffuses to the surface and mixes with the flammable gases
6. Flammable gases ignite and combust at the surface as follows:
   $C_xH_y + O_2 \rightarrow CO_2 + H_2O + CO + C$ ΔH_C^o (polypropylene) = - 11,099 kcal/kg
   ΔH_C^o (polystyrene) = - 9,900 kcal/kg
   ΔH_C^o (wood) = - 4,900 kcal/kg
7. Calcium chloride is heated
8. Calcium chloride decomposes and releases chlorine as follows:
   $CaCl_{2\,(g)} + \frac{1}{2} O_{2\,(g)} \rightarrow CaO_{(s)} + Cl_{2\,(g)}$ $\Delta H_R^o = 48.1$ kcal/gmol
9. Chlorine is converted to hydrogen chloride as follows:
   $Cl_{2\,(g)} + H_2O_{(g)} \rightarrow 2HCl_{(g)} + \frac{1}{2} O_{2\,(g)}$ $\Delta H_R^o = 13.7$ kcal/gmol

3.2 Fire Spread Assumptions

The following simplifying assumptions have been made in order to make it amenable for calculating the rate of fire spread:

- The fire starts in a row of pallets located along one wall
- The adjacent second row is exposed to the fully developed fire of the first row of pallets
- In a fully developed fire, the products of combustion reach a maximum estimated fire temperature assuming 30% of the heat of combustion is emitted as radiant energy
- The first set of pails in the adjacent second row is heated, the combustibles ignite and the calcium hypochlorite decomposes explosively
- Since polypropylene makes up most of the combustibles, the time to ignition is based on polypropylene
- The sprinkler system is not activated in time to stop the spread of the fire
- The first set of pails in the adjacent row (which is immediately exposed to the fire) ruptures and the calcium hypochlorite spills
- The first set of pails burns and reaches the maximum estimated fire temperature
- Once the maximum estimated fire temperature is reached, the combustibles continue to burn at a steady rate and the temperature remains constant at the estimated fire temperature. The reactions proceed to equilibrium during this phase. Since the materials in the fire reach such elevated temperatures, it is assumed that the kinetics of all reactions is very fast and therefore all the reactions proceed to equilibrium.
- As the first set of pails burns, the second set of pails in the row of pallets is exposed to the fire and is heated until the combustibles ignite and the calcium hypochlorite decomposes explosively
- Each set of pails in all the remaining rows are exposed to this line of fire which moves from one end of the warehouse to the other
- The fire is homogenous across the vertical plane
- The line of fire is assumed to spread at an equal rate within pallets and between pallets

3.3 Fire Development and Release Rates of Fire Products

The rate of development of the fire and the source term were estimated using basic principles of heat transfer, thermodynamics and chemistry.

The fire, once started, can be characterized in four stages:

Stage #1: Heating of the adjacent row of pails to ignition.

Stage #2: Calcium hypochlorite decomposition reaction takes off and the pail ruptures spilling its contents. All of the combustibles in the row are engulfed in flames. The maximum estimated fire temperature is reached.

Stage #3: At the maximum estimated fire temperature, all reactions proceed to equilibrium. This stage ends when the combustibles stop burning.

Stage #4: The mass of material cools down.

Three fire scenarios were investigated, to examine the uncertainty associated with the amount of air available for combustion:

1. No air: all the oxygen for combustion is supplied by the decomposition reaction
2. Some air
3. Stoichiometric air: air sufficient to complete the combustion of the plastic and wood is allowed to take part in the fire

The extent of calcium chloride decomposition was not calculated since not all the required thermodynamic data was available. Based on literature values (Gray and Halliburton, 2000), approximately 22% of the calcium chloride produced by the decomposition of calcium hypochlorite was assumed to decompose further into chlorine or hydrogen chloride.

At the expected conditions of the fire, the ratio of the partial pressures of hydrogen chloride to chlorine is 186. Therefore, the conversion of chlorine to hydrogen chloride will be large, approximately 99%.

The details of calculations for the maximum estimated fire temperature, chlorine/hydrogen chloride equilibrium, radiant flux on adjacent row of pails, ignition and mass burning rate of polypropylene, ignition of calcium hypochlorite, rate of fire spread, duration of the fire, and chlorine and hydrogen chloride release rates are presented in the Appendix.

3.4 Modeling Potential Impacts – Dispersion and Thermal Radiation

The hazard end-points of the primary decomposition and combustion products of interest are shown in Table 3.

Table 3 Hazard end-points of the primary decomposition and combustion products of interest

Pollutant	Toxic End-points			
	Toxic end-point (ERPG-2)	Toxic end-point (ERPG-2)	IDLH	IDLH
	(ppm)	(mg/m^3)	(ppm)	(mg/m^3)
HCl	20	30	50	77
Cl_2	3	8.7	10	30
CO	350	406	1,200	1,392
CO_2	NA	NA	40,000	73,200
C (soot)	-	NA	-	1,750

The AERMOD dispersion model (US EPA, 2001) was used to investigate the size of the hazard zones for the various fire scenarios, to establish also the worst case meteorological condition that would lead to the worst case hazard zone. AERMOD is becoming the regulatory air quality model in the US and Canada, replacing the ISC series of models. It's dispersion formulations based on surface similarity models are more sophisticated than the classical Gaussian formulations used in the older models, especially for unstable and stable atmospheric conditions.

This model requires separate surface and upper air data files, and does not use the traditional Pasquill-Gifford stability classes. Instead, atmospheric stability is characterized by a combination of surface radiative flux, wind speed and surface roughness, in the form of a dimensionless parameter called the Monin-Obukhov Length.

The model was run for each hour using a five-year meteorological data set from the nearest airport. The conditions that resulted in the largest hazard zones for each of the three fire scenarios were established as the extreme-high-wind conditions. Other more common wind conditions (medium and low wind speed conditions), in combination with a range of stability conditions characterized by the main stability parameter surface radiative heat flux, were also selected after a detailed examination of the data set. These other conditions were then used for establishing the credible worst-case meteorological conditions for use in determining the corresponding hazard zones.

Source characteristics such as shape and orientation of the building, wake effects, and plume rise were taken into account in the modeling.

4 RESULTS, CONCLUSIONS AND RECOMMENDATIONS FOR FURTHER STUDY

The calculated fire temperatures for the three scenarios vary from approximately 950 C (no air) to 1180 C (stoichiometric).

The fire durations and source terms for the primary pollutants of interest for the three fire scenarios investigated are given in Table 4.

Table 4 Source term estimates for the three fire scenarios

Pollutant	Fire Scenario					
	Stoichiometric air		Some air		No air	
	Total Mass Emitted	Emission Rate	Total Mass Emitted	Emission Rate	Total Mass Emitted	Emission Rate
	kg	kg/s	kg	kg/s	kg	kg/s
HCl	101,478	8.6	101,478	6.1	101,478	5.4
Cl_2	996	0.085	996	0.060	996	0.053
CO	0	0.0	21,748	1.3	18,520	1.0
CO_2	312,942	26.6	215,500	13.1	183,520	9.8
C (soot)	0	0.0	6,378	0.4	5,431	0.3
Fire duration:	3.3 hr		4.6 hr		5.2 hr	

The modeling indicates that the limiting product of fire for determining the largest hazard zone is hydrogen chloride. Table 5 presents the hazard zones for this chemical for the three fire scenarios.

Table 5 Worst imaginable case hazard distances for the three fire scenarios

Worst Case Fire Scenario	Emission Rate	HCl Toxic Hazard Zone Distance (to ERPG-2) Worst case metereology	Thermal Hazard Zone Distance
	kg/s	m	m
Stoichiometric Air	8.6	300	31
Some Air	6.1	280	24
No Air	5.4	290	22

These results further indicate that the reduction in the pollutant emission rate for the cases where less air is available for combustion is somewhat compensated by the smaller plume rise resulting from lower fire temperatures. Thus, all the hazard distances for these fire scenarios are very similar to each other at approximately 300 m.

For the first credible worst case, where the fire conditions are the same as the worst imaginable case, but with more common weather conditions, the hazard zones in a sustained manner are estimated to be not appreciable, due to the strong plume rise expected from the high fire temperatures.

It should be noted that this result is for one-hour average concentrations in air, which the ERPG-2 limit is based on. Concentration fluctuations on shorter time scales may lead to higher shorter term concentrations locally. Also, the edges of the plume (which can cool much faster than the bulk of the plume) could get caught in the building wake, thus again resulting in higher shorter term concentrations. Further fine-tuning of concentration estimates is recommended using more sophisticated concentration fluctuations modeling to take into account these shorter term variations which can lead to higher peak concentrations than modeled here.

The effect of activation of the fire suppression system was also investigated as part of the study. This resulted in some cooling of the fire temperature. However, it was estimated that, unless the fire is extinguished within the first forty five seconds or so of its initiation, and the fire proceeds to become fully developed, the sprinkler system would not be effective in controlling the fire.

Modeling of warehouse fires, even for fires involving relatively simple chemicals like calcium hypochlorite, is complex at best. Many simplifying assumptions are required to make the modeling tractable. However, with careful consideration of the details of the physical situation expected in the fire, it is possible to develop reasonable solutions for order of magnitude estimates of hazard zones, which in turn can be used for emergency planning purposes.

In a real situation, the development of a fire and the subsequent dispersion of the plume from the fire are complicated by actual storage conditions in the warehouse at the time of the fire, the inherent variability of building structures in terms of their structural collapse or damage in response to the fire, and the inherent variability of wind conditions and atmospheric turbulence. Therefore, actual sampling of air quality around the facility is necessary for on-the-ground decision making.

APPENDIX MODELING DETAILS

Maximum Estimated Fire Temperature

The maximum estimated fire temperature is calculated as follows:

$\Sigma\ \Delta H$ (products) = 0.7 [$\Sigma\ \Delta H$ (reactants) - ΔH_R^o]

where $\Sigma\ \Delta H$ (products) is the change in enthalpy of the products from standard state to the adiabatic fire temperature.

$\Sigma\ \Delta H$ (reactants) is change in enthalpy of the reactants from the initial state to the standard state. Since the initial state is assumed to be 25 C, this value is zero.

ΔH_R^o is the enthalpy change at standard state (25 C and 1 atm) for the overall reaction.

0.7 represents a 30% assumed loss due to radiation emitted

$\Sigma\ \Delta H$ (products) = $\int \Sigma\ (n_i\ Cp_i)\ dT$

$\Delta H_R^o = \Delta n_{Ca(OCl)2}\ \Delta H_R^o$ [$Ca(OCl)_2$ decomposition] + $\Delta n_{polypropylene}\ \Delta H_C^o$ [polypropylene combustion] + $\Delta n_{polystyrene}\ \Delta H_C^o$ [polystyrene combustion] + $\Delta n_{wood}\ \Delta H_C^o$ [wood combustion] + $\Delta n_{CaCl2}\ \Delta H_R^o$ [$CaCl_2$ decomposition] + $\Delta n_{Cl2}\ \Delta H_R^o$ [Cl_2 conversion to HCl]

Maximum estimated fire temperatures were calculated for the following three oxygen supply scenarios:

1. No additional air is added into the warehouse. The oxygen supplied for combustion comes primarily from the oxygen released during calcium hypochlorite decomposition and some from the air in the warehouse. Partial combustion (66 %) occurs and uses all the oxygen available. For underventilated fires, carbon monoxide is generated at a rate of 0.3 g per gram of fuel consumed (Pitts, 1996). Carbon particles, as smoke, is produced at a ratio of 0.03 g per gram of carbon dioxide produced (Peacock et al., 1993). The calculated maximum fire temperature for this scenario was 957 C (compared to an adiabatic flame temperature of 1396 C).

2. Some air leaks into the warehouse during combustion but it is less than the stoichiometric amount. Partial combustion (77%) occurs and uses all the oxygen available. The calculated maximum fire temperature for this scenario was 1018 C (compared to an adiabatic flame temperature of 1478 C).

3. Enough air leaks into the warehouse to provide the stoichiometric amount of oxygen for complete combustion. Complete combustion for polypropylene is estimated to be 88% combustion (Chung et al., 1999). The calculated maximum fire temperature for this scenario was 1179 C (compared to an adiabatic flame temperature of 1700 C).

The extent of calcium chloride decomposition was not calculated since not all the required thermodynamic data was available. The extent of calcium chloride oxidation has been reported to be 21.6% when calcium chloride is exposed to moist air at 1200 C (Gray and Halliburton, 2000). Therefore, 21.6% of the calcium chloride produced by the decomposition of calcium chloride was assumed to decompose further to chlorine or hydrogen chloride.

Chlorine / Hydrogen Chloride Equilibrium

At elevated temperatures with the presence of water vapor, the formation of hydrogen chloride from chlorine is favoured.

$\Delta G^o = \Sigma\ \Delta G_f^o$ (products) - $\Sigma\ \Delta G_f^o$ (reactants)

where ΔG^o is the Gibbs Free Energy of change for the reaction at standard state (25 C and 1 atm)

$\Sigma\ \Delta G_f^o$ (products) is the summation of the Gibbs free energy of formation of the products at standard state

$\Sigma\ \Delta G_f^o$ (reactants) is the summation of the Gibbs free energy of formation of the reactants at standard state

K (at 25 C) = exp [$\Delta G^o/RT$]

where K (at 25 C) is the equilibrium constant for the reaction at 25 C

R is the universal gas constant

T is the temperature (298 K)

K = K (at 25 C) exp [$-\Delta H_R^o/R$ (1/T – 1/298 K)]

where K is the equilibrium constant at the adiabatic fire temperature

$K = [(P_{HCl})^2\ (P_{O2})^{1/2}]\ /\ [(P_{Cl2})\ (P_{H2O})]$

where P_{HCl} is the partial pressure of hydrogen chloride

P_{O2} is the partial pressure of oxygen

P_{Cl2} is the partial pressure of chlorine

P_{H2O} is the partial pressure of water vapor

At the expected conditions of the fire, the ratio of the partial pressures of hydrogen chloride to chlorine is 186. Therefore, the conversion of chlorine to hydrogen chloride will be large, approximately 99%.

Radiant Flux on Adjacent Row of Pails

$E = \varepsilon \sigma T^4$

where E is the surface emitted flux. If the transmissivity and the view factor for the receptor is assumed to be unity, the surface emitted flux is equivalent to the radiant flux received by the row of pails.

ε is the emissivity (0.2 for smoky fires; Gray and Halliburton, 2000)

σ is the Stefan-Boltzmann constant (5.67 x 10^{-11} kW/m^2K^4)

At the estimated maximum fire temperatures of 1179 C, the radiant flux on the adjacent row of pails is 50 kW/m^2.

Ignition and Mass Burning Rate of Polypropylene

For a radiant flux of 50 kW/m^2, the time to ignition for polypropylene has been measured to be 41 seconds (CCPS, 2000). Therefore, Stage #1 (heating of the pails) is expected to last for 41 seconds until the polypropylene ignites.

The mass burning rate for polypropylene is reported to be 25 g/m^2-s for an external radiance of 70 kW/m^2 (Babrauskas and Grayson, 1992). Therefore, a mass burning rate of 20 g/m^2-s was taken for an external radiance of 50 kW/m^2.

$t_c = w_{pp} / (b_{pp} S)$

where t_c is time for combustion of polypropylene (and other combustibles) in one row of pails

w_{pp} is the weight of polypropylene for the row of pails

b_{pp} is the mass burning rate of polypropylene

S is the surface area of the row of pails

The duration of combustion (Stage #3) is estimated to be 205 seconds.

Ignition of Calcium Hypochlorite

Time to ignition of calcium hypochlorite was estimated as follows using a computer spreadsheet to solve the ordinary differential equation. Ignition is defined as the point at which the there is a runaway of the calcium hypochlorite decomposition reaction.

Energy balance for a pail of calcium hypochlorite:

[Accum] = [heat in] – [heat out] + [heat generated] – [heat consumed]

[heat consumed] = 0

[heat in] = $[E\ A_e + h_{fl}\ A_e\ (T_{fl} - T)]\ dt$

where E is the radiant flux received by the pail from the previous row which is engulfed in fire

A_e is the vertical surface of the pail exposed to the fire

h_{fl} is the flame convective heat flux coefficient

T_{fl} is the adiabatic fire temperature of the previous row which is engulfed in fire

T is the average temperature of the calcium hypochlorite in the pail

[heat out] = $[h_c\ A_n\ (T - T_a)]\ dt$

where h_c is the convective heat flux coefficient to the ambient air

A_n is the area of the pail which is not exposed to the fire

T_a is the ambient temperature in the warehouse (25 C)

[heat generated] = $[n\ (-\Delta H)\ A\ \exp(-E_A/RT)]\ dt$

where n is the number of moles of the calcium hypochlorite in the pail

ΔH is the heat of reaction of the calcium hypochlorite

A is the Arrhenius pre-exponential factor for calcium hypochlorite decomposition, 8.74 x 10^{13}/min (Uehara et al., 1978)

E_A is the energy of activation for calcium hypochlorite decomposition, 29.5 kcal/gmole (Gray and Halliburton, 2000; Uehara et al., 1978)

For the stoichiometric air case, the time to reach ignition was calculated to be 147 seconds and the temperature when this occurs is approximately 200 C. This is considered to be the maximum time to reach the fully developed fire at the end of Stage #2. The minimum time would be the time to reach ignition of the polypropylene. The actual time was estimated to be at the midway point, (41 s + 147 s)/2 = 94 s.

Duration of Warehouse Fire

Therefore the following four stages are envisioned:

Stage #1: Heating of the adjacent row of pails to ignition , 41 seconds

Stage #2: Calcium hypochlorite decomposition reaction takes off and the pail ruptures spilling its contents. All of the combustibles in the row are engulfed in flames. The maximum fire temperature is reached in 94 seconds.

Stage #3: At the maximum fire temperature, all reactions proceed to equilibrium. This stage ends when the combustibles stop burning, 205 seconds.

Stage #4: The mass of material cools down.

# of Rows in Warehouse

# of pallets = W_{CAPO} / $w_{CAPO, pallet}$

where W_{CAPO} is the total weight of CAPO in the warehouse

$w_{CAPO, pallet}$ is the weight of CAPO in a pallet

# of rows (both columns) = # of pallets / [2 (# of pallets per row)]

where (# of pallets per row) is [# of pallets in each row (long)] x [# of pallets in each row (high)] x [# of pallets in each row (wide)]

Rate of Fire Spread

For the stoichiometric air case, each row of pails must be heated for 94 seconds before it reaches fully engulfed fire conditions.

[time for fire to reach last row] = [# of pails (wide) in each row] x [# of rows (both columns)] x [time duration to reach beginning of Stage #3]

[duration of fire] = [time for fire to reach last row] + [Stage #3 length of time]

[duration of fire] = 3 x 41 x 94 seconds + 205 seconds = 11,767 seconds

Chlorine and Hydrogen Chloride Release Rates

$\dot{M}_{Cl2} = n_{Cl2}\, MW_{Cl2}$ / [duration of fire]

$\dot{M}_{HCl} = n_{HCl}\, MW_{HCl}$ / [duration of fire]

$n_{Cl2} = n^{o}_{Ca(OCl)2}\, y_{CaCl2}\, y_{Cl2}$

$n_{HCl} = 2\, n^{o}_{Ca(OCl)2}\, y_{CaCl2}\, y_{HCl}$

where $\dot{M}_{Cl2}$ is the rate of release of chlorine during the fire

n_{Cl2} is the total number of moles of chlorine generated by the fire

MW_{Cl2} is the molecular weight of chlorine

$\dot{M}_{HCl}$ is the rate of release of hydrogen chloride during the fire

n_{HCl} is the total number of moles of hydrogen chloride generated by the fire

MW_{HCl} is the molecular weight of hydrogen chloride

$n^{o}_{Ca(OCl)2}$ is the original number of moles of calcium hypochlorite in the warehouse

y_{CaCl2} is the fraction of $CaCl_2$ which undergoes decomposition

y_{Cl2} is the fraction of chlorine which is not converted to hydrogen chloride

y_{HCl} is the fraction of chlorine which is converted to hydrogen chloride

The release rate of chlorine is estimated to be 0.085 kg/s. The release rate of hydrogen chloride is estimated to be 8.6 kg/s.

REFERENCES

Babrauskas, V. and Grayson, S.J. (1992), "Heat Release in Fires", New York, Elsevier Applied Science.

CCPA *Site Acute Risk Assessment Implementation Aid* (Second Edition).

CCPA (1996) *Risk Assessment Guidelines* (May).

CCPA (1995) *Guidelines for Site Risk Communication – Attachment 2 Technical Basis of Calculation of Worst Case Scenario* (5 October).

CCPS (2000) *Guidelines for Chemical Process Quantitative Risk Analysis*, American Institute of Chemical Engineers; Center for Chemical Process Safety, , New York.

Chung, Soo Hyun, Park, Jong Jin, Jeon, Sang Goo, Kim, Dong Chan (1999) "Pyrolysis of Waste Plastics Using Synthesized Catalysts from Fly Ash", Thirteenth US – Korea Joint Workshop on Energy and Environment Proceedings, Reno, USA.

Gray, Brian F. and Halliburton, Brendan, (2000) "The Thermal Decomposition of Hydrated Calcium Hypochlorite (UN 2880)", Fire Safety Journal, 35, pp. 223-239.

Hopkins, Donald Jr., "Predicting the Ignition Time and Burning Rate of Thermoplastics in the Cone Calorimeter", University of Maryland M.Sc. Thesis, 1995.

NIOSH (1994) *Pocket Guide to Chemical Hazards*. US Department of Health and Human Services, National Institute for Occupational Safety and Health. June.

Peacock, R. D.; Forney, G. P.; Reneke, P. A.; Portier, R. W.; Jones, W. W. (1993) "CFAST, The Consolidated Model of Fire Growth and Smoke Transport," *NIST TN 1299* , National Institute of Standards and Technology, 246 p., February.

Perry, R.H., Green, D.W. (1997) *Perry's Chemical Engineers' Handbook*. McGraw Hill, New York, New York. ISBN 0-07-049841-5.

Pitts, William M. (1996), "Carbon Monoxide Formation Algorithm", Thirteenth Meeting of the UNJR Panel on Fire Research and Safety, March 13-20. Volume 2.

Uehara, Yoichi et al.. (1978) "Thermal Ignition of Calcium Hypochlorite", Combustion and Flame 32, pp. 85-94.

US EPA (2001) *User's Guide for the AMS/EPA Regulatory Model – AERMOD*. United States Environmental Protection Agency, Office of Air Quality Planning and Standards, Emissions, Monitoring and Analysis Division. Revised Draft. Research Triangle Park, North Carolina 27711. August.

US EPA (1999a) *Risk Management Program Guidance for Offsite Consequence Analysis*. United States Environmental Protection Agency, Office of Solid Waste and Emergency Response. EPA 550-B99-009, April.

20th Annual CCPS International Conference
Risk Management: The Path Forward
Atlanta, Georgia
April 11 – 13, 2005

Session
DUST EXPLOSIONS

Title
TANTALUM DUST EXPLOSIBILITY CHARACTERISTICS

Authors

C. James Dahn
Safety Consulting Engineers, Inc, 2131 Hammond Drive, Schaumberg, Il, USA 60173
Ashok Ghose Dastidar
Safety Consulting Engineers, Inc, 2131 Hammond Drive, Schaumberg, Il, USA 60173
Jacob Finkelshtein
Safety Consulting Engineers, Inc, 2131 Hammond Drive, Schaumberg, Il, USA 60173
Abdullah Kashani
Safety Consulting Engineers, Inc, 2131 Hammond Drive, Schaumberg, Il, USA 60173

Phone: (847) 925-8100 Fax: (847) 925-8120 Email: sceinc@sceinc.com

Abstract

Dust explosions can occur with nearly any finely divided powders in industry. Typically, in the electronic industry, most engineers address the preparation, processing and handling aspects of tantalum powders and are aware of the spontaneous fire hazards of heated material. This study addresses the dust explosion hazard of tantalum powder in the electronics and medical industries. Other elemental metals can be a significant dust explosion hazard in many industries. Thus it is important to properly characterize their dust explosion hazards

INTRODUCTION

Tantalum which occurs principally in the mineral called *"Columbite-Tantalite"* is in the earth in less than 1 part per million. When it is pure, it is ductile and can be formed into fine wire that can be used for evaporating other metals such as aluminum. Tantalum is almost immune to chemical attack at temperatures below 150 °C. At high temperatures, tantalum becomes much more reactive. Alkalis attack it only slowly. Its melting point exceeds only tungsten and rhenium. Tantalum is used to make electrolytic capacitors and vacuum furnace parts which account for over 60 % of it use. Tantalum found wide use in making surgical appliances since it is completely immune to body fluids and non-irritating. In powder form, it costs over $40 per pound.

In some processes, the metal is ground up into powder, agglomerated, and sintered for the intended use. Fires have been known to occur after heating and sintering. Usually, little consideration for its dust explosibility is displayed since it was thought to have a very mild dust explosion output. Its material properties are as follows:

Atomic Weight	180.88 g/mol
Melting Point	3027 °C
Boiling Point	4100 °C
Metal Density	16.6 g/cm³
Powder Density	14.5 g/cm³

In processing tantalum powders, mixing and agglomeration usually occur in closed vessels. Normally, the powder dust during mixing is in high concentrations depending on the mixer type. Absent of ignition sources, most engineers tend to neglect the dust explosion potential since fires are to be expected.

The present study was related to determining the maximum dust explosion output and ignition energy of tantalum. The maximum dust explosion pressure is essential to evaluate survivability of closed or partly closed vessels. The rate of pressure rise in the dust explosion will dictate the degree of vessel venting to prevent structural damage. It is used to determine the dust-deflagration suppression requirements to prevent vessel damage.

DUST EXPLOSIVITY AND IGNITION CHARACTORISTICS

The published dust explosivity properties for tantalum are as follows:

Explosion Pressure (psig)	Rate of Pressure Rise (dP/dt) (psig/s)	Minimum Ignition Energy (mJ)	Cloud Ignition Temperature (°C)	Layer Ignition Temperature (°C)	Minimum Explosible Concentration (g/m³)
55	4400	120	630	300	< 200

The explosion output and minimum ignition energy testing was conducted in this study on a pure tantalum powder to validate the literature values.

Dust Explosion Output Testing

The dust explosion output was conducted on tantalum per ASTM E 1226 using a Bureau of Mines 20-liter dust explosion chamber. Lycopodium powder was tested as a reference. The test setup, as shown in Figure 1, utilizes a pressure transducer to monitor the dust explosion pressure. The output is recorded on a digital computer for data reduction and analysis.

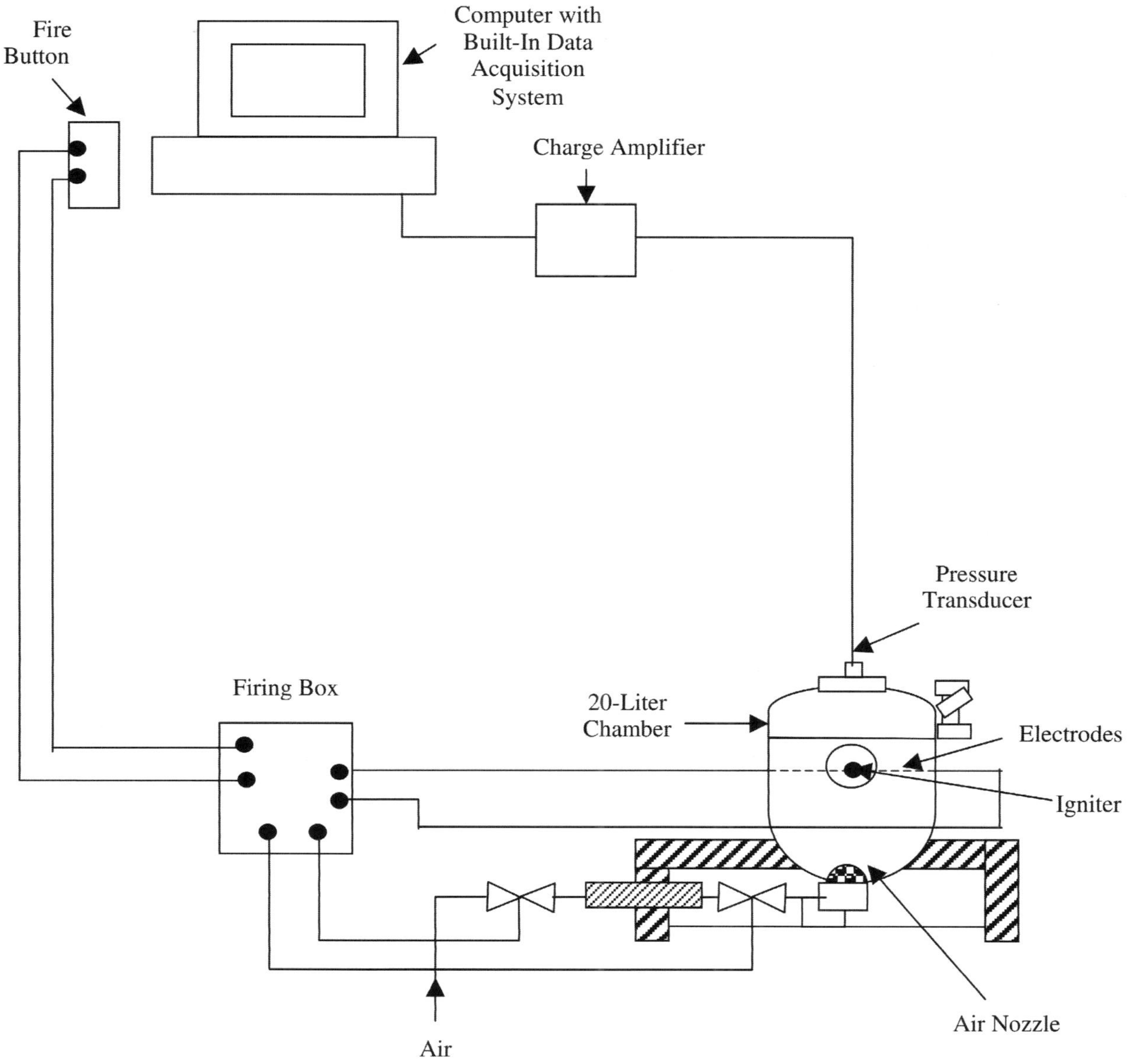

Figure 1 - Dust Explosion Test Setup

The desired amount of the dust (-200 mesh size) is introduced into the dust-well of the chamber. A chemical igniter is positioned at the center of the vessel, the chamber cover is closed. An air pulse from a pressurized canister disperses the dust uniformly into the chamber. At a pre-described time interval, the chemical igniter is activated causing the dust to ignite. Photographs of the chamber are shown in Figures 2 and 3 below.

Figure 2. 20-liter dust chamber in the closed position.

Figure 3. Chamber in open position.

To determine the maximum dust-explosion pressure and rate of pressure rise, successive tests are run at increasing dust concentrations while using the same strength of chemical igniter (10,000 Joules). As increased dust concentrations are used, the pressure and rate of pressure rise increase until they drop off at high concentrations. Typically, this is in the range of 1250 to 1500 g/m³ for organic powder/dusts.

The pressure rise due to the chemical igniter is subtracted from the total pressure output. The maximum pressure rise rate is determined by differentiating the pressure-time curve. The deflagration index (K_{St}), maximum pressure rise rate normalized to a 1-m³ volume was also determined at the optimum dust concentration.

Results

The explosion severity test results and deflagration index (K_{St}) for the tantalum powder and the reference material, Lycopodium, are shown in Table 1 below.

TABLE 1

EXPLOSION SEVERITY TEST RESULTS

Material	Dust Concentration (g/m³)	Maximum Pressure Output (psig)	Maximum Pressure Rise Rate (psi/s)	K_{St} Deflagration Index (bar m/s)
Tantalum Powder	2000	106	8640	162
Lycopodium Dust	750	94	8177	152

The explosion pressure and rate of pressure rise as a function of the dust concentration tested are shown in Figures 4 and 5 below.

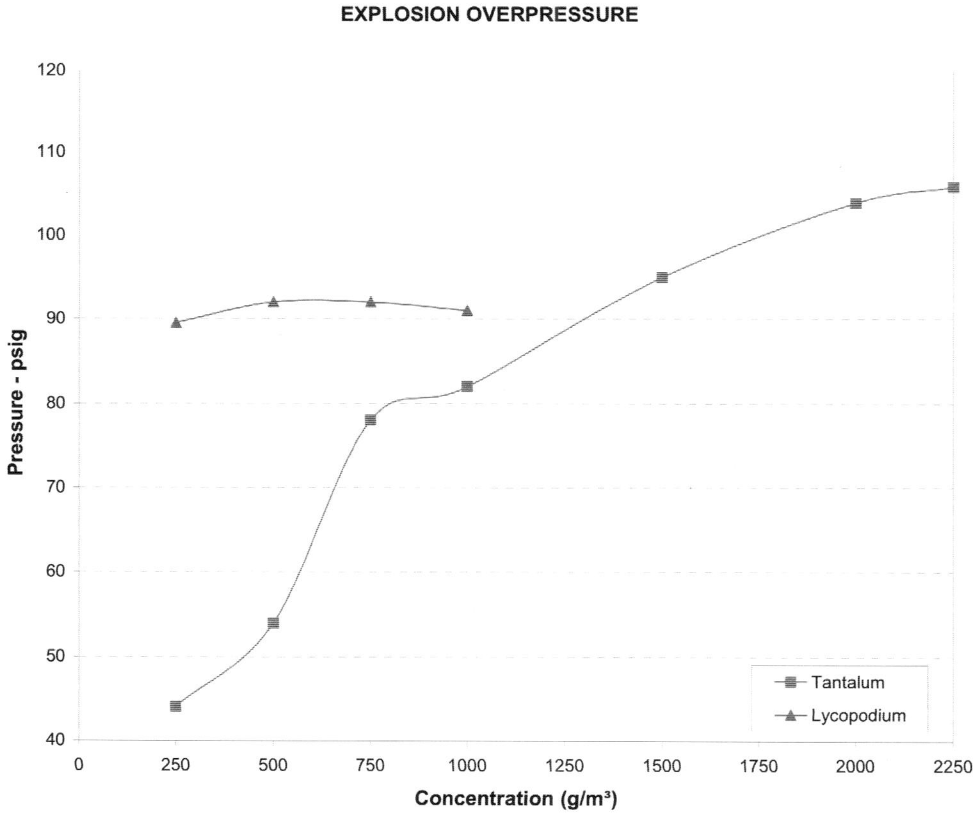

Figure 4. Pressure output variation with concentration.

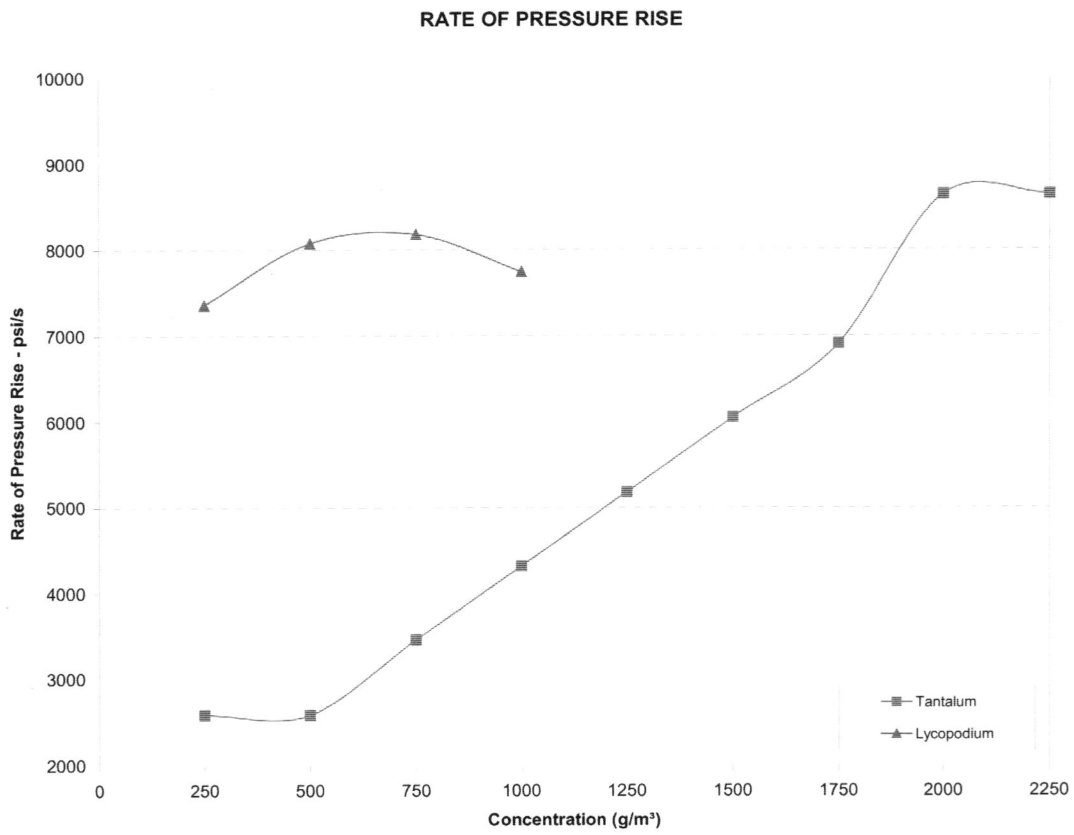

Figure 5. Rate of pressure rise vs. nominal concentration of powder.

From these graphs, tantalum dust deflagration pressure output and rate of pressure output is lower at low dust concentrations as compared to lycopodium dust thus giving one a false sense of safety in operations. Normal dust testing is conducted in concentrations up to 1,000 g/m³ in air. Most metal dusts have deflagration pressures and rates of pressurization in these ranges with exception of energetic metals such as aluminum and magnesium. As can be seen here, further testing at higher concentrations of dust has resulted in explosion severity data that is greater than that stated in literature.

MINIMUM IGNITION ENERGY

The purpose of the minimum ignition energy testing is to determine how easy it is to ignite the dust cloud by an electrostatic discharge. The dust is lofted into a 1.2-liter cylindrical dust chamber where it is exposed to a spark discharge from a known electrical spark discharge circuit. This circuit supplies electrical energy to two electrodes in the chamber from one of several electrical capacitor sizes and applied voltages via a high-speed switch, see Figure 6. The ignition energy is measured by monitoring spark gap voltage and current.

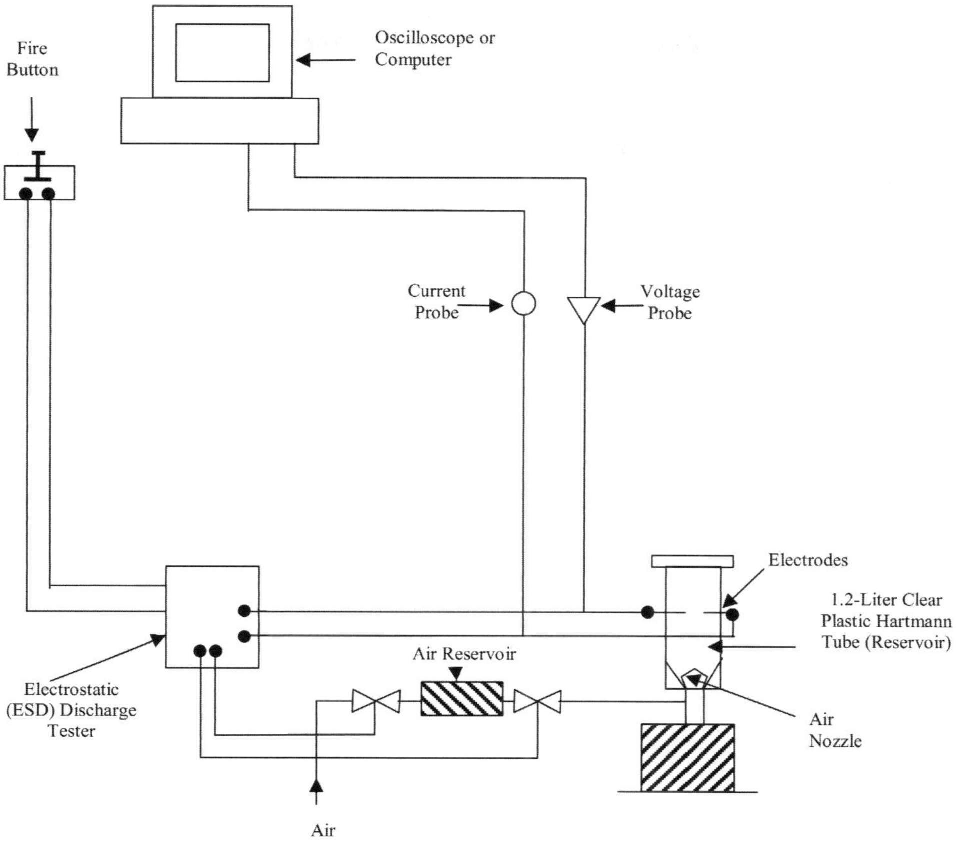

Figure 6. Minimum ignition energy test setup.

The minimum ignition energy was determined by performing successive tests on the tantalum dust material (- 200 mesh sieve size) by varying energy (varying capacitor size and voltage on the capacitor) and dust concentrations. Lycopodium powder was used as reference for testing.

Results

The minimum ignition energy of tantalum powder was found at a dust concentration of 2000 g/m³. In contrast, the minimum ignition energy of lycopodium dust was found at a dust concentration of 750 g/m³. The spark gap ignition energy values are listed as follows:

Material	Minimum Ignition Energy (mJ)
Tantalum	14.4
Lycopodium	11.5

The minimum ignition energy for tantalum, determined in this study, is relatively low. It is lower, by an order of magnitude, than the value stated in literature. It should be noted that under worse case conditions, one could develop electrostatic discharge energies in this range (up to 20 mJ). With the MIE of tantalum dust being ~14 mJ, a lofted dust cloud may be ignited by an electrostatic discharge from plant personnel. Certainly, isolated conductors with stored electrostatic energies can ignite the dust in a cloud form depending on the level of stored energy.

CONCLUSIONS AND RECOMMENDATIONS.

From this study, one can appreciate the danger of handling tantalum in a dust cloud form since it produces high deflagration pressures and rates of pressure rise at high dust concentrations. Beware that this dust is very energetic at high dust concentrations. If the dust contacts water at high temperature, hydrogen gas can form which compounds the hazard by producing a hybrid dust/vapor hazard at vapor concentrations below the lower flammable limit for hydrogen. Thus, the deflagration output can be greatly increased above what was characterized in this study. Its so easy to get wrapped up in the processing and handling of the powder that one can lose sight of its dust explosion potential.

It should also be noted that relaying solely on literature values for the explosion sensitivity and characteristics of a dust might lead one into a false sense of security. The literature data should be examined in context and should not be used unless test specifics (e.g. test equipment, concentration range tested, dust particle size, etc.) are known.

The 27 March 2003 Billy-Berclau Accident: A Technical And Organisational Investigation

Jean-christophe Lecoze, Nicolas Dechy, Samantha Lim, Emmanuel Leprette, Ruddy Branka
INERIS (Institut National de l'Environnement Industriel et des Risques)
Parc Alata, 60550 Verneuil-en-Halatte, FRANCE
Tel: 33 3 44 55 66 77, ineris@ineris.fr

Abstract

Following a dynamite explosion in a pyrotechnic plant in Billy Berclau, France, which caused the death of 4 persons and injured 13 persons, the Environmental Ministry appointed INERIS (National Institute For Environmental Safety) to carry out an investigation. The purpose of the investigation was to consider technical and organisational issues related to the origins of the accident as well as the damages of the accident. This paper will introduce the results of the investigation with a focus on its organisational side.

The organisational requirement within that investigation along with the technical dimension is a first in France in the chemical industry and was enforced partly by the SEVESO II European Union directive context. The context of the Toulouse disaster played its role also. Indeed this directive requires the implementation of a safety management system, dedicated to major hazard prevention for certain plants, according to their risk level (based partly on the quantity of hazardous materials stored). The plant where the accident occurred was under that regulation and had to implement such a management system. The need for an understanding of the management system's underlying root causes in that accident was therefore high.

INERIS has been developing research programs on the organisational sides of major hazard prevention for few years and could therefore use its expertise in that field for the investigation.

In a first part of the paper, the accident will be presented, through a chronology of the events, damages and through the technical causes that have been identified and retained as most probable during the investigation. This first part is a preliminary step for understanding the case itself and to get into the second part of the paper.

The second part will discuss the organisational investigation, from a methodological point of view and then from the findings. The methodological aspect will question the ability to get into the complexity of the organisational life. Organisations are complex systems that need to be looked at with appropriate lenses. These lenses are influenced by the investigator experience and background and it is part of the methodology to address this issue. The level of opening of the organisational investigation is another strong methodological issue. As organisations are complex systems, they interact at all levels, and their understanding implies opening the system to many individuals involved inside and outside the boundaries of the organisation. That is another important issue of the investigation. The outcomes of the investigations are presented according to different levels of the system and are discussed.

1. Introduction And Context Of The Investigation

Following a dynamite explosion in a pyrotechnic plant in Billy Berclau, France, which caused the death of 4 persons and injured 13 persons, the Environmental Ministry appointed INERIS (National Institute For Environmental Safety) to carry out an investigation. The purpose of the investigation was to consider technical and organisational issues related to the origins of the accident as well as the damages of the accident. This paper proposes to describe some related issues around the investigation, with a focus on its organisational side.

1.1 Context Of The Investigation

A dynamite explosion, in a pyrotechnic plant in Billy Berclau, France, occurred on 27 march 2003 at 6h16, and caused the death of 4 persons and injured 13 persons. Fires resulting from the explosion were quickly extinguished by emergency services.

After the event, several investigations were launched by Authorities:

- the litigation inquiry carried out by the Police and the Justice,
- the Labor Ministry carried out its investigation,
- the investigation of the company Nitrochimie S.A. that merged with the Health and Safety committee: their focus was to find out the causes of the explosion,
- the administrative inquiry of IGE-CGM-IGAS (Inspection Générale de l'Environnement, Conseil Général des Mines and Inspection Générale des Affaires Sociales) set up by four Ministries: Environment, Defence, Labor and Industry; the investigation had to focus on circumstances of the accident, causes (technical, human and organisational), safety measures, recommendations that could lead to regulation changes, and should use expertise of INERIS and IPE (Inspection des Poudres et Explosifs); the report would be made publicly available (on the web in April 2004),
- the investigation of INERIS, launched by the Environment Ministry, focusing on technical and organisational dimensions of the accident.

The general objective of administrative inquiries carried out by IGE, IGAS and CGM, that are organisations within their Ministries, is to focus on the circumstances (in case of accidents, pollution,…) and on the local application of the law by companies and local control authorities. The investigation of INERIS brought (final report was issued in September 2003) the parallel inquiry of the IGE-CGM-IGAS, a technical and expertise point of view, independent of the other investigations (litigation, company…). It was also used as a support for the local control authority (DRIRE) to require implementation of safety measures from the company to get authorization to re-start.

This investigation occurred in a broader context following the Toulouse disaster that occurred the 21st September 2001. As a reminder, several investigations were launched by stakeholders (a few more than in this case), and provided numerous findings, lessons and recommendations for changes (Dechy et al, 2004). Disasters are usually seen as catalyst for changes and the Environment Ministry was designing a new law on industrial risk prevention (Law of 2003-699 of 30th of July 2003) and discussing it with stakeholders at the time of this event.

After Toulouse disaster, other decisions of the Environment Ministry were taken to upgrade the existing funding for carrying research at our Institute on the learning from experience, accident investigation methodologies and the integration of human and organisational factors into the major hazard prevention. This investigation was the first one to use most of these developments. As a reminder, INERIS has activities for certifying explosives and hazardous materials (for transportation) in France and EU.

1.2 Objectives, Contents And Management Of The Investigation

The objectives of the investigation were set up by the risk bureau on chemical risk (SEI/BRTICP) and the BARPI (that runs the ARIA database on industrial accidents) within the Environment Ministry and were accepted by the scientific director of INERIS.

The objectives were to:

- estimate the effects of the explosion (inside the site and in the environment), its strength, its eventual domino effects, the differences with the safety studies scenarios,
- find the origin, the circumstances and the possible technical and organisational root causes that should be investigated; for the technical causes, the nature, quantities, reactions of material involved, interactions with the process have to be investigated; organisational factors should be investigated also with regards to the Seveso II regulation and its safety management system requirement,
- formulate technical recommendations on process, workshop design and organisational recommendations such as cleaning of units, risk analysis, learning from experience, etc.

To fulfil these requirements, INERIS management appointed a multidisciplinary team of 5 main investigators (the Institute has 150 people working in accidental risk prevention) that were competent in the field of pyrotechnic process, the pyrotechnic and Seveso II regulations and its safety management requirements, the inspection process, technical investigation, materials sensitivity testing, explosion and pressure waves effects, human and organisational factors. The reports were reviewed by two top level managers (technical director of the Accidental Risk Division and the technical director of the Certifying Division) and approved by the scientific director.

The first visit consisted in collecting facts, photos of the damages, material samples, and documents and were followed, during 3 months, by several interviews of process operators, frontline managers, maintenance technicians, safety managers, top management of plant, top management of the group and control authorities.

2. The Context Of The Accident

2.1 The Company

The company, Nitrochimie S.A., is owned by a French group (SAEPC, Société Anonyme d'Explosifs et de Produits Chimiques) quoted to the stock exchange market of Paris and by the Davey Bickford Smith English group. The turnover was 120 M€ in 2001. The company Nitrochimie S.A. owns two sites in France: the Billy-Berclau site in north of France that produces explosives such as dynamites, blasting gelatine, powder dynamites, and ammonium nitrate fuel oil (ANFO) with 80 to 90 people, and another site (with the Research and Development laboratories) that manufactures blasting emulsions, expansive cement, and monomethylamine nitrate.

The plant is on a site of 750 000 m², on Billy-Berclau and two other cities and its first neighbours are at 700 m. The plant was created a hundred years ago in 1904 to produce nitroglycerine based explosives for the coal mines nearby. It was destroyed during WWI and rebuilt in 1920 with two areas for different manufacturing purposes: one for ANFO and one for dynamites. In 1991, the nitrate activity was stopped and the number of employees was reduced from 170 to 117. Ten years later, in 2003, they were 84 permanent workers and 4 to 6 interim workers.

The explosives manufacturing rates per year were roughly 6 500 tons of dynamites, 8 500 tons of ANFO and requires the consumption of 4 000 tons of sulfonitric mix, 240 tons of DinitroToluène, 10 700 tons of AN and 550 tons of light fuel. These manufacturing explosives to be sold on the EU market needed to be approved according to EU Directives specifications. INERIS does certify explosives in the EU and since 1997 certified those of Nitrochime S.A. and the last time was in May 2001 for 15 different dynamites and 2 ANFO.

2.2 The Safety Related Regulations

These manufacturing activities were under two main type of safety-related regulations:

- The labor regulation and its pyrotechnic regulation: a specific regulation from the labor Ministry was set up in 1979 (Decree n°79-846) and implemented with positive results measured by reduced rates of accident. It required specific prevention measures to be implemented. An emphasis of the investigation was on a comparison between observed effects and estimated effects from the safety cases,
- The environment regulations and its Seveso II regulation (96/82/EC) transposed in 2000 in France: indeed this Directive requires the implementation of a safety management system, dedicated to major hazard prevention for certain plants, according to their risk level (based on the quantity of hazardous materials stored). The plant where the accident occurred was under that regulation for 3 years and had to implement such a management system. The need for an understanding of the management system underlying root causes in that accident was therefore high.

2.3 The Process

There exist several kinds of dynamites that differ in rate of nitroglycerine or nitroglycol. To manufacture dynamite gelatine, four main steps are briefly described:

- Nitration (in a workshop, continuous process) of a mix of glycerine/glycol by a mix of nitric and sulfuric acid to get nitroglyceroglycol (NCG)
- Paste or gelatine manufacturing: it can be done in two different ways: an automated mixing bath holder called Tellex operated at distance in a safe place - or by batch in a tank called Guedu where NCG and adsorbents are poured by operators and the blast gelatine is removed after it has been automatically mixed. The Tellex produces 3 tanks in a row (that are made of polyethylene) for a total amount of 135 kg of dynamite gelatine. These tanks are equipped with caps and 4 wheels. The Guedu produces 3 tanks for a total amount of 120 kg. Then the tanks are driven on a trolley to an intermediate workshop storage n°49 that has a storage capacity of 1 300 kg. From this storage, tanks are automatically driven following a cable, to the processing workshop n°50 to manufacture cartridges of dynamites. These workshops are all separated by barricades of 5 to 6 meters high.
- The automated processing of gelatine to dynamite cartridges at workshop n°50: The inlet of the tanks into the feeder (see figure 1) is automated when the process is started. After the extrusion of the gelatine into the die (photo 1 and 2) and its cut at the roller, the gelatine is rolled in a wax paper before introducing it in the cartridge. The cartridge, processed in the Rollex machine (rate of 900 kg/hour), are transferred on a conveyor to be packed in boxes in a neighbouring workshop n°52. This neighbouring workshop has the video surveillance reporting screen (and recorder) for operators to monitor the process once it is started.

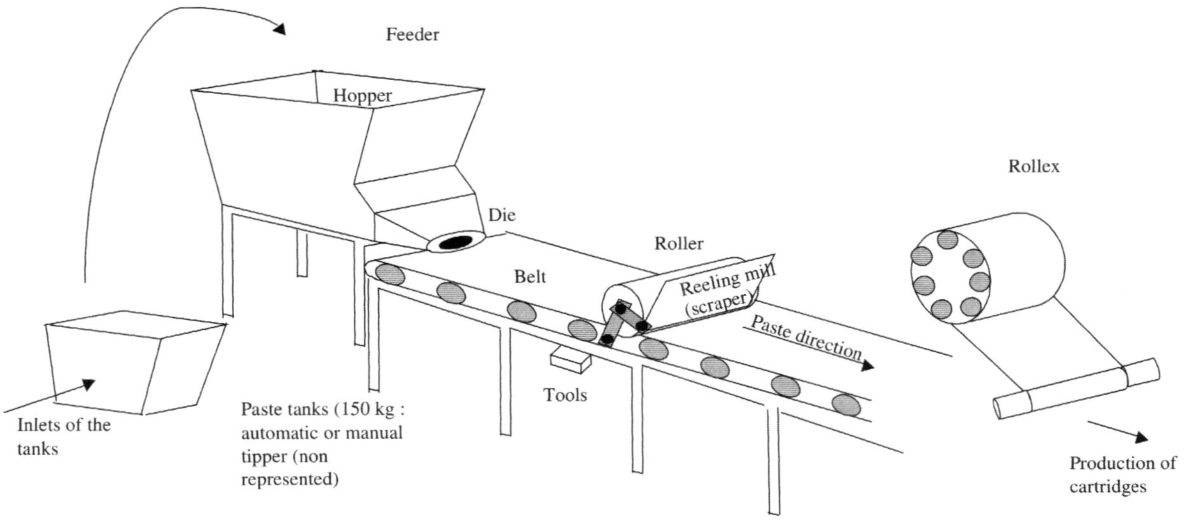

Figure 1: The Rollex Machine And The Feeder

Photo 1: View Of The Feeder

Photo 2: Inside The Feeder (Post-Accident Experiments By The Company)

2.4 Organisation To Operate The Process Involved In The Explosion

Several people are involved on site in the operation and management of the high risk process of dynamites cartridges manufacturing. The key actors were:

- the operator of the Rollex machine: its role is required on the Rollex at the starting of the process; to prepare the automated processing state, its actions are to pour the tanks into the feeder, realising the checks and mechanical adjustments, verifying the size of the cartridge, the level of the conveyor belt, the level of the roller before moving to workshop n°52 to monitoring the process on the video screen; the starting of the automated process is known as a critical operation for the operator's life due to the vicinity with the paste processing into the mechanical machine;
- the assistant of the operator of the Rollex, that helps but does not make the adjustments
- the cleaner of the pyrotechnics waste for the all site,
- a mechanical technician of the maintenance team,
- 3 frontline managers and 1 production manager

The 3 frontline managers and the production manager are responsible for scheduling the production on a monthly basis and then on a weekly basis. The shifts are from 5h-12h and to 12h-19h or from 7h-15h.

2.5 A Summary Of The Chronology

In the days before the accident, a decision was taken to recycle a specific blast gelatine (27 tons of Supedopex) that was stored in UK after the Irish subsidiary decided to cancel the production of these explosives. The gelatine would be recycled in the process for the manufacturing of a specific type of dynamite called Dynaroc 6 after a testing phase. These tests started at the beginning of March 2003.

Because it is not a current product in Billy-Berclau, a new recipe had to be created. The results then have to be approved by the internal laboratory to confirm the quality and safety specifications. The production started three days before the accident.

In the following chronology, there are two frontline managers (A and B): A is responsible for the nitration and mixing process steps and B is responsible for the processing of gelatine to cartridge manufacturing.

The day before the accident, some problems of homogeneity of the mix of Superdopex gelatine and other reactants had delayed the production. The frontline manager B decided to start the shift work at 5 a.m. instead of 7 a.m. to process stocks of paste. In parallel, some testings for a new dynamite called Frangex, have resulted in 8 boxes of 20 kg to be recycled in the production. The operator of the Rollex and its frontline manager B, are unaware of these testings.

So, the day of the accident, the rollex operator arrived at 5h15 in the workshop n°50 (notice that some steps of the chronology have been identified with the video recorder). He looked for gelatine to be processed into the pyrotechnic area and met some operators. From 5h20 to 5h30, he started the rollex machine, made the adjustments, processed some gelatine and packed two boxes (50 kg of cartridges were found after on the workshop 52). There were two other operators scheduled but they were not there that day: one did not wake up, and another switched with a colleague and was not aware of the schedule change. During a half hour some discussions and actions involving the rollex operator and the frontline managers are performed to find two other operators.

After being informed that he will be helped, the rollex operator came back at 6 a.m. to start the process. He faced problems with processing the gelatine through the die. It required him to do a lot of operations (mechanical, cleaning) on the feeder, the plasticating screw, the die, the belt, the roller, etc.

A few minutes before the accident, the assistant joined him. A few seconds before it, the cleaner came into the workshop and took the pyrotechnic paper garbage. He was supposed to start at 7 a.m. but showed up early that day, and with the agreement of frontline manager B, was authorised to start his work of unit cleaning. Also the mechanical technician for maintenance, for some operating reasons (he had forgotten a tool) was walking and passing in front of the tunnel linking workshop n°49 and n°50 when the explosion occurred at 6h16'34.

3. The Effects Of The Accident

3.1 Damages To Structures

The explosion produced extensive damages in the plant and in the nearby village of Billy-Berclau. The methodology of data collection and damage assessment was the same as followed for the investigation of the Toulouse accident by INERIS for IGE (Mouilleau and Dechy, 2002). Some principles of theories on shock wave propagation (Wright J.K., 1961, and Baker W.E. et al, 1983) were used for damages assessment. As a reminder, this damages investigations involves uncertainties and methodological biases that lead, for example, to several TNT equivalent estimates for the explosion of Toulouse (Dechy and Mouilleau, 2004).

Workshop n°50, where the explosion took place, was totally destroyed (photo 3).

Photo 3: Workshop 50, where the explosion occurred $\Delta P > 100$ kPa

Photo 4: Crater in Workshop 50 $\Delta P > 100$ kPa

Photo 5: Damages in workshop 49 (30 m from Workshop 50) - $\Delta P = 10 - 15$ kPa

Photo 6: Damage to a building, 370 m from Workshop 50. (dislodged tiles and broken panes) - $\Delta P = 1\text{-}2$ kPa

The heavy Rollex parcelling machine was thrown against a barricade, barricades were partially collapsed, and two small craters (about 1 m diameter) were dug in the concrete at two locations were dynamite charges were stored on the ground (photo 4). Nearby workshops within about 100 m from the explosion suffered extensive damages (photo 5), but mainly to light structures (roofs, sidings, wood beams). Barricades isolating each workshop from the others proved once again to be an efficient protection safety barrier against the explosion. In the far field, minor damages to light structures were observed (photo 6).

Most buildings in the plant suffered from dislodged tiles and asbestos-cement panes, broken windows and panes. Similar effects were noticed in Billy-Berclau village, about 1000 m distant from the explosion epicentre.

It was not mentioned any domino effect. The whole dynamite charge present in workshop n°50 exploded, but the explosion did not propagate to nearby workshops where high quantities of explosives were also stored.

3.2 Analysis Of The Damages

Overpressure estimates were made on 34 damage points related to typical overpressure level or ranges available in the literature (Mouilleau and Dechy, 2002; Cozzani and Salzano, 2004). The ranges of estimated overpressures recorded are plotted versus distance from the explosion epicentre (figure 2).

The screen effect provided by barricades between workshops was taken into account in the assessment methodology, according to the following rules recommended by the French Pyrotechnic Regulations (figure 3):

- within a distance twice the height of the barricade, the overpressure level at ground is divided by 4 to 6, compared with the overpressure that would be measured without barricade.
- At a distance between twice and four times the height of the barricade, the overpressure level at ground is divided by 2 to 4, compared with the overpressure that would be measured without barricade.

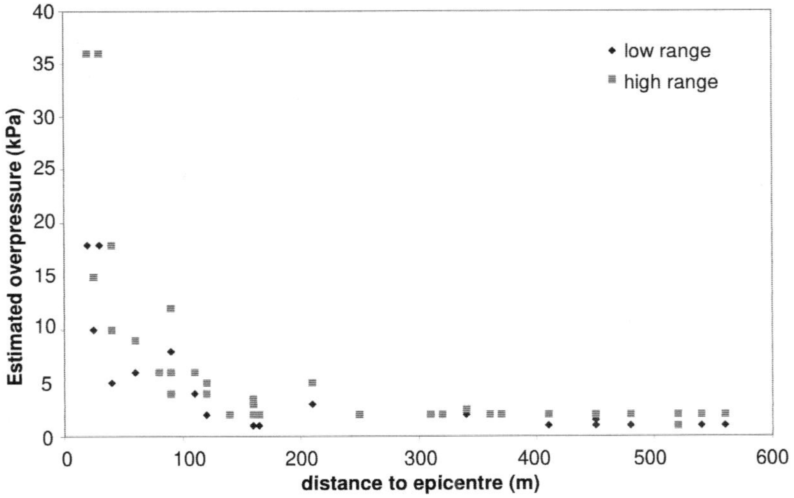

Figure 2: Overpressure Estimations Vs Distance To The Epicentre

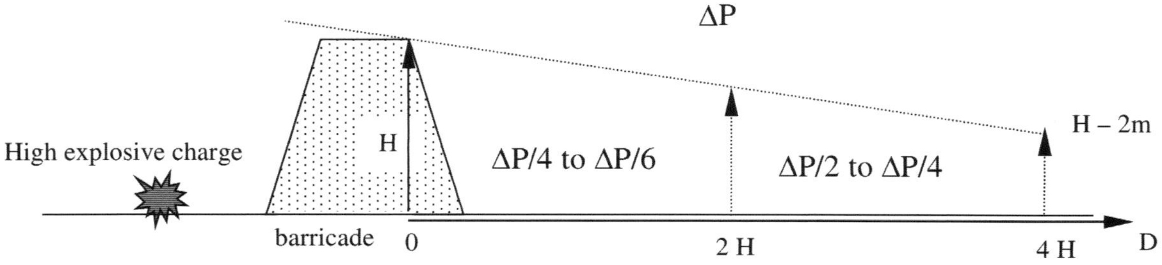

Figure 3: Screen Effect Of A Barricade (Adapted From The French Pyrotechnic Regulations)

As one would expect, the estimates show a decrease in the overpressures the greater the distance measured from the epicentre. In the pyrotechnic area, the shock wave propagation was clearly disturbed by reflection and diffraction phenomena in crossing the barricades, that complicates the damage interpretation (see Wenzel A.B. and Bessey R.L., 1969). After correction by the screen effect, a mean overpressure level from 5 to 15 kPa seems plausible. Outside the pyrotechnic area (beyond 250 m from workshop n°50), the overpressure level seemed about 1 to 2 kPa.

From the very first analysis of the damage points, it has been deduced that the TNT equivalent of this explosion turned out to be between 70 to 130 kg. These values correspond to the medians of the weight values calculated from the overpressure estimates respectively on the low side and on the top side. Considering an energetic equivalent of 1.1 kg TNT for 1 kg dynamite, that would provide a dynamite charge from 60 to 120 kg. However, a discrepancy was noticed in TNT equivalents from inside and outside the pyrotechnic area. The analysis of the inside damage points would suggest a dynamite charge from 50 to 100 kg, whereas the outside damage points would give 100 to 300 kg.

This disparity should be explained by considering:

- in the pyrotechnic area, the overpressure wave disturbance caused by the barricades;
- in the far field, the difficulty to quote accurately damage to light structures, and the possible disturbance by external phenomena, such as the wind speed and the wave reflection by the lower atmospheric layers (Grant et al, 1967, and Reed, J.W., 1973).

Those estimates can be compared with the real explosive charge present in workshop n°50 when the explosion occurred (figure 4). This has been assessed to 580 kg, distributed into:

- two dynamite tanks on hold at the entrance of Workshop 50: 260 kg,
- dynamite in the feeder: 150 kg,
- one tank discharge near the belt: 90 kg,
- 4 boxes of Frangex dynamite: 80 kg

The first explosion has probably occurred in the feeder containing 150 kg of dynamite. Applying the French Pyrotechnic Regulations rules of calculation, the detonation in the feeder would have been transmitted by pressure wave to other close dynamite charges within a distance of no more than:

$$D = 0.5.Q^{1/3}$$

where Q is the explosive mass.

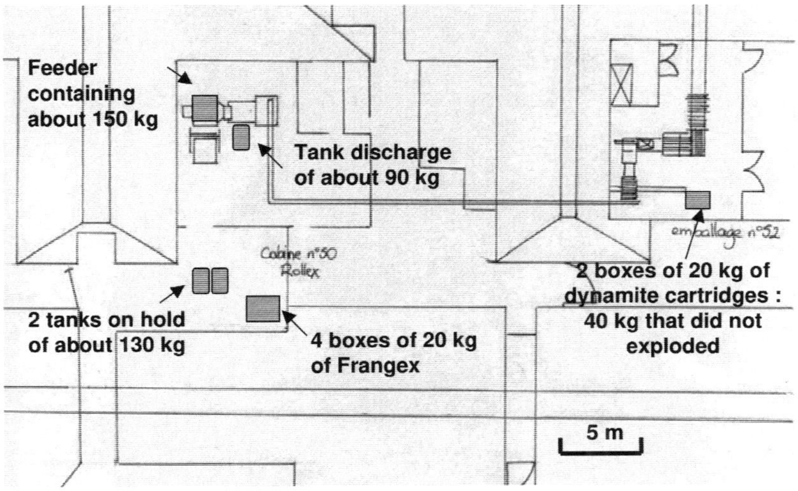

Figure 4: Gelatine Quantities At The Moment Of The Explosion

Thus, it could have been transmitted by pressure wave to the tank discharge closed to the feeder. Regarding the other charges distant from about 10 m from the feeder, the detonation would have rather been propagated by missile projection. Barricades must have prevented the explosion transmission by missile projection from workshop n°50 to nearby workshops.

However, the real total dynamite amount in workshop n°50 turns out to be quite larger than the one estimated by the damage analysis. This gap could be explained considering that several successive detonations generated independent overpressure peaks which did not cumulate. Many witnesses mentioned a "long, dull noise" which could emphasise this assumption.

It could be admitted that two main charges were located in Workshop 50:

- 240 kg near the feeder (feeder + tank discharge)
- 340 kg at the entrance (2 tanks + 4 Frangex dynamites boxes).

If the explosion was transmitted by a missile impact from the first charge to the second one, the speed of transmission would have been about 1000 m/s, meaning that the second charge would have detonated about 10 ms after the first one. As the overpressure pulse of a detonation has a typical period of 1 ms, the global pressure wave signal would have been a succession of independent overpressure peaks that would not have cumulated, at least in the near field. Therefore, the overpressure damages would have been caused by the detonation of the biggest charge (340 kg) rather than the total charge (580 kg).

One can also consider that not two but four independent dynamite charges, as identified on figure 3, generated four successive detonations and four independent overpressure peaks. Thus, the overpressure damages in the near field would have been caused by the detonation of 150 kg of dynamite in the feeder. This hypothesis would agree with the TNT equivalent estimate of 50 – 100 kg from the near field damages interpretations.

Beyond the first barricades, very complex waves diffraction and reflection phenomena should have occurred, and pressure signal should have partially recombined. That could explain the discrepancy between TNT equivalent estimates from the damage analyses in the pyrotechnic area and in the far field.

As a conclusion, it seems that the overpressure effects were not generated by the massive explosion of the whole dynamite charge in workshop n°50, but were most probably caused by several successive and independent detonations whose overpressure signals might have partially recombined in the far field, according to shock waves propagation theories.

3.3 Effects To People

Fourteen accident at work notifications were registered by the company during the days following the explosion. Casualties are located on figure 5. Among them are:

- 4 deaths,
- 2 minor injuries by blasting onto the ground,
- 1 minor injury to ears,
- 1 minor injury by ceiling fall,
- 5 psychological shocks,
- 1 disease,

Notice that five people located within 100 m from the explosion epicentre were not injured.
The four deceased people were located in workshop n°50 and for one of them, in the communicating tunnel between workshops n°49 and n°50. Other people located in the nearby workshops suffered minor injuries, highlighting the protection efficiency against pressure waves offered by barricades between workshops.

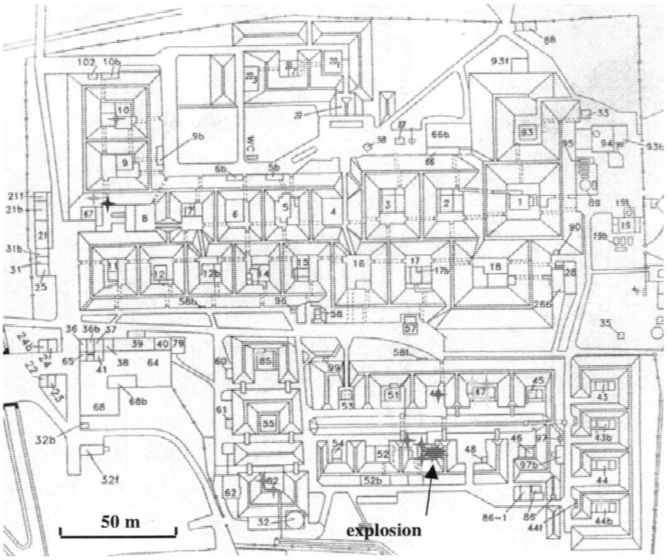

Figure 5: Victims location

death, injury by blasting onto the ground, injury to ears injury by ceiling fall, psychological shock, disease, no injury

4. The Technical And Direct Causes Investigation

The technical analysis of the investigation was performed by using an analytical tool (fault tree) to identify in a structured way the probable causes of the explosion of the workshop n°50. The analysis with a fault tree is an inductive-type method used to determine the technical causes and some human actions leading to an accident.

The building of this fault tree was based on:

- the information gathered during the interviews,
- the tests on the samples of the explosive that was packed and stored in the workshop n°52 at the time of the accident,
- on the analysis of the damage caused by the explosion,
- on the expertise of INERIS in its knowledge about the phenomena of explosives explosion.

The possible location of the explosion had to be determined as a starting point in order to be relevant in our assumption in the remaining analysis. Thus, based on data from the damage plotting, the chronology and the video recorded file, the feeder was identified as the most probable location of the explosion.

Then, the investigation focused on the nature of the blasting material in the feeder. The material in hand here is obviously dynamite but many questions arose (see chronology):

- What kind of dynamite was it ?

- Did the dynamite meet the required technical specifications ?
- Or was it more sensitive than the required technical specifications ?

About the type of dynamite, there were two possibilities: the Dynaroc 6 made on a regular mode at the Tellex or the Dynaroc 6 made with the Superdopex gelatine on a manual mode.

Two charges were put into the hopper: the first one was some Dynaroc 6 made of Superdopex gelatine produced the day before the accident. But, the origin of the second container is still unknown. Because of the sequence of events, it certainly came from one of the storage buildings, but the inventory made after the explosion did not help much to draw clear conclusions.

Concerning the suitability of the dynamite to the technical specifications, the tests carried out at INERIS on the samples collected from the company, if one considers them as representative of what was in the feeder at the time of the explosion, helped to conclude that the explosive was adequate and not more sensitive to shock or to friction than the dynamite usually produced by Nitrochimie S.A..

Yet, even if the explosive was adequate, its sensitivity can be questioned as, for example, the gelatine had to remain a long time (one night) in the feeder and the die. The gelatine was hard to process (see chronology) that had an effect on the exposure of operator to risk. Many factors may have increased the sensitivity of dynamite:

- the formulation of the gelatine can be inadequate,
- the viscosity of the gelatine can be off-specification,
- the absorbents can have a high acidity rate,
- exudation can occur,
- the dynamite can be unstable,
- the gelatine can have an inadequate homogeneity.

Consequently, except the issue of acidity, all the factors were kept for different reasons. For example, an exudation could have occurred in the die due to the long stay of the paste in it. Besides, even if it is difficult to set a relation between an inadequate homogeneity of the gelatine and its level of explosibility, this factor cannot be turned down.

There is a minimum amount of energy required to trigger the explosion and thus knowing the paste sensitivity appears to be of high interest. Some of the ignition sources could have been present but not powerful enough. Several ignition sources were identified:

- heat action by an external fire,
- electrical, electrostatic arc,
- external pressure wave,
- missile impacts,

- mechanical action from the process or something else.

The approach consisted in making an inventory for each type of ignition source of the situations or circumstances where they could have appeared given the equipment, the labor organization, the task performance itself in the workshop, etc.

According to the available data, only the mechanical action was kept as probable. Among the possible causes of a mechanical ignition, there were possibilities of friction due to different metallic parts between the container, the feeder, the hopper or the tipper or the friction of a foreign matter on the feeder. This last option was kept as the most probable.

Because not enough evidence was collected, many options are still conceivable. The foreign matter could have been introduced:

- in the hopper by the operator, notably when he threw a ball of gelatine (that fell into or close to a tool box placed along the Rollex belt) to the feeder a few seconds before the explosion,
- in the hopper by the fall of an object of the tank when unloaded
- in the hopper by the fall of an object from above the feeder
- in the tank during the manufacturing of the dynamite, the transportation of tank on the trolley to the workshop n°50 or in the workshop n°50
- in the hopper before the paste was loaded in.

As a preliminary conclusion of that step of the investigation, there are still some uncertainties about the exact technical causes of the explosion. It would have been interesting to continue the tests on the gelatine used at the moment of the explosion to clarify the link between its properties (viscosity, exudation, ...) and its sensitivity to explosion. Besides, there is still this uncertainty about the origin of the second tank.

The most probable scenario is the presence of a foreign matter in the feeder either brought inadvertently by an operator or due to the fall of objects from the container or the hopper. But the timing and the circumstances where it could have been introduced are still unspecified. The internal company investigation concluded also that the most probable scenario is a foreign matter in the feeder.

However, this was not the only issue as a matter of a fact that four people died in the explosion, introducing the risk factor of the exposure of operators. Limiting exposure is a key thing in the pyrotechnic industry. In that accident, the circulation issue appeared therefore as a key element of the understanding of the severity of the human losses. The cleaner, the maintenance operator as well as the assistant of the main operator were not supposed to be where they were in these particular circumstances, at the time of the explosion.

With this most probable technical scenario assumption and the circulation problem, we could get into the issue of why the organization was not able to prevent this scenario. Safety barriers and controls to prevent this scenario and its severity, which was a known risk by the company, were analysed in their organizational context.

5. The Organisational Investigation

Three main points are discussed in the following part: the methodological aspects of the organisational investigation and the background models, the outcomes and a discussion around them.

5.1 Methodological Issues

5.1.1 Rationale For Investigating Organisations

Organisations are complex systems. They imply a high number of interactions, at all levels, from the shop floor to the board committee but also the outside of the organisation, with individuals located in its environment (regulators, economic partners). These interactions are rich; they involve the nature of the social world. This social world is made of situation of power between people, of cultural dimensions, of social construction of reality, of norms, of values, of rule. Organisations also can be seen through several lenses like the financial one, the human resources one, the economical one, the institutional one, the historical one, the societal one… Organisations are therefore a subject of research for various disciplines like economy, politics, sociology, anthropology, psychology, psychosociology, all of them applying different filters. This makes the organisational life a quite dense subject of study (J.Rasmussen and I.Svedung (1997, 2000) suggested a representation of the socio-technical system where many of these features are introduced and linked).

Qualitative studies investigating on large organisations but even on small organisations deal with the problem of collecting and analysing data for interpretation, and also deal with the ability to consider various possible angles and dimensions, across levels.

However, with investigations, one helping thing is that the focus following an accident is given. Indeed, the accident sequence reveals some inadequacies that are the starting point of the investigation. The aim of the investigator is therefore to give a context to the events. This implies to go back into the history of the organisation in order to better describe in which circumstances the decisions and actions of the individuals were taken. The idea of going back in history is to find out more about the conditions that "*shaped the landscape in which the accident unfold itself*" (J.Rasmussen I. Svedung, 2000*)* further back from the close events, like human errors and technical failure leading to the accident.

The purpose of an organisational investigation could then be said to describe the events with a broad scope, to bring some explanation (defining some causal or rather conditional links between the events and a context) and sometimes to build a predictive model from it. The latter has not been performed so far many times. It is the work of researchers, social scientists, who conceptualised further their interpretation of the organisational dimensions, applying social science techniques of theorising from grounded data, and single cases.

What is actually performed is represented in this figure n°6.

As said, it is nevertheless not such a common thing to theorise from accidents. Thus, when we go through the existing literature on the organisational side of accidents, it shows that there are actually several types of possible approaches for investigating.

We classified them in three types (JC Lecoze 2005).

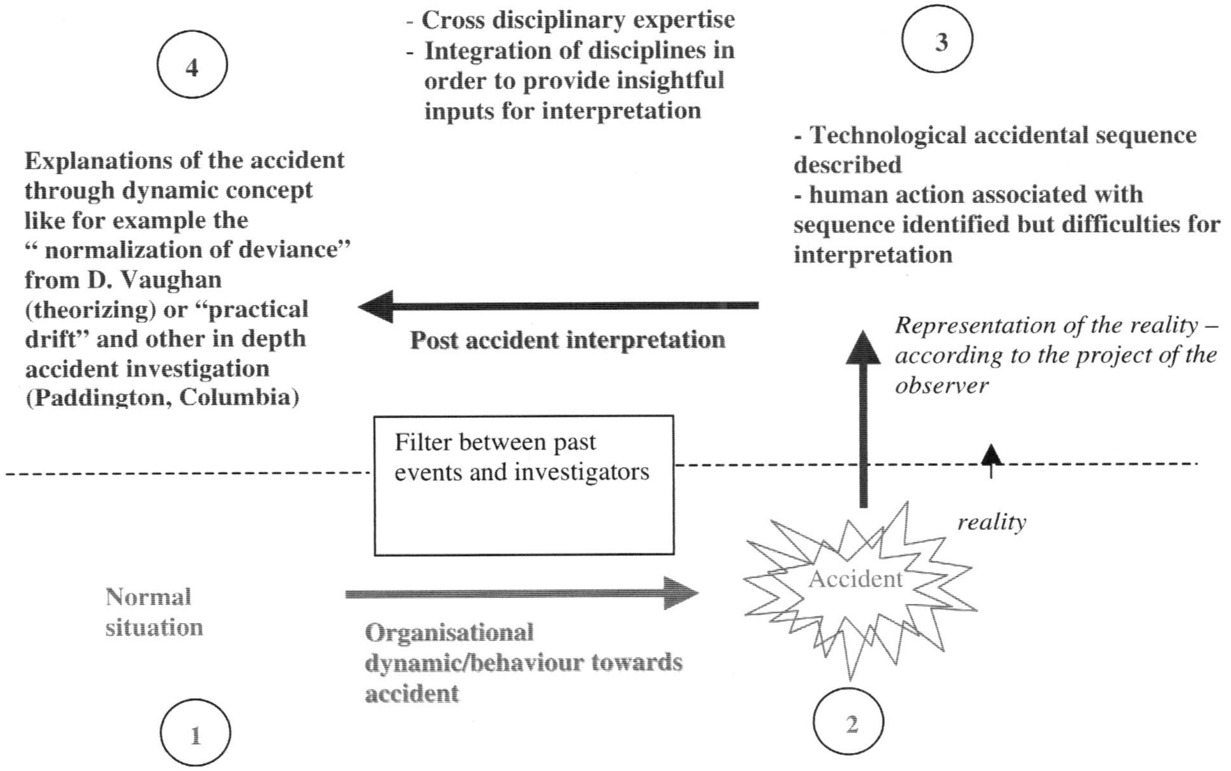

Figure 6: Purpose And General Process Of Investigating (Lecoze Et Al 2002)

5.1.2 Three different approaches of investigations

First, these works already mentioned with research and theorizing perspectives (i.e D.Vaughan, 1996, S.Snook, 2001), secondly the commissions set up after disasters (i.e: Columbia, 2003, Ladbroke grove, 2000) and then the structured root causes analysis (i.e MORT, STAMP).

All of these three types must face the bias of forcing the data.

The "forcing the data" concept comes from one of the difficulty of the social science theorizing process. In social science, theories can be grounded, generated and inferred from the data collected through fieldwork. These are not to be generated purely by a logico-deductive approach. However, in front of the richness of the organisational life, theories can be validated through strategy consisting in forcing the data to fit the theory. This is a symptom that also could be called the "data that should fit the model" rather than the "model that should fit the data" approach (from Glaser and Strauss, 1967).

The three types of investigating techniques deal with it differently.

Pre defined models, in structured root causes analysis, apply a "data that should fit the model" approach but are made for systematic recommendation purposes and are useful in that sense. Commission and research work are more descriptive and explanatory, and do not specifically rely on pre defined model. However, they also introduce underlying models and theories to bring some light on the data extracted from the investigation.

These three types involve different relationship with the underlying models that they use for treating the data, but also different time constraints, techniques, resources (number of people, competences, expertise etc), scope (micro-meso-macro levels) and objectives (describing, explaining, making recommendation).

5.2 Methodology Used For The Case

5.2.1 The Underlying Models

In the investigation performed for the Billy Berclau accident, the approach was based on an investigating process that was half way between a structured root cause analysis (i.e MORT, implying focused questions and pre-defined safety management system) and a more descriptive approach based on open interviews and openness of the system with micro-meso-macro perspective as well as a historical approach.

The investigation report made clear in a introductory part that several organisational dimensions would be investigated, with references to works in the field, like the High Reliability Organisation (i.e Roberts 1990, Rochlin 1996, 1999, Weick, 1987), Normal Accident theory (C.Perrow, 1999), Incubation period (B.Turner, 1978), Normalisation of deviance (D.Vaughan, 1996), Latent and active failure (Reason, 1990, 1997) as well a the idea of organisational behaviour towards accidents (Rasmussen and Svedung, 2000).

Also, it has been stated that some dimensions were investigated and articulated with the help of a type of matrix as suggested by Glandon and Waring (1998). This matrix contains the following dimensions:

Risk contexts
Organisational environments
Economies and market
Public policy, legislation and regulation
Social and political climate

	Technology
	History, operating territories and conditions
Human Factors	
	Culture
	Power relations, political processes and decision making
	Perception, cognition and meanings of success
Formal coping arrangements	
	Risk management
	Risk assessment
	Management systems
	Approaches to change

Table 1: Thematic table linking theoretical aspects of risk with empirical studies (A.Waring and A.I. Glendon, 1998)

All of the dimensions can be investigated with related theories in the background, models or works in the safety field (like in perception, power in companies, safety management systems etc).

The idea was not to validate any of the frameworks suggested by any theories and concepts but rather to explicitly mention that they were in the mind of the investigator when collecting and analysing the data. In her article dealing with the heuristic of case analysis (1992), D. Vaughan has debated this point of the pre-existence of models in mind when investigating and theorizing.

The importance of the references and underlying models in organisational investigation is high. It has been quite rightly stated that an investigator could find what he/she wanted to find in accidents.

5.2.2 The link between technical and organisational factors

The technical anchoring of the organisational investigation has been performed through the barriers and controls revealed through the technical sequence. Twenty three barriers and controls were identified based on the most probable scenario retained. The barriers from B1 to B12 were aimed at decreasing the likelihood of foreign object presence in the feeder, B13-B16 were designed for preventing the explosion of the feeder in case of a foreign object friction and the B17-B23 decreased the severity of the human loss in case of accident. Each of the barriers were investigated in detailed through interview and then replaced in a broader context.

To illustrate it in a simple way, the investigation has been performed like the next figure n°7 is suggesting it (adapted from a representation used for MORT, Johnson, 1980 and mixed with Waring and Glendon, see table 1).

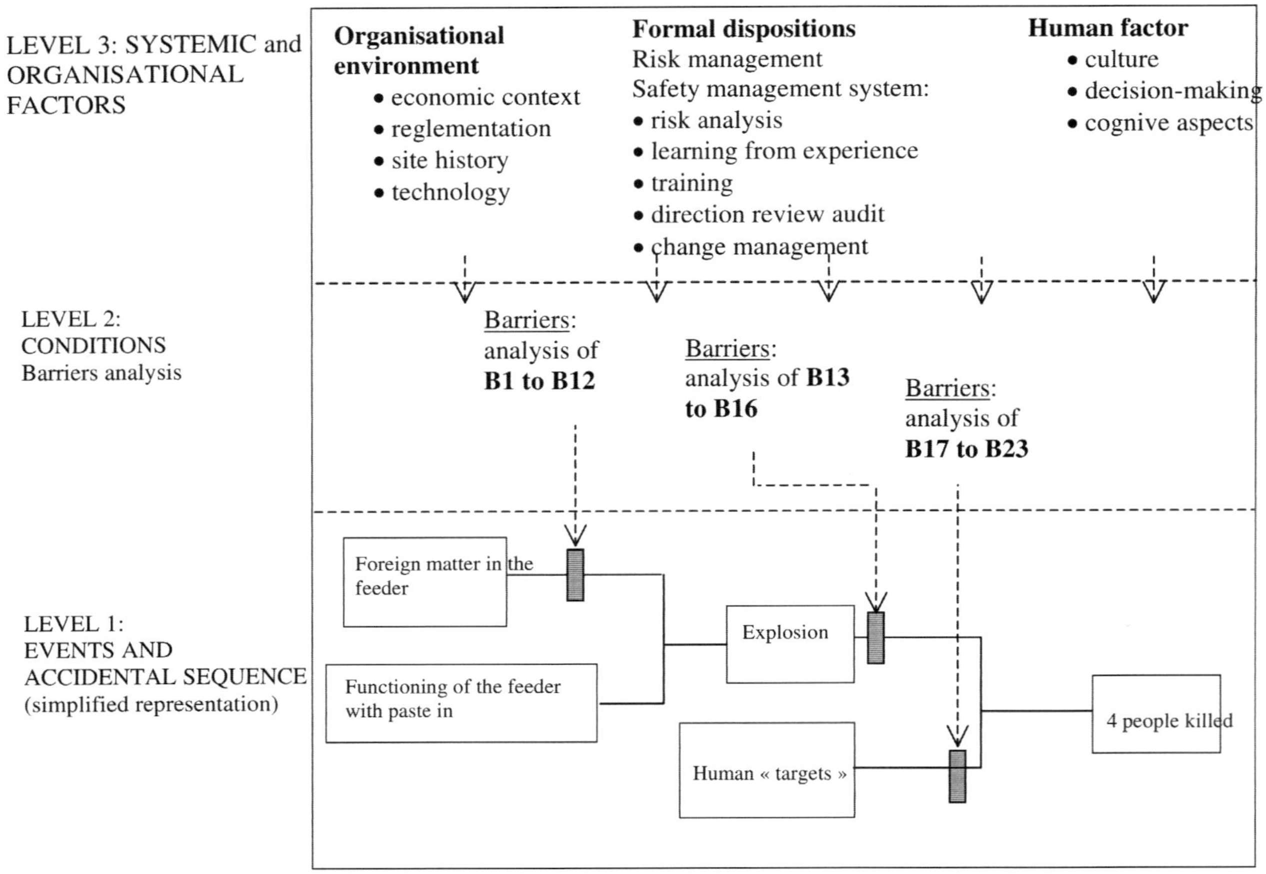

Figure 7: Link Between Technical And Organizational Factors In The Billy Berclau Case

The first level is the accident sequence with the inadequacies of the barriers and controls, the second deals with the conditions for these barriers and control not to fulfil their functions and the third level, which brings some insights on the overall context influencing the conditions.

In fact, this investigating process shares some similarities with what has been described for example in Fortune and Peters (1995:110):

- Pre-analysis
- Identification of significant failures (similar to level 1)
- System selection (similar to level 2)
- System modelling (similar to level 2)
- Comparisons (similar to level 2 and level 3)
- Further analysis (similar to level 3)
- Synthesis

The investigation consisted in a series of 30 interviews (70% for the levels 1 and 2, 30% for the level 3) with people from the shop floor to the board committee at the corporate level. An extensive investigation focusing on the documents provided by the company describing quality and safety management systems as well as safety cases, and the records of the activity of learning from experience, safety meeting, risk analysis, etc was also performed. The use of a book about the culture and history of the company celebrating a century of existence was used and very helpful for getting to know better the company and its history.

5.3 Outcomes Of The Investigation, Lessons To Learn

A certain numbers of findings made are quite consistent with some already mentioned influencing organisational factors, in other investigations like in A. Hopkins, (2001) but also general lessons learnt throughout accident (Llory et al., 2004) and were similar to accident models mentioned earlier in this paper. Ninety recommendations were made for the company to improve the level of safety. They were made at several levels of the system, from the technical level to the management and regulatory level.

Here is a selection and discussion of relevant aspects that were identified.

Some lessons at the level of the authorities and regulations (risk contexts Public policy, legislation and regulation, in regard with table 1):

The regulations are increasingly dense for industries, as much from their contents as from their overlapping. The pyrotechnical site was liable to many regulations concerning safety under the pyrotechnics regulation depending on the health and safety regulations but also under the environmental regulation (the classified installations in France and Seveso II Directive in EU) dealing with major hazards. The site was classified high threshold concerning the transposition of the SEVESO II directive. In this case, there is an overlapping of two regulations which however both deal with the prevention of major hazard.

These regulations have nevertheless basic differences. One of these difference relates to prescriptive nature for one (the law on pyrotechnics safety) and the other by objectives (in particular the SMS within the framework of the application of the SEVESO II directive).

Consequently, a time for integrating the new regulation was required. This adaptive time is all the more important than the culture of the industry has been shaped by the regulations applied for more than two decades. In the case of pyrotechny, the decree of 1979, very detailed and seen as an effective regulation, which allowed notable improvements of safety on sites at the time of its implementation, had somehow played this shaping role.

Another one of these differences relates to the risk assessment practices and calculations of safety distances, which is not without causing administrative issues. Indeed, the two regulations have different thresholds, different classifications. This brings confusion and a difficulty of treatment of the safety reports and a difficulty in meeting regulations requirements. Concretely, at the level of the plant, this also resulted in an important volume of administrative work which has been regularly growing since the nineties with the increase of the environmental regulation and requirements. This risk context took part in diverting the safety department from its operational activities.

The diversity of the regulations implies also the diversity of controls. This can be a good thing, but can also be a bad thing when the links between the parts have consequences for the consistency of the whole. In particular, it is necessary to take into account that this company faced variations in terms of number, objectives and quality of inspection. An inspectorate like the IPE (Inspection des Poudres et Explosives - Explosive and Powder Inspectorate) has the means and experience for understanding well the operations of the installations and can bring a useful and constructive help for risk management, while keeping a control and regulative function. This is therefore, not a surprise to find clearly stated in the regulation the possibility for the assistance for in site inspection and safety reports by the IPE for the DRIRE and the labour inspectorate (in charge of work safety regulation as well labour regulations). Consequently, in case of numerous updating of safety reports, the company has different strategies available for dealing with the submission of the reports, to one of the control authorities. In that case, this precise situation occurred and revealed consequently several non-conformities.

This general context of regulation overlapping with sometimes different objectives and contents, with multiple control authorities, creates a complex regulatory situation for which the company must organise its resources in order to cope specifically with it. These remarks do not, of course, justify non-conformity, but remind about the sometimes dense and complex regulatory contexts that organizations are facing, and shed some light on the difficulties that they meet through their attempt to comply fully with them. The pyrotechnical site faced a situation from this point of view for which a great part of its safety operational activity was diverted to treat these administrative tasks involved.

Authorities in France work actively on this issue of overlapping regulations.

At the level of the organization of the company and site (human factors and formal coping arrangement, in regards to table 1)

- ✓ *Changes in organizations*

Changes in organizations must be done with a careful attention given to the potential impact of these on the safety practices at the operational level. Usually, the impact of changes is approached for technical issues but much less in terms of organizational issues. It is clear that the impact can’t be measured the same way. The impact of organizational changes is clearly not easy to assess, but some measures should be taken before and after. In this accident, important changes were done at the level of the site top management and at the level of the technical corporate department. These two changes impacted on the operational safety of the site. Indeed, following the changes, the safety department was not able anymore to play its safety role against the production pressure, as it had been working for years. Thus, it is believed that any changes that imply an evolution in the way the system sustains its level of safety through a balance between safety against the natural production pressures, increases the risk of operating outside the boundaries of safe practices. The evolution that was identified through the investigation in this case concerned the way information was available at the level of the board committee. This information was important for high levels in order to hear about the level of safety at the operational level. This situation was not without reminding the "trickle down effect" described by D.Vaughan about the Challenger accident (1997) when she describes the effect of political, economical and organisational arrangements made at the top of the system that slowly and gradually spread across the lower levels.

✓ *The cultural dimension*

Culture is seen as an important concept today for explaining how companies operate. However it is still a scientific concept that is strongly debated. For the purpose of the investigation, it appeared however relevant for some aspects. We retained about culture that, according to E.Schein, (1992), it is:

" *a pattern of shared basic assumptions that the group learned as it solved problems of external adaptation and internal integration, that has worked well enough to be considered valid and, therefore, to be taught to new members as the correct way to perceive, think, and feel in relations to this problems"*

The history of the company is hundred years old and the site is one of the oldest. A long tradition of operators has been working on the site, a dynasty as they would call them (a term that was found in the document celebrating the century of the company). This long history, through the events and the technology used, shaped the practices and maintained a tradition in the way the blasting gelatine was handled. Despite the new technology, requiring less manual handling of the paste, some remaining practices described as "pratique à l'ancienne" equivalent for "old time practices" were still in use. They consisted in regularly manipulating the paste with bare hands. This kind of old practices interacting with more recent practices required by the new technology can interact inappropriately with safety principles introduced with the technology. In the accident, the video revealed a behavior of the operator taking a piece of gelatine that fell on a tool box and putting it directly back in the feeder. This is not without reminding an old way of dealing with the paste.

It was also found interesting to talk about safety culture in our case as it is today a widely shared concept (or rather assumption) in the research and practice though also still very debated. There is again no consensual definition among scientists. Here, for practical purposes, we can retain the HSE (UK Health and Safety Executive) definition about safety culture as being:

"The product of individual and group values, attitudes, competencies, and patterns of behaviour that determine the commitment to and the style and proficiency of, an organisation's Health and Safety programmes. Organisations with a positive safety culture are characterised by communications founded on mutual trust, by shared perceptions of the importance of safety, and by confidence in the efficacy of preventive measures"

As a guidance through the concept of culture the HSE gives these items related to the dimensions of safety cultures for which some are relevant in this accident case:

- **Organizational commitment and communication**
- **Line management commitment**
- Risk taking behaviour and some contributory influences

- Personnel role
- Workmates' influence
- Competence
- Supervisors' role
- Permit to work systems
- **Reporting of accidents and near misses**

In bold the dimensions that appeared illustrative, as items representative of aspects of safety in the site. The line management commitment and organisational commitment appeared through the interview and documents analysis to be less than adequate to maintain the activities against natural and expected drifts of safety practices (this point is highlighted below). However, the concept of culture is not used further here as it tends to give a deterministic vision of the organisation. A vision which implies that culture creates the conditions for people to behave the way they behave although individuals have also a strategic and adaptive dimension, with their own intentions, that are never fully determined by outside constraints (it is one of the difficulties for predicting organisational behaviour as a whole).

✓ *Safety management (formal coping arrangements in regards with table 1)*

From a more structural point of view of the organisation, it was clear that the SMS required by SEVESO II directive was currently being under implementation, but had some shortcomings in the way it was thought and designed. A context for this difficulty to fully apply the safety management principles has already been stressed with the existence of a prescriptive regulation.

Some are the shortcomings pointed out during the investigation:

- the principle of defences in depth (also called the barrier and control approach) extracted from the risk analysis was not fully applied and only limited to a small number of defences. This strategy was implemented in order to focus only on the last defence, called the ultimate defence principle. The idea for the company was not to identify too many defences which is very heavy afterwards in terms of management, maintenance, inspecting and auditing. Moreover only technical defences were included and not behavioural ones (like the circulation issue). It is a problem not to make barriers and controls more explicit for a risk control perspective. This was a serious flaw in the way the system could have been thought. Also, it appeared that the risk analysis sessions were not opened to operators. This can be surprising as it is a widely known practice within industries but was eventually less surprising in a prescriptive regulatory context where everything was already codified and written, ready to be applied.

- The learning from experience is a key activity that must help operators and managers to think and work on the conditions leading to the incidents. These conditions can be technical as well as organisational. In this case, the organisational dimensions of the incident were not so much analysed, as the safety department was no longer involved into the learning sessions following the recent changes in the management of the plantAS a consequence, there was more focus more on technical rather than organisational issues. A long list of warnings concerning foreign objects in the last 3 years triggered some concerns about the cleaning within the installations. The management had therefore planned to improve cleaning practices.

- The training was also a concern for the company who had to manage in a flexible manner the production for coping with the low visibility of the market. This implied recruitment of temporary workers, which created the problem of the following up on training. Training must also be used as a way of reminding people about safety principles, to prevent the work routine from becoming unsafe, according to the law. The company had not been very cautious on the follow-up for all operators on the refresher sessions.

- This aspect of routine introduces the idea of dynamic drifting (as for example described in S.Snook, 2000). This consists in shifting at the level of the operators from a rule based approach to a more task based approach, where the individual can locally cross the safe boundaries of the system (J.Rasmussen, 1997), and therefore can create risk, but also being exposed to risks. Micro deviations can install in the organisation with no impact on the production as a whole but create a kind of deviance within the system in terms of safety (D.Vaughan, 1996, 1999). As dynamic and evolving, it is of no surprise to discover that operational practices defined by rules, confronted with multiple dimensional and sometimes conflicting aspects, are moving towards adaptive tasks. It appeared, with retrospect, that what was observed in this accident, had obvious common points with what J. Rasmussen (1997) described:

- "In any well designed work system, numerous precautions are taken to protect the actors against occupational risk and the system against major accidents, using a 'defence-in-depth' design strategy. One basic problem is that in such a system having functionally redundant protective defences, a local violation of one of the defenses has no immediate, visible effect and then may not be observed in action. In this situation, the boundary of safe behaviour of one particular actor depends on the possible violation of defenses by other actors. Therefore, in systems designed *according to the defence-in-depth strategy, the defenses are likely to degenerate systematically through time, when pressure toward cost-effectiveness is dominating."*

  Although the pressure towards cost effectiveness didn't show to be so important through the analysis of the economical constraints of the company, an internal pressure for production was however clearly stated through the interviews. In the accident the various drifts around the rules involved the shift schedule that was modified without front line managers knowing about it but also the early start of the installations without the required number of people, the issue of circulation in the pyrotechnic area and the cleaning of installations. They all appeared, with hindsight all to be outside the rules defining the way things should have been performed. However, they couldn't be seen as individual and isolated mistakes of operators in a traditional sense, but had to be seen as a general behaviour moving outside expected practices and safe boundaries, and involving all levels within the organisation, that allowed this situation to occur.

  K.Weick (1987) gave an insightful comment about the problem of drifting and crossing the safe boundaries. He suggests an interesting explanation for the reasons why safe boundaries are crossed. It also had some quite strong echoes with our case, particularly with the issue of circulation, with the installations cleaning but also the specific morning situation:

"Reliability is invisible in the sense that reliable outcomes are constant, which means there is nothing to pay attention to. Operators see nothing and seeing nothing presume that nothing is happening. If nothing is happening and they continue to act the way they have been, nothing will continue to happen. This diagnosis is deceptive and misleading because dynamic inputs create stable outcomes"

Operators regularly do things to cope with situations in order to be able to make their way out of conflicting goals but also unexpected situations, and therefore can cross, without totally noticing it, the boundaries of safe operations. In this accident for example, without any adaptation and with a strict respect of the rules, the operations wouldn't have started that day as two operators were missing at the beginning of the shift.

Deviance or drifts can sometimes be sustained in the long run, as the normal(ized) practices, in some specific organisational, economical and historical contexts. Companies should acknowledge this situation and work on these micro adaptations through the design of organisational means where people can talk about them and find way to overcome the risky ones. These kind of dynamic aspects have clearly to be addressed specifically with an appropriate organisational support. This would need to be experienced and developed in the future, for safer operations.

- The safety indicators related to occupational health and safety (LTI for Lost Time Injury) were very good. The company had a past records of no accidents for four years. This led to the conclusion, following the accident, that these types of indicators do not make clear how good a company manages the major hazard prevention. This is also very consistent with what has been identified by A. Hopkins in his investigation of the Longford plant gas accident (2001). Indicators of these types can be misleading and give a wrong picture of the safety state of a plant, in terms of major hazard prevention. Efforts should therefore be made to indicate the level of safety through the concept of barriers for example, where specific links with the major hazards are possible. A quality type of approach for indicators can be used for that purpose, which could be seen maybe as a bit expensive but that is believed to be worthwhile implementing. The limits of such systems should however also be stated, and complemented for example with softer approach where people can deal with deviation from expected practices which have no direct and measurable impact on safety, and are sometimes necessary in light of conflicting or constraining goals.

- The last point discussed here from this case is the usefulness of cross sites auditing. In the company, this practice had been gradually decreased throughout the years. It seemed that this decrease has revealed to be detrimental to the level of safety. It is believed, that quite consistently with some ideas derived from the normalisation of deviance concept, that external auditing from competent individual is a great help for getting out and pointing out established practices at odds with what is being done outside or known outside as being safe. Control authorities can of course fulfil this mission but internal cross auditing between plants make the process of exchanging maybe sometimes easier, and more open to discussion.

6. Conclusion

The multidisciplinary approach developed in this accident investigation covered a wide range of issues from technical to organisational ones. Damages analysis, technical cause analysis, material tests, chronology, barriers analysis and organisational investigation all gave complementary insights, in order to better understand the accident. The damages analysis, chronology and material test supported the technical cause analysis and the barriers approach helped to create a link with the organisation. There are of course many aspects that could be developed further. But the time spent on the case helped to get deep enough to generate insights.

The methodological approach developed for the organisational investigation has been described as half way between a type of structured root cause analysis and a more descriptive and contextual approach looking for conditional links between facts, using the existing literature on the subject as the background models for guiding the interpretation. The methodology has the potential for being applied again in other accident investigations with organisational dimension, but will be also practicable for smaller investigations.

The findings about the organisational side of this accident have some similarities with in depth investigations dealing with this level of analysis. Some of them have been described in the outcomes discussed in this paper.

7. References

Baker W.E., Cox P.A., Westin P.S., Kulesz J.J., Strehlow R.A., 1983: Explosion hazards and Evaluation, Elsevier, New York.

Columbia Accident Investigation Board, 2003, Vols. II-VI of Final Report.

Cozzani and Salzano, 2004, The quatitative assessment of domino effects caused by overpressure Part I, Probit models, Journal of Hazardous Materials A107 (2004) pp 67-80.

Cullen, Lord. 2000 The Ladbroke Grove Rail Inquiry Part 1 and 2 reports, HSE books, her majesty stationery Office, Norwich, U.K.

Dechy N. and Mouilleau, Y. 2004, « Damages of the Toulouse disaster, 21st september 2001 », Proceedings of the 11th International Symposium Loss Prevention 2004, Praha, 31 May - 3 June 2004.

Dechy N., Bourdeaux T., Ayrault N., Kordek M.-A., Le Coze J.-C., 2004, First lessons of the Toulouse ammonium nitrate disaster, 21st september 2001, AZF Plant, France, Journal of Hazardous Materials 111 - July 2004, (special issue on JRC-ESReDA seminar on Safety accident investigation, Petten, the Netherlands, 12-13 May 2003)

Fortune J. Peters G. (1995) Learning from failure, the system approach. John Wiley and Sons.

Glaser B. G. Strauss A. L. (1967) The discovery of grounded theory: strategies for qualitative research. Aldine de Gruyter.

Grant, R.L., Murphy J.N., Bowser M.L., 1967: Effect of weather on sound transmission from explosive shots, Bureau of Mines Report of Investigation 6921

Hopkins A. 2000. Lessons from Longford: the Esso Gas Plant Explosion. Sydney. CCH Australia.

Lecoze J-C., Lim S., Intégration des aspects organisationnels dans le retour d'expérience: l'accident majeur, un phénomène complexe à étudier, pour le Ministère de l'Ecologie et du Développement Durable, 2002, downloadable on http://www.ineris.fr

Lecoze J-C, 2005, Organisation and disaster: from lessons learnt to theorising, forthcoming

Llory.M, Dien Y, Montmayeul R. (2004) Organisational investigation methodology and lessons learned. Journal of Hazardous Materials 111 – July 2004

Mouilleau Y. and Dechy N., 2002. « Initial analysis of the damage observed in Toulouse after the accident that occured on 21st of september on the AZF site of the Grande Paroisse company », *International ESMG Symposium, Nürnberg, Germany 8-10 October 2002, on Process safety and industrial explosion protection*

Perrow C. (1999) Normal accident theory, living with high risk technology, second edition

Rasmussen, J. (1997) Risk management in a dynamic society: a modelling problem, in safety science vol 27 n°2/3, p 183-213.

Rasmussen J. Svedung I. (2000). Proactive risk management in a dynamic society. Swedish Rescue Service Agency. Karlstad.

Reason.J. L'erreur humaine. Presse Universitaire de France. Collection Le Travail Humain. 1993.

Reason J. 1997. Managing the risk of organisation accident. Ashgate.

Reed, J.W., 1973: Distant Blast Predictions for Explosions, Minutes of the Fifteenth Explosives Safety Seminar, Dpt of Defense Explosives Safety Board, Washington D.C., vol. II, pp. 1403-1424

Roberts K.H. (1990) Some characteristics of one type of high reliability in organisation. Organisation Science. 1(2). 160-176.

Rochlin G.I. (1996) Reliable organisations: Present research and future directions, journal of contingencies and crisis management, volume 4, june.

Rochlin G.I. (1999) Safe operation as a social construct. Ergonomics. Volume 42. N° 11. 1549-1560.

Schein.E, Organisational culture and leadership, second edition, Jossey-bass, 1992.

Snook, S.A. (2000) Friendly fire, the accidental shootdown of US black hawks over northern Iraq, Princeton university press.

Turner B. Pidgeon N. (1997) Made-man disaster: the failure of foresight. Second edition, Butterworth-Heinmann.

Vaughan D. (1992) Theory Elaboration: The Heuristics of Case Analysis, in Charles Ragin and Howard S. Becker (eds.), What is a Case?: Exploring the Foundations of Social Inquiry. Cambridge University Press.

Vaughan D. (1996) The Challenger launch decision: risky technology, culture and deviance at NASA, University of Chicago Press, Chicago.

Vaughan, D. (1997) The trickle down effect: policy decisions, risky work and the challenger tragedy, California Management Review. vol 39. n°2.

Weick K. (1987) "Culture as a source of high reliability" , California Review of Management.

Waring A, Glendon A.I. (1998) Managing risk, critical issues for survival and success into the 21st century. Thompson Learning.

Wenzel A.B., Bessey R.L., 1969: Barricaded and Unbarricaded Blast Measurements, Contract n° DAHC04-69-C-0028, Subcontract 1-OU-431, Southwest Research Institute, San Antonio, Texas

Wright J.K., 1961: Shock tubes, John Wiley & Sons Inc., New York

Reports of INERIS on the Billy-Berclau accident (to download on http://www.ineris.fr)

R. Branka, N. Dechy, J.-C. Lecoze, E. Leprette, S.Lim, Intervention de l'INERIS après le sinistre survenu le 27 mars 2003 sur le site de Billy-Berclau de la Société Nitrochimie, Rapport final, Septembre 2004, 30 pages. (*global summary report*)

S. Lim, J.-C. Lecoze, Chronologie de l'accident du 27 mars 2003 survenu à Billy-Berclau sur le site de Nitrochimie, Rapport Annexe 1, MEDD/DPPR, août 2003, 15 pages. (*chronology*)

R. Branka, Examen de l'explosif en cause dans l'accident du 27 mars 2003 survenu à Billy-Berclau sur le site de Nitrochimie, Rapport Annexe 2, MEDD/DPPR, Juin 2003, 8 pages. (*material tests*)

N. Dechy, E. Leprette, Analyse des dommages observés à Billy-Berclau (62) après le sinistre du 27 mars 2003 survenu sur le site de la société Nitrochimie, Rapport Annexe 3, MEDD/DPPR, Août 2003, 63 pages. (*Damages investigation*)

S. Lim, J.-C. Lecoze, Détermination de l'arbre des causes de l'accident du 27 mars 2003 survenu à Billy-Berclau sur le site de Nitrochimie, Rapport Annexe 4, MEDD/DPPR, Août 2003, 24 pages. (*technical analysis*)

J.-C. Lecoze, S. Lim, Analyse organisationnelle de l'accident du 27 mars 2003 survenu à Billy-Berclau sur le site de Nitrochimie, Rapport Annexe 5, MEDD/DPPR, Août 2003, 133 pages. (*organisational investigation*)

Reports of IGE-IGAS-CGM on the Billy-Berclau accident (to download on http://www.ladocumentationfrancaise.fr/brp/notices/044000226.shtml)

Y. Lecointe, J.-P. Chassine, H. Legrand, P. Monsel, Accident survenu à l'usine Nitrochimie de Billy-Berclau (Pas-de-Calais), Inspection Générale de l'Environnement, Conseil général des Mines, Inspection Générale des Affaires Sociales, Mars 2004.

When Risk Becomes Reality:
Formosa Plastics' Response to a Plant Explosion

Robert Thibault
Formosa Plastics
9 Peach Tree Hill Road
Livingston, NJ 07039
rthibault@fpcusa.com

Prepared for Presentation at
American Institute of Chemical Engineers
2005 Spring National Meeting
39th Annual Loss Prevention Symposium
Atlanta, Georgia
April 11-13, 2005

UNPUBLISHED

When Risk Becomes Reality: Formosa Plastics' Response to a Plant Explosion

Robert Thibault
Formosa Plastics
9 Peach Tree Hill Road
Livingston, NJ 07039
rthibault@fpcusa.com

Abstract

At 10:45 p.m. Central Time on Friday, April 23, 2004, an explosion ripped through Formosa Plastics' polyvinyl chloride plant in Illiopolis, IL. The explosion effectively destroyed the plant, leaving five people dead, several others injured, caused the evacuation of thousands of people for two square miles and generated headlines in media around the world.

Formosa's response to the incident has been called a "textbook example of how to handle a crisis." This presentation will outline the steps Formosa Plastics took before, during and after the crisis to help reduce the impact on victims and their families, the community, the company and its employees and customers.

Handling Chemicals in Small Containers

Albert Ness
Rohm and Haas
3100 State Road
Croydon, PA, 19021
aness@rohmhaas.com

Richard Gibson
Rohm and Haas
4800 State Route 12
Elma, WA. 98541
rgibson@rohmhaas.com

Prepared for Presentation at
American Institute of Chemical Engineers
2005 Spring National Meeting
39th Annual Loss Prevention Symposium
Atlanta, Georgia
April 11-13, 2005

January, 2005

UNPUBLISHED

Handling Chemicals in Small Containers

Albert Ness
Rohm and Haas
3100 State Road
Croydon, PA, 19021
aness@rohmhaas.com

Richard Gibson
Rohm and Haas
4800 State Route 12
Elma, WA. 98541
rgibson@rohmhaas.com

Abstract

This paper describes three incidents involving buckets or other small, open containers that have occurred at Rohm and Haas facilities. In one, a reaction between Sodium and water propelled a bucket with considerable force, resulting in an injury. In another, a fire occurred when the lid a plastic container containing a liquid peroxide became contaminated during filling and was blown off. In the last, a solid peroxide and several different initiators and catalysts were staged on a pallet for use in batch reactors. A fire was caused by mixing of small spills from the containers on or under the pallet used to hold them. The fire caused severe damage and temporary shutdown of a reactor train.

1. Introduction

Trevor Kletz once wrote that, in a chemical plant, buckets are more dangerous than a book of matches![1] This paper describes three incidents involving buckets or other small, open containers that have occurred at Rohm and Haas facilities.

In the first, an operator was injured when a bucket previously containing sodium had not been handled and cleaned by the normal procedures. The bucket was then exposed to water from a heavy rain, and the resulting reaction propelled the bucket into the operator's leg with considerable force.

In the second, a fire occurred when the lid of a plastic container containing a Cumene Hydroperoxide (CHP) was blown off. A contaminated container was used, resulting in a reaction of the contaminant with CHP.

In the third, a fire started on or under a pallet used to stage small containers of several different initiators and catalysts. Over time small spills of a solid peroxide and other materials from the containers accumulated on a cable tray located under the pallet and the resulting mixture finally caught fire. The fire caused severe damage to electrical and instrumentation wiring, resulting in a temporary shutdown of a reactor train.

---

[1] Kletz, Trevor, "Improving Chemical Engineering Practices; a new look at old myths of the chemical industry" second edition, Hemisphere Publishing Corporation, New York, 1990.

2. Bucket Propelled by Sodium - Water Reaction

2.1 Incident Description

At a plant that uses sodium, small pieces of waste metallic sodium are generated during line removals and clean-outs, and each time an empty sodium rail car is disconnected. The waste sodium pieces were kept in the sodium handling building in a stainless steal bucket filled with mineral oil until enough pieces were generated for reclamation. The reclamation process involved dissolving the pieces of metallic sodium in clean, dry methanol, with a water analysis of less than one percent. This operation was done in a stainless steel bucket. When the sodium had fully reacted in the methanol, as determined by the absence of sodium "chunks", the mix was transferred from the bucket to a sample recovery pot.

Earlier in the week of the incident, an operator successfully dissolved sodium in methanol in two other buckets, and noticed nothing unusual. On the day of the incident, the operator came across a bucket that contained 2-3" of a sludge material. He asked co-workers about the material and what to do with it. The co-workers were not sure where the sludge material in the bucket came from so they collectively recommended adding methanol to ensure that there was no sodium remaining in the sludge. The operator added methanol to the bucket, placed a lid on it, and placed it on a steam-out pad. About 6-7 hours later the operator returned to the bucket and assumed that, if there was any sodium in the sludge, it would have reacted and transferred the contents of the bucket into the recovery pot. After completing the transfer, the operator noticed there was still residual material left in the bucket. Thinking that all the sodium had reacted, he proceeded to an effluent building to dump the sludge into an effluent pit and wash the bucket out. The operator did not see any sign of reaction when the sludge from the bucket was dumped into the pit. The operator placed the bucket down on the walkway and went to get a hot water hose to rinse the bucket out. At this time an extremely heavy rain started, igniting the contents of the bucket. While attempting to cover the bucket, he knocked it over and its contents spilled onto the wet pad. The resulting reaction propelled the bucket into the operator's leg, resulting in an injury.

2.2 Findings

The investigation found that the bucket still contained sodium even after a 7 hour hold with methanol to recover the sodium, because the sodium was under a layer of dolomite and did not contact the methanol.

The sodium - dolomite mixture was left over from an earlier maintenance activity, and the bucket did not get properly labeled.

Although the operators discussed what to do with the "unknown" material, an adequate EHS management of change review was not conducted for this unusual situation.

2.3 Actions Taken

The sodium reclamation procedure was halted immediately.

The plant contacted the sodium supplier for advice on reclamation procedures, and discovered that the supplier sent their waste to a hazardous waste treatment facility. After that, the plant also started to do this and stopped doing the reclamation process.

For the possibility that some sodium reclamation may have to be done, the procedure was revised to include the following learning's: ensure all containers are properly labeled, do not perform the procedure outdoors when it is raining, ensure all the sodium chunks are dissolved by stirring until no more chunks are left, and, if any unusual situation arises, ensure that an EHS review is conducted.

Several training sessions were held on how to properly label the buckets, how to prepare the sodium waste for disposal and on the importance of management of change and EHS reviews.

3. Container Fire due to Contamination

3.1 Incident Description

This incident involved two operating areas in a plant which we will call Area 1 and Area 2 for convenience. Area 1 supplies an intermediate product to Area 2. Cumene Hydroperoxide (CHP) is a raw material for the intermediate. Storing and handling of CHP, as well as some other small charges, are done by Area 2. The procedure is to have the operators from Area 2 weigh up the small charge of CHP into a labeled plastic container with a hand pump and transfer the containers to Area 1 when needed.

At about 8:00 PM, the Area 1 operator located the plastic containers labeled for CHP use and contacted the Area 2 operator to fill the containers for a batch to be run early the next morning. The hand pump normally used was broken, so another one was located and washed out with water from a flush drum. The containers were covered and stored at Area 2. At approximately 11:15 PM, one of the CHP containers was found smoking. Pressure had blown the top off the container. The fire brigade was called and directed water onto all of the containers. The containers and the drum from which they were pumped were all moved to a safe storage area outside and the top bungs loosened to safely vent any possible reaction. No further reactions occurred.

3.2 Findings

The Area 2 operator noted that he had observed dark debris on the outside of one of the containers, and that it looked like Ferrous Sulfate. After the containers were filled, the operator also noted that the contents of the container looked darker than the other containers, but believed it was an effect from the debris on the outside of the container.

The MSDS notes that the CHP is incompatible with acids, oxidizing agents, reducing agents, oxidation catalysts, copper, lead and iron. Additionally, when CHP and Ferrous Sulfate are entered into the NOAA *Chemical Reactivity Worksheet*[2], it returns the entries; "Explosive when mixed with oxidizing substances", "Heat generated from chemical reaction may initiate explosion", "May cause fire" and "Heat generation by chemical reaction, may cause pressurization" for the combination.

The investigation also found that when the handling of CHP containers by the two areas began, clear responsibility for maintaining the containers was not established.

The investigators concluded that the most probable cause was contamination with Ferrous Sulfate from the outside of the container. A possible cause was contamination from residual chemicals in the flush drum, which is used to flush lances for the drum pump used for monomers, but the investigators considered this less likely.

3.3 Actions Taken

Clean replacement containers were obtained.

A dedicated storage area for the empty and full CHP containers that protects them from contamination from other chemicals was established.

A new pump dedicated to CHP service was obtained.

The operator training was updated and repeated to ensure all operators have an understanding of the hazards of CHP and to emphasize that containers for storing chemicals should be free of contamination inside and out.

The process operating procedures were updated to establish clear ownership of the procedural steps in making this product, including the maintaining and cleaning of the containers.

The process operating procedures were updated to limit the time frame for staging containers of CHP.

4. Fire from Oxidizer Spills

4.1 Incident Description

Small charges for several batch processes were weighed into small containers and staged near the reactors on a pallet. Among the materials handled in these containers were solid organic peroxide, a persulfate and other small charges. Figure 1 shows a re-creation of the staged pallets.

A fire occurred in the process building on the pallet of oxidizers containing the peroxide and persulfate. The sprinkler system activated, but its effectiveness was limited because a cable tray was between the sprinklers and the pallet. The fire caused extensive damage to electrical power,

---

[2] *Chemical Reactivity Worksheet*, U. S. National Oceanic and Atmospheric Administration, http://response.restoration.noaa.gov/chemaids/react.html

control and instrumentation wiring, resulting in considerable downtime in a reactor train. Figure 2 shows the actual pallet after the fire.

Figure 1: Re-Creation of Pallet with Pre-Weighed Materials in Process Area

4.2 Findings

The investigation revealed that, in the past, there had been spills of the different materials because buckets tipped over and cracks had been reported in the buckets. The floor was an open grate and a cable tray was located underneath the staging area, creating a location for the spills to collect because it was difficult to clean up.

The investigators concluded that the most likely cause of the fire was a spill of the peroxide from a bucket onto a cable tray below the grating which contained other, incompatible materials, particularly some activators that had spilled over time. The peroxide reacted with other materials starting a fire.

The location of the pallet near cable trays and junction boxes led to the extensive damage.

Contributing to the incident was a lack of detailed knowledge of how the specific materials would react if mixed. There was a general understanding that the materials should not be mixed, but not in-depth knowledge of what would happen. This led to a practice of staging incompatible materials needed for several batches on the same pallet.

Figure 2: Remains of pallet after fire

4.3 Actions Taken

All chemical residues on pipes and equipment below the grating on the floor of the process building were cleaned up.

All existing containers for pre weighed materials were inspected and replaced as needed. Procedures were established to inspect all containers within one year and adjust the frequency of the inspection or implement a replacement policy based on the results.

Separate storage areas for of catalysts and activators, away from critical equipment and potential contamination sources were set up.

Separate, dedicated, color coded pallets for storing incompatible materials were obtained and used.

Spill containment was provided for all pre-weighed, packaged materials staged on the process floor.

Limits on the amount of materials that could be pre weighed and staged on the process floor were established.

The peroxide containers were replaced with new, color-coded buckets with lids with gaskets.

5. Conclusions and Recommendations

These incidents have several things in common:

- Open buckets or other small containers were used for handling reactive chemicals.
- Contact occurred between the material in the containers and other, incompatible material(s) through spills or contamination.
- The corrective actions rely heavily on administrative procedures.

The incidents show that whenever small containers or buckets are used to handle small charges, the potential for contamination or inadvertent mixing exists. To prevent these kinds of incidents, look for ways to avoid use of small containers. If this is not possible:

- Determine the potential hazards of contamination. Constructing a reactivity matrix is a good way to do this (see the reference to the Chemical Reactivity Worksheet for a good source for beginning to construct a reactivity matrix).
- Provide training to personnel on the potential hazards of contamination and ignition of the materials being handled.
- Provide separate areas for weighing and staging incompatible materials such as activators and initiators.
- Use dedicated containers and handling equipment, such as pumps and pallets.
- Clearly label the containers, and consider color coding them.
- Provide lids for containers.
- Establish procedures for cleaning and inspecting of containers.